Computational Microelectronics

Edited by S. Selberherr

Compact Transistor Modelling for Circuit Design

H. C. de Graaff
F. M. Klaassen

Springer-Verlag Wien New York

Dr. Henk C. de Graaff
Dr. François M. Klaassen
Philips Research Laboratories
Eindhoven, The Netherlands

This work is subject to copyright.
All rights are reserved, whether the whole or part of the material is concerned, specifically those of translation, reprinting, re-use of illustrations, broadcasting, reproduction by photocopying machine or similar means, and storage in data banks.
© 1990 by Springer-Verlag/Wien
Softcover reprint of the hardcover 1st edition 1990
Typeset by Asco Trade Typesetting Ltd., Hong Kong

With 184 Figures

ISBN-13: 978-3-7091-9045-6   e-ISBN-13: 978-3-7091-9043-2
DOI: 10.1007/978-3-7091-9043-2

Preface

During the first decade following the invention of the transistor, progress in semiconductor device technology advanced rapidly due to an effective synergy of technological discoveries and physical understanding. Through physical reasoning, a feeling for the right assumption and the correct interpretation of experimental findings, a small group of pioneers conceived the major analytic design equations, which are currently to be found in numerous textbooks.
Naturally with the growth of specific applications, the description of some characteristic properties became more complicated. For instance, in integrated circuits this was due in part to the use of a wider bias range, the addition of inherent parasitic elements and the occurrence of multi-dimensional effects in smaller devices. Since powerful computing aids became available at the same time, complicated situations in complex configurations could be analyzed by useful numerical techniques. Despite the resulting progress in device optimization, the above approach fails to provide a required compact set of device design and process control rules and a compact circuit model for the analysis of large-scale electronic designs.
This book therefore takes up the original thread to some extent. Taking into account new physical effects and introducing useful but correct simplifying assumptions, the previous concepts of analytic device models have been extended to describe the characteristics of modern integrated circuit devices. This has been made possible by making extensive use of exact numerical results to gain insight into complicated situations of transistor operation. Of course, in some cases the result is a compromise between accuracy and complexity of description.
In this way we hope to serve three categories of readers: professionals involved in the development of IC processes, experts in the design and analysis of integrated circuits and students of applied physics or electrical engineering with an interest in the above subjects.
Generally the material has been split into three main parts. Chapters 1, 2, 3 and 6 give the more general aspects of analytic device modelling (methods, approaches etc.). Then in chapters 4, 5, 7, 8 and 9 specific compact models

are discussed. Finally, the last chapters are devoted to the practical and experimental aspects of device modelling. Since the field covered is rather broad and the authors have expertise in different subjects, we decided to split up the tasks. While chapters 1, 2, 10 and 11 have been jointly written, the first author is responsible for chapters 3, 4 and 5 and the second author for chapters 6, 7, 8 and 9.

At the end of a period of collecting and selecting abundant material, solving almost hopeless problems and setting down what we feel to be the main points of this subject, we would like to express our gratitude to all those who made this book possible:

— The Directors of Philips Research Laboratories, Eindhoven, namely Dr. K. Bulthuis and Mr. W. G. Gelling, who gave us the opportunity to combine our daily work with the activities necessary for preparing the manuscript.
— A large number of our colleagues within the Philips organization. We would like to express our thanks in particular to Ing. W. J. Kloosterman who contributed so much to the numerical development of new device models and their parameter extraction methods. Ir. H. R. Claessen took a very active part in the development of process blocks and statistical modelling. Lic. R. M. D. Velghe's work on MOSFET parameter extraction is gratefully acknowledged. Mr. T. N. Jansen and Ir. R. Dekker did a great amount of work on the comparison of various bipolar and MESFET models, respectively. Dr. J. J. H. van den Biesen did essential work on remedying the deficiencies of existing models for lateral bipolar devices. Ir. T. Smedes' work on MOSFET charge modelling and ECL gates is gratefully acknowledged. Mr. P. Leduc pointed out to us the importance of the unity parameter concept. Ing. R. Petterson and Ing. J. H. A. van der Wielen made several numerical calculations to support the MOSFET models presented.
— Mrs. M. U. A. Jansen, Mrs. M. van Hoof, Mrs. A. F. Loos and Mr. H. J. M. Alblas, who with considerable skill and accuracy took care of the typing and artwork.
— The numerous readers of our manuscript, who by their comments and criticism, improved the initial text considerably. Ir. P. T. J. Biermans, Dr. P. A. H. Hart, Ir. M. C. A. Koolen and Dr. M. M. J. Pelgrom should be mentioned in particular. In addition our thanks are due to Dr. M. J. Bolt, Dr. J. Engel, Ir. A. J. Linssen, Prof. J. A. Pals, Dr. M. Rocchi, Ir. M. F. Sevat, Prof. M. F. Schuurmans, Ir. J. A. van Steenwijk, Ir. H. P. Tuinhout, Ir. M. van der Veen and Mr. C. van Winsum.

Eindhoven, June 1989   H. C. de Graaff and F. M. Klaassen

Contents

1 Introduction 1
   1.1 Compact Models 2
      1.1.1 Models Based on Device Physics 2
      1.1.2 Numerical Table Models 3
      1.1.3 Empirical Models 4
   1.2 Compact Models and Simulation Programs 4
   1.3 Subjects Treated in This Book 5
   References 6

2 Some Basic Semiconductor Physics 7
   2.1 Quantum-Mechanical Concepts 7
   2.2 Distribution Function and Carrier Concentration 9
   2.3 The Boltzmann Transport Equation 13
   2.4 Bandgap Narrowing 19
   2.5 Mobility and Resistivity in Silicon 21
   2.6 Recombination 25
   2.7 Avalanche Multplication 27
   2.8 Noise Sources 30
      2.8.1 Shot Noise 30
      2.8.2 Diffusion Noise and Thermal Noise 31
      2.8.3 Flicker Noise 32
   References 32

3 Modelling of Bipolar Device Phenomena 35
   3.1 Injection and Transport Models 35
      3.1.1 Solution of the Continuity Equations 35
      3.1.2 Injection Model 38
      3.1.3 Transport Model 38
   3.2 The Quasi-Static Approximation and the Charge Control Principle 40

3.3 Collector Currents and Stored Charges 42
    3.3.1 General Relation Between Collector Current and Charges 42
    3.3.2 The Integral Charge Control Relation 45
    3.3.3 Current, Charges and Minority Carrier Concentrations 48
        3.3.3.1 The Low-Injection Case: $n(x) \ll N_a(x)$ 50
        3.3.3.2 The High-Injection Case: $n(x) \gg N_a(x)$ 52
        3.3.3.3 The General Case 54
3.4 Base Currents 56
3.5 Depletion Charges and Capacitances 58
    3.5.1 Influence of Current on Q_{T_c} 60
3.6 Early Effect 62
3.7 Quasi-Saturation, Base Widening and Kirk Effect 65
    3.7.1 The Charge Storage in the Epilayer 65
    3.7.2 Influence of I_c: Ohmic and Hot Carrier Behaviour (Kirk Effect) 67
    3.7.3 Inverse Mode of Operation 71
3.8 Avalanche Multiplication 72
3.9 Series Resistances 75
    3.9.1 Emitter Series Resistance 75
    3.9.2 Base Resistance 76
    3.9.3 Collector Series Resistance 79
3.10 Time- and Frequency-Dependent Behaviour 80
    3.10.1 Charge Control and Quasi-Static Approach 80
    3.10.2 Exact One-Dimensional Solution 81
    3.10.3 Time Delays 84
    3.10.4 Base Charge Partitioning 85
    3.10.5 Second-Order Differential Operators 86
3.11 Transit Time and Cut-Off Frequency f_T 87
3.12 Noise Behaviour 91
3.13 Temperature Dependences 93
References 96

4 Compact Models for Vertical Bipolar Transistors 99

4.1 Ebers-Moll-Type Models 99
    4.1.1 Basic Ebers-Moll Model 99
    4.1.2 Extensions of the Basic Ebers-Moll Model 102
    4.1.3 Temperature Dependence of the Parameters 104
    4.1.4 Typical Results 105
4.2 Gummel-Poon-Type Models 107
    4.2.1 Basic Gummel-Poon Model 107
    4.2.2 Extensions 110
    4.2.3 Full Quasi-Saturation Model 111
    4.2.4 Typical Results 112

Contents

- 4.3 The MEXTRAM Model   114
  - 4.3.1 Main Currents and Stored Charges   117
  - 4.3.2 Quasi-Saturation and Hot-Carrier Effect in the Epilayer   119
  - 4.3.3 Depletion Charges   123
  - 4.3.4 Base Currents   124
  - 4.3.5 Series Resistances   125
  - 4.3.6 Modelling the Inactive Part and Substrate   125
  - 4.3.7 Typical Results   126
- 4.4 Short Review   128
  - 4.4.1 Basic Ebers-Moll Model   128
  - 4.4.2 Extensions to the Ebers-Moll Model   128
  - 4.4.3 Basic Gummel-Poon Model   129
  - 4.4.4 Extensions to the Gummel-Poon Model   129
  - 4.4.5 Mextram Models   129
- References   130

5 Lateral *pnp* Transistor Models   132

- 5.1 Model Definitions   133
  - 5.1.1 Lateral *pnp* Models of the Ebers-Moll Type   133
  - 5.1.2 Lateral *pnp* Models of the Gummel-Poon Type   136
- 5.2 Results   138
- 5.3 Shortcomings of Existing Models   140
- References   142

6 MOSFET Physics Relevant to Device Modelling   144

- 6.1 Formation of the Inversion Layer   144
  - 6.1.1 Qualitative Discussion   144
  - 6.1.2 Quantitative Analysis   147
- 6.2 The Ideal MOS Transistor Current   151
- 6.3 The Threshold Voltage   156
  - 6.3.1 The Body Effect   156
  - 6.3.2 Effect of Implants Additional to the Substrate Doping   157
  - 6.3.3 Effect of Implants of Opposite Type to the Substrate Doping   160
  - 6.3.4 Temperature Dependence   164
  - 6.3.5 Short-Channel Effect   165
  - 6.3.6 Narrow-Width Effect   169
- 6.4 Carrier Mobility in Inversion Layers   170
  - 6.4.1 Bias Dependence of the Carrier Mobility   170
  - 6.4.2 Temperature Dependence   173
  - 6.4.3 Modelling of Effects Other than Mobility Via the θ-Parameters   174

6.5 Saturation Mode 175
   6.5.1 Static Feedback 175
   6.5.2 Channel-Length Modulation 176
6.6 Dynamic Operation 178
   6.6.1 Quasi-Static Operation 178
   6.6.2 Charges, Charge Distribution and Capacitances in the Active Region 181
   6.6.3 Charges in the Off-State Region 183
   6.6.4 Parasitic Contributions 184
6.7 Intrinsic Parasitics 185
   6.7.1 Series Resistance 185
   6.7.2 Gate-Junction Capacitance 188
References 190

7 Models for the Enhancement-Type MOSFET 195

7.1 Long-Channel Models 195
   7.1.1 The Drain Current of Transistors in Uniformly Doped Substrates 196
   7.1.2 The Drain Current of Transistors with Threshold Adjustment Implant 200
   7.1.3 Charges and Capacitances 203
   7.1.4 Effect of Velocity Saturation on the Drain Current 211
7.2 Small Transistor Models 213
   7.2.1 The Drain Current in Small MOSFETS 213
      7.2.1.1 The Threshold Voltage 213
      7.2.1.2 The Substrate Effect 216
      7.2.1.3 The Drain Saturation Voltage 218
      7.2.1.4 Static Feedback and Channel Length Modulation 222
      7.2.1.5 The Subthreshold Mode 223
   7.2.2 Charges 225
      7.2.2.1 Strong-Inversion Region 225
      7.2.2.2 Capacitances 228
      7.2.2.3 Charge in the Subthreshold Region 229
   7.2.3 Effect of Series Resistance on the Drain Current 229
   7.2.4 The Substrate Current 233
7.3 Models for Analog Applications 235
   7.3.1 Review of Existing Models 235
   7.3.2 Improved Description of the Drain Current 238
   7.3.3 Capacitances 241
   7.3.4 Noise 243
      7.3.4.1 Thermal Noise 243
      7.3.4.2 Flicker Noise 246
References 247

8 Models for the Depletion-Type MOSFET 251

- 8.1 Long-Channel Model 252
  - 8.1.1 Mobile Charge Density 252
  - 8.1.2 Threshold and Saturation Voltages 253
  - 8.1.3 Channel Current 255
- 8.2 Short-Channel Model 258
  - 8.2.1 Specific Problems 258
  - 8.2.2 Depletion-Mode Channel Conductance for a Linear Doping Profile 259
  - 8.2.3 The Drain Current of a Short-Channel Depletion MOSFET 262
- 8.3 Charges and Charge Distribution 265
- References 265

9 Models for the JFET and the MESFET 267

- 9.1 The Drain Current of the Junction-Gate FET 268
  - 9.1.1 The Classical Description 268
  - 9.1.2 A Model for Short-Channel Transistors 270
- 9.2 The Drain Current of the MESFET 273
  - 9.2.1 Review of Empirical Models 273
  - 9.2.2 An Improved Model 275
- 9.3 Charges and Capacitances 277
- References 279

10 Parameter Determination 281

- 10.1 General Optimization Method 281
  - 10.1.1 The Linear Case 284
- 10.2 Specific Bipolar Measurements 285
  - 10.2.1 Measurements of Series Resistances 286
  - 10.2.2 Measuring the Cut-Off Frequency f_T 292
- 10.3 Example of Parameter Extraction for a Bipolar Transistor Model 292
  - 10.3.1 The Depletion Capacitances 293
  - 10.3.2 Early Effects 294
  - 10.3.3 The Gummel Plots 295
  - 10.3.4 The Quasi-Saturation 297
  - 10.3.5 The Cut-Off Frequency f_T 297
  - 10.3.6 Concluding Remarks 298
- 10.4 Parameter Determination for MOSFETs 299
  - 10.4.1 Enhancement Devices 299
  - 10.4.2 Depletion Devices 308
- 10.5 Specific MOSFET Measurements 309
- References 312

11 Process and Geometry Dependence, Optimization and Statistics of Parameters   314
   11.1 Unity Parameters and Geometrical Scaling in Bipolar Modelling   316
      11.1.1 Geometry Dependence   316
      11.1.2 Process Dependence of Unity Parameters   318
   11.2 Bipolar Process Blocks and Circuit Optimization   322
   11.3 Geometry- and Process Dependence of MOSFET Parameters   325
      11.3.1 Geometry Dependence   325
      11.3.2 Process Dependence   329
   11.4 Statistics: Definitions and Formulas   331
   11.5 Bipolar Statistical Modelling   333
      11.5.1 Process Blocks and Statistical Models   334
      11.5.2 Correlation Between Compact Model Parameters   337
      11.5.3 Correlation at the Process Level   339
   11.6 MOS Statistical Modelling   340
      11.6.1 Mismatch in MOSFETs   340
      11.6.2 Parametric Yield Estimation in MOS VLSI   344
   References   346

Subject Index   348

Introduction 1

In the early days of transistor applications in electronic circuits the design of these circuits relied heavily on empirical methods; with a certain concept in mind the designer actually built his circuit on a (printed-)circuit board with discrete elements (capacitors, resistors, inductors, transistors, etc.) and checked its electrical performance. If the result was not satisfactory, the discrete elements were changed in value and/or type, until the electrical specifications were met. This empirical procedure was called breadboarding. With the advent of the integrated circuit this method was no longer appropriate, mainly for two reasons:

— discrete elements have properties different from their integrated counterparts. In the first place the integrated circuit has a substrate, common to all elements, that has to be taken into account (e.g. the parasitic *pnp* transistor, accompanying each vertical *npn* transistor, and the coupling between the various devices via the substrate). Furthermore, integrated resistors have rather large parasitic capacitances, etc.
— the distances between the discrete elements of a breadboarded circuit are much larger (tens of millimetres) than those in integrated circuits (tens of microns) and so are the interconnections. Moreover, discrete elements are encapsulated. This means that the parasitic capacitive coupling between the elements in a breadboarded circuit is quite different from that in an integrated circuit and the self-inductances of the interconnection lines in a breadboarded circuit are much larger.

Nowadays the designer of integrated electronic circuits makes ample use of numerical simulation of the electrical circuits [1.1]: it goes faster than breadboarding, it is more flexible and allows for components that are difficult to breadboard. It also enhances the understanding of how a circuit operates. To this end the designer uses circuit analysis programs like SPICE, SLIC, PHILPAC etc. These programs contain mathematical models for the quantitative description of the terminal behaviour of the elements. Modern integrated circuits may consist of hundreds to thousands of elements (or even more), so it is of paramount importance that the mathematical model of each element is as simple as possible, otherwise the required CPU time will be prohibitive. The use of analytical functions has advantages in this respect over a purely numerical approach. On the other hand, the model must also be reasonably accurate to make the circuit simulation reliable.

This book will deal with the mathematical models of active devices (bipolar, MOS and other field-effect transistors) for circuit analysis purposes, as used in computer-aided circuit design methods. We will call these models "compact models", as distinct from the more elaborate numerical device simulation models that are used for the study of device physics and for device design [1.2].

The accuracy of a compact model does not only depend on the set of mathematical equations describing the terminal behaviour, but also on the accuracy of the numerical constants in these equations. Such constants are called parameters; examples of parameters are, among many others, the threshold voltage (V_T) of a MOS transistor and the saturation value (I_s) of the collector current in a bipolar transistor.

Parameter values are obtained from measurements or, if no measurements are available, from numerical device simulation results. So, the development of a compact model must not only deal with defining the model equations, it must also pay attention to methods for extracting the various parameter values.

In this book we have restricted ourselves to active devices; models for passive elements (resistors and capacitors) are beyond its scope. Nearly all models described here refer to silicon devices, with one exception: in chapter 9 a compact model for the GaAs MESFET is given.

Modelling the heterojunction bipolar transistor [1.3] is still in the research phase and a mature compact model for that type of device has not yet emerged. For that reason we have decided to omit that subject from this book.

1.1 Compact Models

We can distinguish several types of compact models for circuit design.

1.1.1 Models Based on Device Physics

If the defining model equations are directly derived from device physics we speak of physical device models. They consist of a set of analytical functions, preferably in explicit form. That the functions be analytical is of great importance, because discontinuities in the functions and their derivatives (up to the third derivative) may cause numerical instabilities. Moreover, in a.c. small-signal applications harmonic-distortion calculations are often needed and this too requires the availability of the derivatives. A well-known problem here is that the analytical expressions following from the device physics, are only valid in a certain range of bias conditions and that outside that range other expressions must be used. Examples are for MOS transistors the subthreshold region and the strong-inversion region, and for bipolar transistors the nonsaturated region and the (quasi-)saturation region. In such cases smoothing procedures of a curve-fitting nature are applied to ensure a continuous transition from one region to another.

It is also important that the model equations behave "decently" in regions far away from the bias conditions normally encountered, because during the iteration process of the circuit simulator the device model may be subject to unpractical and unphysical conditions, and this must not endanger the convergence.

The advantages of using physical device models are as follows:

— the parameters have physical significance, which can be used as a check on the correctness of the parameter extraction procedures and is helpful in circuit design.
— physical models have a forecasting ability within their own class of devices (e.g. vertical *pnp* transistors, *n*-channel MOS transistors).
— geometrical scaling rules can be applied with confidence.
— the correlation between the model parameters is governed by device physics, which can be used for realistic statistical modelling. However, one must be aware that the parameter extraction can introduce an unphysical interdependence of some parameters (e.g. in bipolar transistor models the values of the series resistances may influence the high injection current parameter).

The drawbacks of physical models are:

— the analytical expressions are often simplifying approximations that may affect the accuracy.
— developing a physical model takes a long time (several man-years) by experienced device physicists.
— new devices or alternative structures demand new models or major modifications.

1.1.2 Numerical Table Models

The use of table models is based on the storage of data of all the measured points of the relevant electrical characteristics of a device [1.4]. The tables may also be filled with the results of numerical device simulations. Suitable interpolation procedures must be available to ensure the continuity; spline functions can be used here [1.5].

The great advantage over physical device models is that table models can be developed in a short time. The many disadvantages are that table models are essentially only valid for the measured range: the model validity is uncertain with extrapolation outside the measured range. We have a very large storage of data, which gives problems with the computer memory size.

The forecasting ability is almost zero and geometrical scaling is not possible (each device needs its own tables).

1.1.3 Empirical Models

The model-defining equations in an empirical model are analytical expressions that are of a curve-fitting nature, and not primarily based on device physics. This curve-fitting nature excludes forecasting abilities, geometrical scaling and proper correlation between parameters. The advantages of empirical modelling are the shorter development time with respect to physical modelling and the smaller data storage with respect to table look-up modelling.

Purely empirical models do not exist in practice, but often in physical compact models, the more delicate device phenomena (e.g. transit times) are described by empirical expressions. Empirical expressions are also used in connection with table models, to partly replace the interpolation procedures.

1.2 Compact Models and Simulation Programs

Compact models are especially meant for use in circuit simulation programs, whereas the device simulation programs are mostly used for device design [1.2]. However, it is possible to couple a device simulator directly to a circuit simulator and to use the output of the device simulator to replace the compact model [1.6]. This then requires a much longer CPU time and can only be used for a strictly limited number of devices. A more common method is to use the characteristics, generated by the device simulator, for the extraction of compact model parameters and to do the circuit simulation in the usual way, with compact device models. In this approach the parameter extraction program and the compact model can then be incorporated in an integrated chain of simulation programs [1.7, 1.8]. Such an integrated chain consists of a process simulator, a device simulator, a parameter extractor and a circuit simulator plus compact models. Fig. 1.1 gives the block diagram.

The process simulator consists of routines for diffusion, implantation, oxidation, etc. [1.9]. The input data are taken from the process flowchart, the output gives the device structure in one- or two-dimensional form: doping profiles, oxide, p and n regions. This serves as input for a one- or two-dimensional device simulator [1.10, 1.11] that generates the electrical device characteristics. These are fed into the parameter extraction for the determination of the compact model parameters. It must be stressed here that the parameter extractor input can also be in the form of measured characteristics. Finally, the set of parameter values obtained is used for the compact models in the circuit simulator. Sometimes a so-called process block is inserted between the parameter extractor and the compact model to enable

1.3 Subjects Treated in This Book

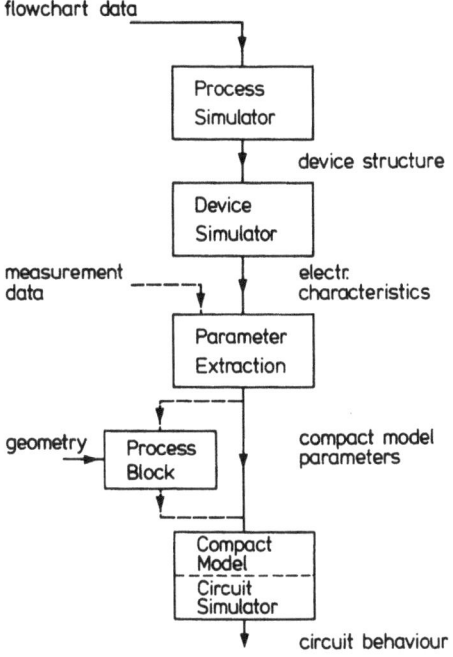

Fig. 1.1 Block diagram of an integrated chain of process, device and circuit simulators

the circuit designer to choose the desired device geometry (see chapter 11 of this book).

1.3 Subjects Treated in This Book

This work deals with compact device models for integrated circuit design. For reasons outlined in section 1.1, our preferred type of compact model is the model based on device physics. Numerical table models will not be dealt with and empirical modelling will be avoided whenever possible, although it cannot be excluded totally.
After this introductory chapter, chapter 2 will consider a branch of basic semiconductor physics, mainly related to charge transport and introducing the concepts of bandgap, mobility, recombination, avalanche multiplication, noise, etc.
Bipolar device physics is discussed in chapter 3, with the emphasis on quasi-static and dynamic behaviour, charge-control concepts, high injection, quasi- and hard saturation. In chapters 4 and 5 specific bipolar transistor models for the vertical *npn* and lateral *pnp* devices will be defined: Ebers-Moll, Gummel-Poon and the recently developed MEXTRAM models. Chapter 6 will deal with the device physics of MOS transistors: inversion

layers, ideal drain current, threshold voltage, surface mobility, saturation, dynamic operation and parasitics.

Actual model descriptions for the enhancement- and depletion-type MOSFETS are to be found in chapters 7 and 8, for both long- and short-channel devices. Models for analogue applications are also considered.

Models for the junction-gate FET (JFET) and the GaAs MESFET are given in chapter 9.

The subject of parameter extraction is dealt with in chapter 10, where the non-linear optimization methods are discussed and a survey of specific measurement methods is given.

The problems of geometrical scaling, process dependence and statistical correlation of model parameters are given in the last chapter (chapter 11), on statistical modelling and design optimization.

References

[1.1] L. W. Nagel: SPICE2—A Computer Program to Simulate Semiconductor Circuits. Electr. Res. Lab. Memo ERL-M520, University of California, Berkeley (1975).
[1.2] W. L. Engl, H. K. Dirks, B. Meinerzhagen: Device Modeling. Proc. IEEE *71*, 10 (1983).
[1.3] J. Tasselli, A. Marty, J. P. Bailbe, G. Rey: Verification of the Charge-Control Model for GaAlAs/GaAs Heterojunction Bipolar Transistors. Solid-St. Electr. *29*, 919 (1986).
[1.4] W. M. Coughran, E. H. Grosse, D. J. Rose: Aspects of Computational Circuit Analysis. Techn. Memo, AT & T Bell Labs, June 26 (1986).
[1.5] J. A. Barby: Multidimensional Splines for Modeling FET Non-Linearities. Thesis UW/ICR 86-01, Inst. Comp. Research, University of Waterloo (1986).
[1.6] W. L. Engl, R. Laur, H. K. Dirks: MEDUSA—A Simulator for Modular Circuits. IEEE Trans. on CAD, CAD-1, (1982).
[1.7] E. J. Prendergast: An Integrated Approach to Modeling. NASECODE IV, 83 (1985).
[1.8] V. Marash, R. W. Dutton: Methodology for Submicron Device Model Development. IEEE Trans. CAD 7, 299 (1988).
[1.9] D. Chin, M. Kump, H. Lee, R. W. Dutton: Process Design Using Two-Dimensional Process and Device Simulators. IEEE Trans. Electr. Dev. *ED-29*, 336 (1982).
[1.10] J. W. Slotboom: Iterative Scheme for 1- and 2-Dimensional dc Transistor Simulation. Electr. Lett. *5*, 677 (1969).
[1.11] S. Selberherr, A. Schutz, H. W. Pötzl: MINIMOS—A Two-Dimensional MOS Transistor Analyzer. IEEE Trans. Electr. Dev. *ED-27*, 1540 (1980).

Some Basic Semiconductor Physics 2

In this chapter we will deal shortly with a number of fundamental concepts of semiconductor physics (distribution functions, doping levels, carrier transport, mobility, etc.). One can also find here a set of formulas that are needed in the description of device phenomena and in the formulation of model equations.

From a viewpoint of theoretical solid-state physics the treatment is certainly not rigorous, the only purpose is to have the basic equations available in this book. Furthermore, we have assumed that the reader is already familiar with the existence of electrons and holes, p- and n type conduction, p-n junctions etc.

2.1 Quantum-Mechanical Concepts

Let us consider a single particle moving in a potential field $V(r)$ with a velocity $\mathbf{v}$ and a momentum $\mathbf{p} = m\mathbf{v}$. Its total energy is then $E = (|\mathbf{p}|^2/2m) + V$. In quantum mechanics the Schrödinger wave equation [2.1] is obtained by putting $\mathbf{p} = -i\hbar \nabla_r$, where $\hbar = h/2\pi$ and h is Planck's constant; ∇_r is a linear differential vector operator:

$$\nabla_r = j_1 \frac{\partial}{\partial x} + j_2 \frac{\partial}{\partial y} + j_3 \frac{\partial}{\partial z}. \tag{2.1}$$

$j_{1,2,3}$ are the unit vectors in the x, y and z directions, respectively. The wave equation becomes

$$\left(-\frac{\hbar^2}{2m} \nabla_r^2 + V\right)\psi = E\psi. \tag{2.2}$$

For $V = 0$ the wave function $\psi(r, t)$ has the form

$$\psi = A \exp\{i(\mathbf{k}\cdot\mathbf{r} - \omega t)\}, \tag{2.3}$$

where ω is the angular frequency, $\mathbf{k}$ is the wave vector, $|\mathbf{k}| = 2\pi/\lambda$ is called the wave number and λ the wavelength.

Applying $\mathbf{p}$ as operator to the wave function ψ gives

$$\mathbf{p}\psi = -i\hbar\nabla_r\psi = \hbar\mathbf{k}\psi$$

or

$$\mathbf{p} = m\mathbf{v} = \hbar\mathbf{k}. \tag{2.4}$$

Thus the wave vector is proportional to the particle velocity **v** and its momentum **p**.

In the expression (2.3) for the wave function the scalar product $\mathbf{k}\cdot\mathbf{r}$ stands for $k_x x + k_y y + k_z z$. Let us suppose that the wave function ψ is confined in space to a cube with sides L. Quantization now requires that $k_{x,y,z} = n_{1,2,3}(2\pi/L)$, where $n_{1,2,3}$ are integers. Each set of allowed $k_{x,y,z}$ values defines a cell in k-space, with a volume $L^3/8\pi^3$.

If the particle is an electron in a semiconductor [2.2], then the density of allowed states is $2*L^3/8\pi^3 = L^3/4\pi^3$; a spin degeneracy of two is here taken into account.

In a crystal lattice the potential V in Eq. (2.2) has the same spatial periodicity as the lattice; the wave function of Eq. (2.3) may still be used, if we consider A a function of position, again with the same periodicity. The particle mass changes into an effective mass tensor with elements $m_{xy}^{-1} = \hbar^{-2}\partial^2 E/\partial k_x \partial k_y$. Even in the isotropic case electrons and holes have unequal effective masses ($m_e \neq m_h$).

The energy of an electron in the conduction band is often written as

$$E(\mathbf{k},\mathbf{r},t) = E_c(\mathbf{r},t) + \frac{\hbar^2}{2m_e}(k_x^2 + k_y^2 + k_z^2). \tag{2.5}$$

For a hole in the valence band we use a similar expression:

$$E(\mathbf{k},\mathbf{r},t) = E_v(\mathbf{r},t) + \frac{\hbar}{2m_h}(k_x^2 + k_y^2 + k_z^2). \tag{2.6}$$

In expressions (2.5) and (2.6) E_c is the bottom of the conduction band and E_v the top of the valence band (see Fig. 2.1).

The distance from E_c to the vacuum level outside the semiconductor is given by the affinity χ. Between E_c and E_v we have a forbidden zone, the bandgap E_{gap}.

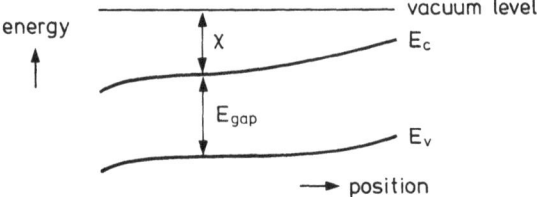

Fig. 2.1 Diagram indicating the vacuum level, the bottom of the conduction band (E_c) and the top of the valence band (E_v). χ is the affinity, E_{gap} is the forbidden zone or bandgap. All these quantities may be position-dependent

2.2 Distribution Function and Carrier Concentration

The distribution function $f(E)$ gives the probability that an electron occupies a state with energy level E [2.3]. The density of states per unit volume is $1/4\pi^3$, so $1/4\pi^3 \, f(E) \, d\mathbf{r} \, d\mathbf{k}$ denotes the number of electrons with energy E that can be found at the time t in a spatial position between $\mathbf{r}$ and $\mathbf{r} + d\mathbf{r}$, with a velocity in the range given by $\mathbf{k}$ and $\mathbf{k} + d\mathbf{k}$.

For electrons in semiconductors the appropriate distribution is the Fermi-Dirac function:

$$f(E) = \frac{1}{1 + \exp\{(E - E_F)/k_B T\}} . \tag{2.7}$$

Here E_F is the Fermi level, k_B is Boltzmann's constant and T is the absolute temperature in Kelvin. For $(E - E_F) \gg k_B T$, $f(E)$ simplifies into the (Maxwell-)Boltzmann function

$$f(E) = \exp\{-(E - E_F)/k_B T\} . \tag{2.8}$$

E_F is only defined in equilibrium situations; it is position-independent. In non-equilibrium situations we use quasi-Fermi levels, one for electrons (E_{F_n}) and one for holes (E_{F_p}).

In high electric fields a shifted Maxwellian distribution function may be used:

$$f(E) = \exp\left[-\left\{E - \frac{\hbar^2}{2m}(\mathbf{k} - \mathbf{k}_0)^2 - E_{F_{n,p}}\right\}\Big/ k_B T\right] . \tag{2.9}$$

The quantities $E_{F_n, p}$, T, E_c, E_v are in general functions of position $\mathbf{r}$ and time t, whereas $(E - E_c)$ is a function of $\mathbf{k}$ only. So we will consider $\mathbf{r}, \mathbf{k}$ and t as the independent variables of the system. The electron concentration follows from

$$n(\mathbf{r}, t) = \frac{1}{4\pi^3} \int f(E) \, d^3k ,$$

where d^3k stands for $dk_x \, dk_y \, dk_z$. Substituting for $f(E)$ the Fermi-Dirac distribution from Eq. (2.7) we get, by using Eq. (2.5),

$$n = \frac{1}{2\pi^2}\left(\frac{2m_e}{\hbar^2}\right)^{3/2} \int_0^\infty \frac{\sqrt{E - E_c} \, dE}{1 + \exp\left(\dfrac{E - E_F}{k_B T}\right)} .$$

Introducing the Fermi integrals [2.3]

$$F_j(u) = \frac{1}{\Gamma(j+1)} \int_0^\infty \frac{x^j \, dx}{1 + \exp(x - u)} \tag{2.10}$$

we get

$$n = \frac{1}{2\pi^2}\left(\frac{2m_e k_B T}{\hbar^2}\right)^{3/2} \Gamma(\tfrac{3}{2}) F_{1/2}\left(\frac{E_F - E_c}{k_B T}\right) , \tag{2.11}$$

bearing in mind that we can write for the gamma function [2.4] $\Gamma(j+1) = j\Gamma(j)$, so that $\Gamma(\frac{3}{2}) = \frac{1}{2}\Gamma(\frac{1}{2}) = \frac{1}{2}\sqrt{\pi}$.
Moreover, with the Boltzmann approximation we have

$$F_{1/2}\left(\frac{E_F - E_c}{k_B T}\right) \approx \exp\{(E_F - E_c)/k_B T\}$$

and

$$n \approx 2\left(\frac{m_e k_B T}{2\pi \hbar^2}\right)^{3/2} \exp\{(E_F - E_c)/k_B T\}$$
$$= N_c \exp\{(E_F - E_c)/k_B T\}. \tag{2.12}$$

In a similar way we can derive for the hole concentration in the valence band

$$p \approx N_v \exp\{(E_v - E_F)/k_B T\} \tag{2.13}$$

with

$$N_v = 2\left(\frac{m_h k_B T}{2\pi \hbar^2}\right)^{3/2}.$$

In undoped material p and n are equal to the intrinsic concentration n_i. The Fermi level in that situation is given by

$$E_F = E_i = \frac{1}{2}(E_c + E_v) + \frac{1}{2} k_B T \ln \frac{N_v}{N_c}. \tag{2.14}$$

From Eqs. (2.12) and (2.13) it follows that, apart from unequal electron and hole masses, the intrinsic Fermi level E_i is roughly in the middle of the bandgap.

The pn product in thermal equilibrium is given by

$$pn = N_c N_v \exp\{-(E_c - E_v)/k_B T\}$$
$$= cT^3 \exp(-E_{\text{gap}}/k_B T), \tag{2.15}$$

and this holds also for non-intrinsic material, because in the pn product only the bandgap and the temperature are involved. For the same reason we can write in general that

$$pn = n_i^2.$$

By using the intrinsic Fermi level we can write Eqs. (2.12) and (2.13) in an alternative way:

$$n = n_i \exp\left(\frac{E_F - E_i}{k_B T}\right) \tag{2.16}$$

and

$$p = n_i \exp\left(\frac{E_i - E_F}{k_B T}\right). \tag{2.17}$$

By adding impurities to the semiconductor we can create energy levels in

2.2 Distribution Function and Carrier Concentration

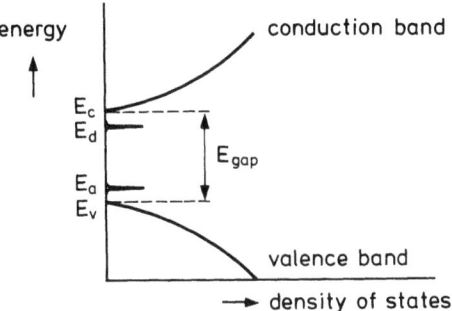

Fig. 2.2 Sketch of the relation between energy and density of states. E_d and E_a are the respective donor and acceptor levels

the forbidden zone and thus manipulate the concentration of electrons and holes.

Donor levels, due to phosphorus for instance, are lying near E_c, whereas boron creates acceptor levels near the valence band; see Fig. 2.2.

A donor becomes ionized by losing an electron to the conduction band, and an acceptor becomes ionized by accepting an electron from the valence band. The occupation statistics for donors and acceptors must take into consideration that most donor and acceptor levels can have only one electron of either spin value. If N_d is the donor concentration, the concentration of occupied, non-ionized donor levels is given by [2.5]

$$n_d = \frac{N_d}{1 + \frac{1}{2}\exp\{(E_d - E_F)/k_B T\}}. \tag{2.18}$$

For the occupied, ionized acceptor levels we get

$$N_a^- = \frac{N_a}{1 + 2\exp\{(E_a - E_F)/k_B T\}}. \tag{2.19}$$

The position of the Fermi level with respect to the band edges is calculated by means of the charge neutrality condition

$$N_d - n_d + p = n + N_a^-. \tag{2.20}$$

Substituting the expressions for n, p, n_d and N_a^-, (2.12), (2.13), (2.18), (2.19) into (2.20) we can solve for E_F.

From Eqs. (2.18) and (2.19) it will be clear that in principle not all the donors and acceptors will be ionized. The number of ionized donors determines the electron concentration in the conduction band and follows from Eq. (2.18):

$$N_d^+ = N_d - n_d = \frac{N_d}{1 + 2\exp\{(E_F - E_d)/k_B T\}}$$

$$= \frac{N_d}{1 + 2\exp\{(E_F - E_c + E_c - E_d)/k_B T\}}$$

or

$$N_d^+ = \frac{N_d}{1 + 2\frac{n}{N_c}\exp(\Delta E_d/k_B T)}. \quad (2.21)$$

Here use has been made of Eq. (2.10); $\Delta E_d = E_c - E_d$ is the ionization energy of the donor level.

Equation (2.21) can be rewritten as a quadratic equation [2.6]:

$$2\left(\frac{N_d^+}{N_c}\right)^2 \exp\left(\frac{\Delta E_d}{k_B T}\right) + \frac{N_d^+}{N_c} - \frac{N_d}{N_c} = 0. \quad (2.22)$$

For given values of ΔE_d and N_d, we can calculate the value of N_d^+. The results are shown in Fig. 2.3.

At 300 K and with $\Delta E_d = 45$ meV we find that $N_d^+/N_d \approx 0.9$ for $N_c/N_d = 100$, that is $N_d \approx 2.5 * 10^{17}$ cm^{-3}. A similar situation exists for acceptors.

According to the results in Fig. 2.3 the percentage of unionized impurity levels increases when the dope concentration N_d increases. However, when N_d increases, the ionization energy of the impurity level decreases,

$$\Delta E_d = \Delta E_d(0) - \alpha N_d^{1/3} \quad (2.23)$$

with $\alpha = 3.1 * 10^{-8}$ eV/cm.

Moreover, at high values of N_d we get a broadening of the donor level into an impurity band, that will overlap with the conduction band.

This effect enhances the ionization in such a way that for large N_d concentrations ($> 3 * 10^{18}$ cm^{-3}) ΔE_d goes to zero and $N_d^+ = N_d$. The phenomenon of unionized impurity dopes is strongest ($\approx 15\%$ at 300 K) for dope concentrations between 10^{17} and $3 * 10^{18}$ cm^{-3} [2.7].

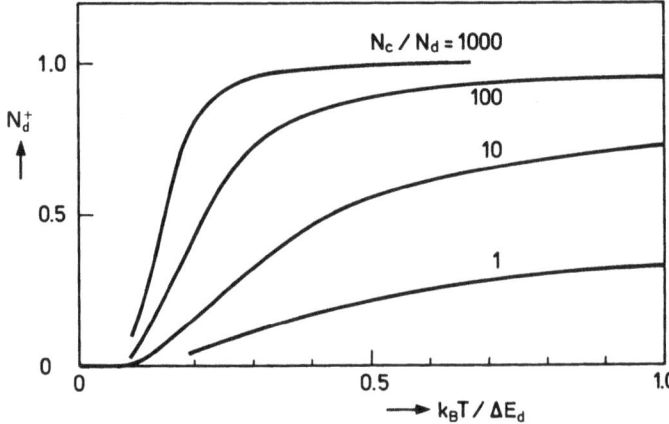

Fig. 2.3 Ionized impurity ratio (N_d^+/N_d) as a function of the normalized ionization energy ($k_B T/\Delta E_d$) for various doping levels (N_c/N_d)

2.3 The Boltzmann Transport Equation

We will consider a distribution function $f(\mathbf{r}, \mathbf{k}, t)$, with the position $\mathbf{r}$, the wave vector $\mathbf{k}$ and the time t as independent variables. If forces are acting on the system of particles (e.g. electrons in the conduction band of a semiconductor) the change of f with time will be [2.8]:

$$\frac{df}{dt} = (\nabla_r f) \cdot \frac{d\mathbf{r}}{dt} + (\nabla_k f) \cdot \frac{d\mathbf{k}}{dt} + \frac{\partial f}{\partial t}. \tag{2.24}$$

Here ∇_k is also a differential vector operator:

$$\nabla_k = \mathbf{j}_1 \frac{\partial}{\partial k_x} + \mathbf{j}_2 \frac{\partial}{\partial k_y} + \mathbf{j}_3 \frac{\partial}{\partial k_z}.$$

The dots in Eq. (2.24) mean the scalar product of the two vectors involved. If we disregard for the moment such effects as recombination and generation of carriers, the counteracting forces, oppposing the change in f, are the collisions. These can be approximated by a relaxation time model, $(f - f_0)/\tau$, where f_0 is the equilibrium distribution function. In the steady state we then have

$$\frac{df}{dt} = \left(\frac{df}{dt}\right)_{\text{coll.}} = -\frac{f - f_0}{\tau(\mathbf{r}, \mathbf{k})}. \tag{2.25}$$

The relaxation time τ is in general a function of $\mathbf{r}$ and $\mathbf{k}$, which means that τ is position- and energy-dependent.
In (2.24) $d\mathbf{r}/dt$ is the velocity of the particles, so $d\mathbf{r}/dt = (\hbar/m)\mathbf{k}$, and $d\mathbf{k}/dt$ represents the acceleration: for electrons in the conduction band we then have $d\mathbf{k}/dt = -(1/\hbar)q\mathbf{F}$. The vector $-q\mathbf{F}$ is the force acting on an electron and is determined by the gradient of the conduction band: $q\mathbf{F} = \nabla_r E_c$. Combining Eqs. (2.24) and (2.25) the Boltzmann equation for electrons in the conduction band becomes [2.9]:

$$\frac{\partial f}{\partial t} + \frac{\hbar}{m_e} \mathbf{k} \cdot \nabla_r f - \frac{q}{\hbar} \mathbf{F} \cdot \nabla_k f = -\frac{f - f_0}{\tau}. \tag{2.26}$$

In the static case $\partial f/\partial t = 0$ and f satisfies

$$f = f_0 - \tau \left(\frac{\hbar}{m_e} \mathbf{k} \cdot \nabla_r f - \frac{q}{\hbar} \mathbf{F} \cdot \nabla_k f\right).$$

If we now replace the function f by f_0 in the terms between brackets, we get the first-order approximation for f [2.10]:

$$f^{(1)} = f_0 - \tau \frac{\hbar}{m_e} \mathbf{k} \cdot \nabla_r f_0 + \tau \frac{q}{\hbar} \mathbf{F} \cdot \nabla_k f_0. \tag{2.27}$$

Here $f^{(1)}$ has only a linear field term, but this is sufficiently accurate if $|\mathbf{F}| \ll k_B T/ql$, where l denotes the mean free path. In silicon $k_B T/ql \sim 10^5$

V/cm at carrier temperatures of 1000 K [2.11]. With the help of the distribution function f or its first-order approximation $f^{(1)}$ we can define several quantities. We have already mentioned the particle concentration

$$n = \frac{1}{4\pi^3} \int f\, d^3k; \qquad (2.28)$$

see also Eq. (2.11).

We can also introduce an energy density W_n for the electrons by [2.12]:

$$W_n = \frac{1}{4\pi^3} \int (E - E_c) f\, d^3k. \qquad (2.29)$$

If f is e.g. a "shifted Maxwellian" (see Eq. (2.9)), then

$$E - E_c = \frac{\hbar^2}{2m_e}\{(k_x - k_0)^2 + k_y^2 + k_z^2\}$$

and

$$W_n = \frac{1}{4\pi^3}\frac{\hbar^2}{2m_e}\left(\int k^2 f\, d^3k + k_0^2 \int f\, d^3k\right).$$

With the help of Eqs. (2.11) and (2.28), and the fact that $\int \mathbf{k} f_0\, d^3k = 0$, W_n finally becomes

$$W_n = \frac{3}{2} k_B T n \frac{F_{3/2}\left(\dfrac{E_F - E_c}{k_B T}\right)}{F_{1/2}\left(\dfrac{E_F - E_c}{k_B T}\right)} + \frac{1}{2} m_e v_0^2 n. \qquad (2.30)$$

The electron current density $\mathbf{J}_n$ is given by

$$\mathbf{J}_n = \frac{-1}{4\pi^3}\frac{q\hbar}{m_e}\int \mathbf{k} f\, d^3k$$

$$\approx \frac{-1}{4\pi^3}\frac{q\hbar}{m_e}\int \mathbf{k} f^{(1)}\, d^3k, \qquad (2.31)$$

where again use has been made of $\int \mathbf{k} f_0\, d^3k = 0$.

We now substitute the expression for $f^{(1)}$ from Eq. (2.27) into the integral in Eq. (2.31), making use of the fact that for Fermi-Dirac functions $f\{E(\mathbf{k})\}$

$$\nabla_r f = T \frac{\partial f}{\partial E} \nabla_r\left(\frac{E - E_F}{T}\right) \qquad (2.32)$$

and

$$\nabla_k f = \frac{\hbar^2}{m}\mathbf{k}\frac{\partial f}{\partial E}; \qquad (2.33)$$

see Eq. (2.7). In the process of evaluating the expression in Eq. (2.31) we

2.3 The Boltzmann Transport Equation

come across integrals of the type

$$I_M = -\frac{\hbar^2}{4\pi^3 m^2} \int \tau k^2 (E - E_c)^{M-1} \frac{\partial f_0}{\partial E} d^3k. \tag{2.34}$$

We then finally get for $\mathbf{J}_n$

$$\frac{1}{q}\mathbf{J}_n = \frac{1}{3}I_1\left\{q\mathbf{F} + T\nabla_r\left(\frac{E_{F_n} - E_c}{T}\right)\right\} + \frac{1}{3}I_2\frac{1}{T}\nabla_r T. \tag{2.35}$$

For an isothermal carrier gas $\nabla_r T = 0$ and Eq. (2.35) reduces to

$$\mathbf{J}_n = -q\mu_n n \nabla_r \varphi_n. \tag{2.36}$$

Here φ_n is the Fermi potential for electrons: $-q\varphi_n = E_{F_n}$ and $q\mathbf{F} = \nabla_r E_c$ has also been used. The mobility μ_n is given by

$$\mu_n = \frac{1}{3}\frac{qI_1}{n} = -\frac{1}{3}q\left(\frac{\hbar}{m_e}\right)^2 \frac{\int \tau k^2 \frac{\partial f_0}{\partial E} d^3k}{\int f_0 d^3k}. \tag{2.37}$$

We may take the relaxation time τ as a function of energy [2.8]:

$$\tau = \tau_0 \left(\frac{E - E_c}{k_B T_0}\right)^s. \tag{2.38}$$

The value of s depends on the type of scattering: e.g. for acoustic phonon scattering $s = -\frac{1}{2}$.
Substituting Eq. (2.38) into Eq. (2.37) leads, after extensive manipulation, to

$$\mu(T) = \frac{q\tau_0}{m_e}\left(\frac{T}{T_0}\right)^s \frac{\Gamma(s + \frac{5}{2})}{\Gamma(\frac{5}{2})} \frac{F_{s+1/2}\left(\frac{E_F - E_c}{k_B T}\right)}{F_{1/2}\left(\frac{E_F - E_c}{k_B T}\right)}. \tag{2.39}$$

$F_{s+1/2}/F_{1/2}$ is a quotient of Fermi integrals (see Eq. (2.10)) that goes to one in the Boltzmann approximation.
It should be stressed here that T in the above-mentioned expressions is the temperature of the carriers (electrons); it might differ from the crystal lattice temperature, e.g. when the carriers acquire high energy in high electric fields ("hot" carriers).
There is an alternative way of writing down $\mathbf{J}_n$: After the substitution of Eq. (2.27) into Eq. (2.31) the ∇_r term is

$$\frac{q}{4\pi^3}\left(\frac{\hbar}{m}\right)^2 \int \tau (\mathbf{k} \cdot \nabla_r f_0)\mathbf{k}\, d^3k$$

$$= \frac{1}{3}\frac{q}{4\pi^3}\left(\frac{\hbar}{m}\right)^2 \nabla_r \int \tau k^2 f_0\, d^3k = q\nabla_r(D_n n).$$

This defines D_n as

$$D_n = \frac{1}{3}\left(\frac{\hbar}{m_e}\right)^2 \frac{\int \tau k^2 f_0 \, d^3k}{\int f_0 \, d^3k}. \tag{2.40}$$

The diffusion constant D_n is closely related to the mobility of Eq. (2.39),

$$\mu_n k_B T / q D_n = \frac{F_{s+1/2}\left(\frac{E_F - E_c}{k_B T}\right)}{F_{s+3/2}\left(\frac{E_F - E_c}{k_B T}\right)}. \tag{2.41}$$

This is the Einstein relation; in the Boltzmann approximation the right-hand side becomes equal to one.

The total current density is now the sum of the field and diffusion terms:

$$\mathbf{J}_n = q\mu_n n \mathbf{F} + q\nabla_r(D_n n). \tag{2.42}$$

Eq. (2.42) is also valid if $\nabla_r T \neq 0$, in contrast with Eq. (2.36), where $\nabla_r T = 0$ is required. If we can put $\nabla_r D_n = 0$, Eq. (2.42) further simplifies in

$$\mathbf{J}_n = q\mu_n n \mathbf{F} + q D_n \nabla_r n. \tag{2.43}$$

The moving electrons give rise to an energy flux $\mathbf{S}_n$, given by [2.11, 2.12]

$$\mathbf{S}_n = \frac{\hbar}{4\pi^3 m_e} \int \mathbf{k}(E - E_c) f \, d^3k. \tag{2.44}$$

Substituting $f^{(1)}$ for f and working out the integral we obtain

$$\mathbf{S}_n = -\frac{1}{3} I_2 \left\{ q\mathbf{F} + T\nabla_r\left(\frac{E_F - E_c}{T}\right) \right\} - \frac{1}{3} I_3 \frac{1}{T} \nabla_r T. \tag{2.45}$$

If we substitute for $\mathbf{F}$ by using Eq. (2.35) we get

$$\mathbf{S}_n = -\frac{I_2}{I_1} \frac{1}{q} \mathbf{J}_n + \frac{1}{3} I_2 \left(\frac{I_2}{I_1} - \frac{I_3}{I_2}\right) \frac{1}{T} \nabla_r T. \tag{2.46}$$

$I_{1,2,3}$ are integrals as given by Eq. (2.34):

$$I_2/I_1 = (s + \tfrac{5}{2}) \frac{F_{s+3/2}}{F_{s+1/2}} k_B T$$

and

$$I_3/I_2 = (s + \tfrac{7}{2}) \frac{F_{s+5/2}}{F_{s+3/2}} k_B T.$$

All the Fermi integrals have $(E_F - E_c)/k_B T$ as argument. In the Boltzmann approximation the quotients go to one and $\mathbf{S}_n$ becomes

2.3 The Boltzmann Transport Equation

$$\mathbf{S}_n = -(s + \tfrac{5}{2})\frac{k_B T}{q}\mathbf{J}_n - (s + \tfrac{5}{2})\frac{k_B^2 T}{q}\mu_n n \nabla_r T. \tag{2.47}$$

We now return to the Boltzmann transport equation (Eq. (2.26)). Multiplying each term by $1/4\pi^3$ and integrating over the total k-space we get:

$$\frac{1}{4\pi^3}\int \frac{\partial f}{\partial t} d^3k = \frac{\partial}{\partial t}\frac{1}{4\pi^3}\int f\, d^3k = \frac{\partial n}{\partial t},$$

$$\frac{1}{4\pi^3}\int \frac{\hbar}{m_e}\mathbf{k}\cdot\nabla_r f\, d^3k = \frac{1}{4\pi^3}\frac{\hbar}{m_e}\nabla_r\cdot\int \mathbf{k} f\, d^3k = -\frac{1}{q}\nabla_r\cdot\mathbf{J}_n,$$

$$\frac{1}{4\pi^3}\int \frac{q}{\hbar}\mathbf{F}\cdot\nabla_k f\, d^3k = 0.$$

Together this becomes the continuity equation for electrons:

$$\frac{\partial n}{\partial t} - \frac{1}{q}\nabla_r\cdot\mathbf{J}_n = -\frac{n - n_0}{\tau_n}.$$

The right-hand side of this equation stems from the collision term in its relaxation time formulation. This relaxation, which is restricted to the system of electrons in the conduction band, is very fast ($1/2\pi\tau_n \sim 1500$ GHz). Usually there is also an exchange of particles between the conduction and valence bands due to generation and recombination, which process is much slower. So for frequencies well below 1500 GHz we can better replace $(n - n_0)/\tau_n$ by the generation rate (G) minus the recombination rate (R):

$$\frac{\partial n}{\partial t} - \frac{1}{q}\nabla_r\cdot\mathbf{J}_n = G - R. \tag{2.48}$$

Multiplying Eq. (2.26) by $(1/4\pi^3)(E - E_c)$ and integrating over k-space gives us

$$\frac{1}{4\pi^3}\int (E - E_c)\frac{\partial f}{\partial t} d^3k = \frac{\partial W_n}{\partial t},$$

$$\frac{1}{4\pi^3}\int \frac{\hbar}{m_e}(E - E_c)\mathbf{k}\cdot\nabla_r f\, d^3k = \nabla_r\cdot\mathbf{S}_n,$$

$$\frac{1}{4\pi^3}\int \frac{q}{\hbar}(E - E_c)\mathbf{F}\cdot\nabla_k f\, d^3k = -\frac{q}{4\pi^3}\frac{\hbar}{m_e}\left\{F_x\int k_x f\, d^3k \right.$$
$$\left. + F_y\int k_y f\, d^3k + F_z\int k_z f\, d^3k\right\}$$
$$= \mathbf{F}\cdot\mathbf{J}_n,$$

$$\frac{1}{4\pi^3}\int (E - E_c)\frac{f - f_0}{\tau} d^3k = \frac{W - W_0}{\tau_{nw}},$$

where

$$\tau_{nw}^{-1} = \frac{\int (E - E_c)\frac{f - f_0}{\tau} d^3k}{\int (E - E_c)(f - f_0) d^3k}.$$

τ_{nw} is the energy relaxation time.
The result of all this is an energy continuity equation for electrons:

$$\frac{\partial W_n}{\partial t} + \nabla_r \cdot \mathbf{S}_n + \frac{W_n - W_{n0}}{\tau_{wn}} = \mathbf{F} \cdot \mathbf{J}_n. \qquad (2.49)$$

The right-hand side is the energy delivered to the electrons and $(W_n - W_{n0})/\tau_{wn}$ represents the energy transfer to the crystal lattice due to the collisions. The energy flux $\mathbf{S}_n$ follows from Eq. (2.46).
So far everything has been derived for electrons in the conduction band, but similar equations hold for holes in the valence band, and the total charge transport in semiconductors is governed by the following set of equations:

$$\left.\begin{aligned}
\frac{\partial n}{\partial t} - \frac{1}{q}\nabla_r \cdot \mathbf{J}_n &= G - R, \\
\frac{\partial p}{\partial t} + \frac{1}{q}\nabla_r \cdot \mathbf{J}_p &= G - R, \\
\frac{\partial W_n}{\partial t} + \nabla_r \cdot \mathbf{S}_n + \frac{W_n - W_{n0}}{\tau_{wn}} &= \mathbf{F} \cdot \mathbf{J}_n, \\
\frac{\partial W_p}{\partial t} + \nabla_r \cdot \mathbf{S}_p + \frac{W_p - W_{p0}}{\tau_{wp}} &= \mathbf{F} \cdot \mathbf{J}_p.
\end{aligned}\right\} \qquad (2.50)$$

Furthermore we have Poisson's equation, relating the electrostatic potential ψ to the space charge:

$$-\nabla^2 \psi = \frac{q}{\varepsilon}(N_d^+ - N_a^- + p - n). \qquad (2.51)$$

The electrostatic potential is defined here by means of the intrinsic level E_i in Eq. (2.14)

$$E_i = -q\psi.$$

Note that the electric field strength $-\nabla_r \psi$ is not necessarily the same as the force $\mathbf{F}$ in Eq. (2.26), because $\nabla_r E_c \neq \nabla_r E_i$ if the bandgap is not constant but position-dependent. The forces acting on electrons and holes may then be different. See also section 2.4.

2.4 Bandgap Narrowing

The additional flux equations are:

$$\begin{aligned}
\mathbf{J}_n &= -q\mu_n n \nabla_r \varphi_n \approx q\mu_n n \mathbf{F} + qD_n \nabla_r n, \\
\mathbf{J}_p &= q\mu_p p \nabla_r \varphi_p \approx q\mu_p p \mathbf{F} - qD_p \nabla_r p, \\
\mathbf{S}_n &= -(s+\tfrac{5}{2})\frac{k_B T_n}{q}\mathbf{J}_n - (s+\tfrac{5}{2})\frac{k_B^2 T_n}{q}\mu_n n \nabla_r T_n, \\
\mathbf{S}_p &= (s+\tfrac{5}{2})\frac{k_B T_p}{q}\mathbf{J}_p - (s+\tfrac{5}{2})\frac{k_B^2 T_p}{q}\mu_p p \nabla_r T_p.
\end{aligned} \qquad (2.52)$$

Note that the right-hand side terms in the continuity equations for electrons and holes are the same $(G - R)$, which implies that carrier trapping is excluded. Note also that the flux equations for $\mathbf{S}_n$ and $\mathbf{S}_p$ contain in principle different temperatures T_n and T_p for electrons and holes. If there is no carrier heating ($T_n = T_p = T_{\text{lattice}}$) and no temperature gradient, we can then simplify the set of equations to the continuity equations for the carriers plus Poisson's equation.

2.4 Bandgap Narrowing

From Eqs. (2.12–13) it is clear that the *pn* product is given by

$$pn = n_i^2 = N_c N_v \exp(-qV_{\text{gap}}/k_B T).$$

In the temperature range 250–450 K the bandgap voltage decreases more or less linearly with temperature: $V_{\text{gap}} = V_{g0} - \alpha T$. Moreover, the quantity $N_c N_v$ increases with T^3, so we can write [2.13]:

$$n_i^2 = \text{constant} * T^3 \exp(-qV_{\text{gap}}/k_B T) \qquad (2.53)$$

$$= 9.61 * 10^{32} * T^3 \exp(-qV_{g0}/k_B T) \qquad (\text{cm}^{-6}\ \text{K}^3).$$

Because the bandgap depends a little bit on the dope concentration, the extrapolated bandgap voltage V_{g0} is also a function of the dope concentration and for silicon is given by

$$V_{g0} = 1.205 - \Delta V_{\text{gap}}(N) \qquad (\text{volts}). \qquad (2.54)$$

For n_i^2 we then have

$$n_i^2 = n_{i0}^2 \exp(q\Delta V_{\text{gap}}/k_B T), \qquad (2.55)$$

where n_{i0}^2 is the equilibrium *pn* product in undoped silicon. There are several empirical formulas known for V_{gap} as a function of the dope concentration:

—the Slotboom-de Graaff formula [2.14]:

$$\Delta V_{\text{gap}} = 9 * 10^{-3} \left\{ \ln \frac{N}{N_{\text{ref}}} + \sqrt{\left(\ln \frac{N}{N_{\text{ref}}}\right)^2 + 0.5} \right\}, \qquad (2.56)$$

$$N_{\text{ref}} = 1 * 10^{17}\ \text{cm}^{-3}.$$

This expression is assumed to be valid for both n- and p-type silicon, although it was only measured for p type. For n-type silicon an alternative is —the del Alamo, Swanson and Swirhun formula [2.15]

$$\Delta V_{gap} = 18.7 * 10^{-3} \ln(N/N_{ref}), \quad (2.57)$$
$$N_{ref} = 7 * 10^{17} \text{ cm}^{-3}.$$

—Finally Bennett [2.16] calculates the bandgap narrowing for heavily doped material ($N > 10^{19}$ cm^{-3}) as

$$\Delta V_{gap} = 2\frac{k_B T}{q} \ln \left\{ 1 + 1.9 \exp\left(-\frac{N - 2.4 * 10^{20}}{1.68 * 10^{20}}\right) \right\}. \quad (2.58)$$

Eq. (2.58) gives much less bandgap narrowing than the other formulas, which can only be compensated by assuming more recombination (section 2.6) together with a higher minority carrier mobility, to explain the observed hole currents in n^+ regions. Fig. 2.4 shows the different results of the various empirical models for ΔV_{gap}.

In non-equilibrium we introduce quasi-Fermi potentials [2.17] for electrons ($\varphi_n = -(1/q)E_{F_n}$) and holes ($\varphi_p = -(1/q)E_{F_p}$); in equilibrium they coincide: $\varphi_n = \varphi_p = -(1/q)E_F$.

The electron and hole concentrations (Eqs. (2.16) and (2.17)) are rewritten as

$$n = n_i \exp\{q(\psi - \varphi_n)/k_B T\}$$

and

$$p = n_i \exp\{q(\varphi_p - \psi)/k_B T\}. \quad (2.59)$$

The electron current density J_n in the isothermal case ($\nabla_r T = 0$) is given by Eq. (2.52):

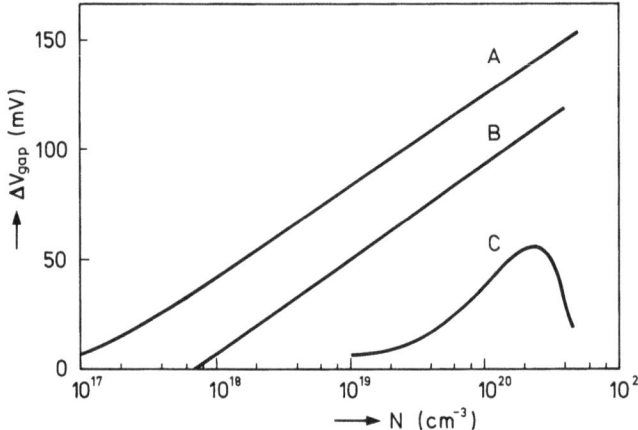

Fig. 2.4 The bandgap narrowing voltage ΔV_{gap} as a function of the dope concentration, according to Slotboom and de Graaff (A) [2.14], del Alamo et al. (B) [2.15] and Bennett (C) [2.16]

2.5 Mobility and Resistivity in Silicon

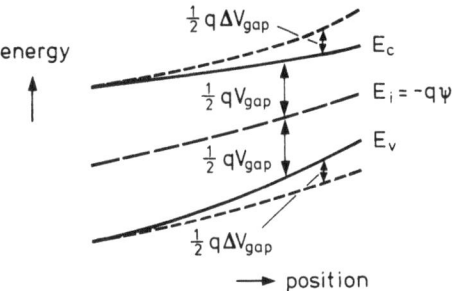

Fig. 2.5 Diagram indicating the position-dependent bandgap (qV_{gap}), the bandgap narrowing ($q\Delta V_{gap}$) and the electrostatic potential ($q\psi$)

$$\mathbf{J}_n = -q\mu_n n \nabla_r \varphi_n.$$

With the help of Eq. (2.59) this can be rewritten as

$$\mathbf{J}_n = q\mu_n n \left(-\nabla_r \psi - \frac{k_B T}{q} \frac{1}{n_i} \nabla_r n_i + \mu_n k_B T \nabla_r n \right). \tag{2.60}$$

The last term represents the diffusion component, the first is the drift component with

$$-\nabla_r \left(\psi + \frac{k_B T}{q} \ln n_i \right)$$

as acting force **F**:

$$\mathbf{F} = -\nabla_r \left(\psi + \frac{k_B T}{q} \ln n_i \right) = -\nabla_r \left(\psi + \frac{1}{2} V_{gap} \right) = \frac{1}{q} \nabla_r E_c.$$

For the hole current density we get a similar result:

$$\mathbf{J}_p = q\mu_p p \nabla_r (-\psi + \tfrac{1}{2} V_{gap}) - \mu_p k_B T \nabla_r p. \tag{2.61}$$

Now we have

$$\nabla_r \left(-\psi + \frac{1}{2} V_{gap} \right) = \frac{1}{q} \nabla_r E_v.$$

The drift of electrons and holes is the result of different acting forces $\nabla_r E_c$ and $\nabla_r E_v$, respectively. These acting forces consist of the electric field $-\nabla_r \psi$ and the influence of the bandgap narrowing ΔV_{gap}. If $\Delta V_{gap} \equiv 0$ we get $\nabla_r E_c = \nabla_r E_v = -\nabla_r \psi$. See also Fig. 2.5.

2.5 Mobility and Resistivity in Silicon

Since the charge carriers are subject to different scattering mechanisms, each with a characteristic relaxation time, the evaluation of the mobility expression (2.37) is cumbersome. Moreover, Eq. (2.37) has been derived with the relaxation time approximation and under the assumption that the distribution function $f \approx f^{(1)}$ has only a linear field term (see section 2.3). Therefore in practice a mixture of semi-theoretical and semi-empirical formulas are used. Generally the carrier mobility depends on temperature, doping level, electric field (to be distinguished in parallel and perpendicular to the current flow) and surface or bulk location.

In principle for nonpolar semiconductors four scattering mechanisms have to be taken into account [2.18].

a. In low-doped bulk material acoustic phonon scattering prevails. In this case the mobility is given by

$$\mu_l = \text{constant} * m_e^{-5/2} T^{-3/2}.$$

However, for the majority of bulk devices the current is determined by higher values of the doping. A special case forms the inversion layer of MOSFET devices. Here the carrier motion is quantized in one direction and for such layers theory predicts.

$$\mu_s = \text{constant} * T^{-1}(Q_d + \tfrac{1}{2}Q_i)^{-1/3}, \quad (2.62)$$

where Q_d and Q_i are the charge in the depletion and inversion layer, respectively. Owing to this relation the mobility in MOSFET devices becomes dependent on the field normal to the surface (compare section 6.4).

b. For more highly doped substrates ($N > 10^{17}/\text{cm}^3$) Coulomb scattering by ionized impurities becomes dominant. At a reference temperature of 300 K the following empirical relation is commonly used [2.19]:

$$\mu_b(300, N) = \mu_{\min} + \frac{\mu_{\max} - \mu_{\min}}{1 + (N/N_{\text{ref}})^\alpha}. \quad (2.63)$$

The parameter values $\mu_{\min}$, $\mu_{\max}$, N_{ref} and α are given in Table 2.1 for electrons and holes. At very high doping levels ($N > 10^{20}$ cm^3) Eq. (2.63) gives too high a value for μ_b and we have to add a correction term to obtain

Table 2.1

	Electrons		Holes
	Arsenic	Phosphorus	Boron
$\mu_{\min}$	52.0	68.5	45.0
$\mu_{\max}$	1415	1415	470
μ_i	43.0	56.0	29.0
N_{ref}	9.70×10^{16}	9.20×10^{16}	2.20×10^{17}
N_1	3.40×10^{20}	3.40×10^{20}	6.10×10^{20}
α	0.68	0.71	0.72
β	2.00	1.98	2.00

2.5 Mobility and Resistivity in Silicon

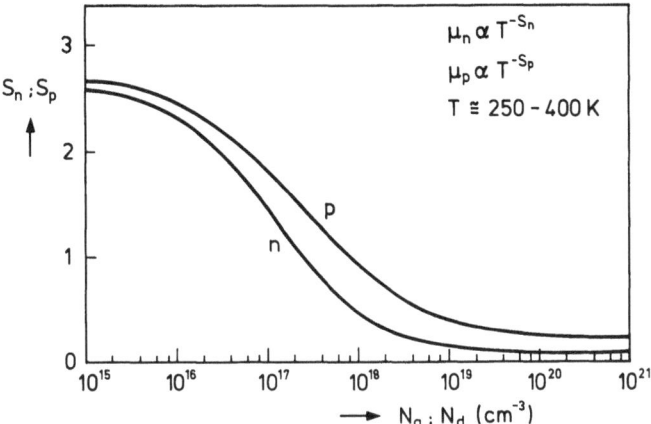

Fig. 2.6 The exponents for the power-law approximations of the mobility-temperature curves, as a function of dope concentration

a good fit with the measured values [2.20]:

$$\mu_b(300, N) = \mu_{min} + \frac{\mu_{max} - \mu_{min}}{1 + (N/N_{ref})^\alpha} - \frac{\mu_i}{1 + \left(\frac{N_1}{N}\right)^\beta}. \tag{2.64}$$

The additional parameters μ_i, N_1 and β are also given in Table 2.1.
In the temperature range 250–450 K the temperature dependence of the mobility can be approximated by (see also Eq. (2.39)):

$$\mu_b(T, N) = \mu_b(300, N)\left(\frac{T}{300}\right)^s. \tag{2.65}$$

The exponent s is different for holes and electrons ($s_n \neq s_p$) and dependent on the dope concentration; see Fig. 2.6.

c. For MOSFETS in strong inversion ($Q_i > 10^{-6} C/cm^2$) another scattering mechanism arises, which is assumed to be caused by surface roughness. In this case the evaluation of Eq. (2.37) predicts [2.18]:

$$\mu_s = \text{constant} * (Q_d + \tfrac{1}{2}Q_i)^{-2}. \tag{2.66}$$

Fig. 2.7 shows that the first and third mentioned scattering effects are present in MOSFET devices of several process generations (to be distinguished by different values of gate insulator thickness) [2.21].

d. The last effect, scattering by optical phonons, occurs when charge carriers gain sufficient energy in a parallel field to be scattered by optical phonons. This finally leads to velocity saturation. For majority carriers in silicon the related decrease of mobility with field is often expressed by the following empirical relation [2.19]:

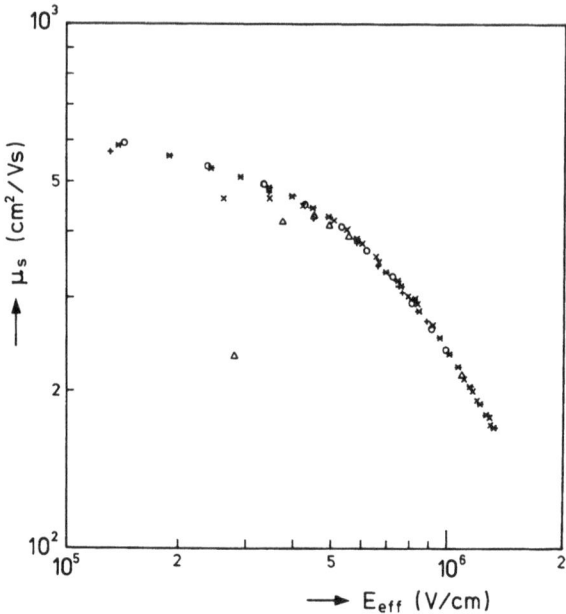

Fig. 2.7 Electron inversion layer mobility as a function of the perpendicular electric field, for n channels. The gate insulator has been varied between 50 nm (○) and 10 nm (×)

$$\mu = \frac{\mu_0(T, N, F_\perp)}{\left\{1 + \left(\frac{\mu_0 F_\|}{v_s}\right)^\beta\right\}^{1/\beta}}, \qquad (2.67)$$

where μ_0 is either μ_b or μ_s, or a combination of both according to Matthiessen's rule. For electrons we have roughly $\beta \approx 2$, for holes $\beta = 1$. v_s is the saturated drift velocity; in silicon $v_s = 10^7$ cm/s at 300 K.

The specific resistance of the material is given by

$$\left. \begin{array}{l} \rho = \dfrac{1}{q\mu_n(N_d^+ - N_a^-)}, \\[6pt] \rho = \dfrac{1}{q\mu_p(N_a^- - N_d^+)}, \end{array} \right\} \qquad (2.68)$$

for n- and p-type material, respectively; N_d^+ and N_a^- are the ionized impurity concentrations. The temperature dependence of ρ is determined by that of μ and by the variation of N_d^+ or N_a^- with temperature, see e.g. Eqs. (2.22) and (2.23).

The resistance of a layer with thickness h is often characterized by the so-called sheet resistance, that is the resistance of one square:

$$\rho_\square = \left\{q \int_0^h \mu(N) N^\pm(x)\, dx\right\}^{-1}. \qquad (2.69)$$

2.6 Recombination

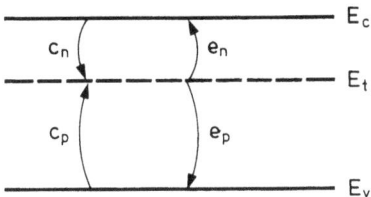

Fig. 2.8 Capture and emission of a trap in the forbidden zone at energy level E_t, interacting with the conduction and valence bands

2.6 Recombination

In silicon we can distinguish between recombination via defects with energy levels in the forbidden zone (Shockly-Read-Hall mechanism) and direct band-to-band recombination (Auger mechanism).

In the SRH mechanism [2.22] we have a defect or trap density N_t at an energy level E_t: see Fig. 2.8.

The trap or defect can capture electrons from the conduction band and emit electrons to the conduction band. The capture rate is proportional to n and $(N_t - n_t)$, the emission rate is proportional to n_t, where n_t is the concentration of trapped electrons. The distribution function for trapped electrons is

$$f_t = \frac{n_t}{N_t} = \frac{1}{\{1 + g_t^{-1} \exp\{(E_t - E_F)/k_B T\}}, \qquad (2.70)$$

where g_t is the trap degeneracy factor: $g_t \geqslant 1$.

The net capture rate of conduction band electrons can now be written as

$$R_n = c_n n (N_t - n_t) - e_n n_t,$$

with c_n and e_n as proportionality constants for capture and emission.

The net capture rate of valence band holes is $R_p = c_p p n_t - e_p(N_t - n_t)$. In equilibrium there is a detailed balance, that is $R_n = R_p = 0$. From this detailed balance it follows that

$$e_n = c_n n \left(\frac{N_t}{n_t} - 1\right) = c_n n \left(\frac{1}{f_t} - 1\right)$$

$$= c_n n \frac{1}{g_t} \exp\{(E_t - E_F)/k_B T\}.$$

In non-degenerate material we can use Boltzmann statistics and put

$$n = N_c \exp\{-(E_c - E_F)/k_B T\}.$$

Instead of c_n (or c_p) the capture cross sections σ_n (or σ_p) are often used, $c_{n,p} = \sigma_{n,p} v_{th}$, where $v_{th} = \sqrt{3k_B T/m}$ is the thermal velocity. We then get

$$e_n = \frac{\sigma_n v_{th} N_c}{g_t} \exp\{(E_t - E_c)/k_B T\}$$

or, assuming that σ_n is temperature-independent,

$$\left.\begin{aligned} e_n &= \text{constant} * T^2 \exp\{(E_t - E_c)/k_B T\} \\ &\text{and for} \\ e_p &= \text{constant} * T^2 \exp\{(E_v - E_t)/k_B T\}. \end{aligned}\right\} \quad (2.71)$$

We assume furthermore that Eq. (2.71) also holds in non-equilibrium situations and that then $R_n = R_p = R \neq 0$.
Finally we get for the net recombination rate

$$R = N_t \frac{c_n c_p pn - e_n e_p}{c_p p + c_n n + e_p + e_n}$$

or

$$R = \frac{pn - n_i^2}{\tau_{n_0}(p + p_1) + \tau_{p_0}(n + n_1)} \quad (2.72)$$

with

$$\tau_{n,p_0} = \frac{1}{\sigma_{n,p} v_{th} N_t},$$

$$p_1 = \frac{e_p}{c_p} = g_t N_v \exp\{-(E_t - E_v)/k_B T\},$$

$$n_1 = \frac{e_n}{c_n} = \frac{1}{g_t} N_c \exp\{-(E_c - E_t)/k_B T\}.$$

The SRH recombination has thus four parameters: σ_n, σ_p, N_t and $g_t \exp(-E_t/k_B T)$. These parameters are usually dependent on the doping levels and may vary with position.

An additional recombination mechanism for an indirect-gap semiconductor is the Auger recombination in which three particles are involved: a hole and an electron recombine directly and the energy release accelerates a third carrier [2.23]. This recombination rate is given by

$$R_{Au} = (C_n n + C_p p)(pn - n_i^2). \quad (2.73)$$

C_n and C_p are constants that vary sightly with temperature. At room temperature $C_n \approx C_p \approx 1.5 * 10^{-31} \text{ cm}^6 \text{ s}^{-1}$.

In heavily doped n-type material we now have as the total recombination rate

$$R = \left\{ C_n n + C_p p + \frac{1}{\tau_{n_0}(p + p_1) + \tau_{p_0}(n + n_1)} \right\} (pn - n_i^2)$$

$$\approx \left(C_n n + \frac{1}{\tau_{p_0} n} \right) pn$$

$$= \left(C_n n^2 + \frac{1}{\tau_{p_0}} \right) p$$

$$= p/\tau_{\text{eff}}.$$

2.7 Avalanche Multiplication

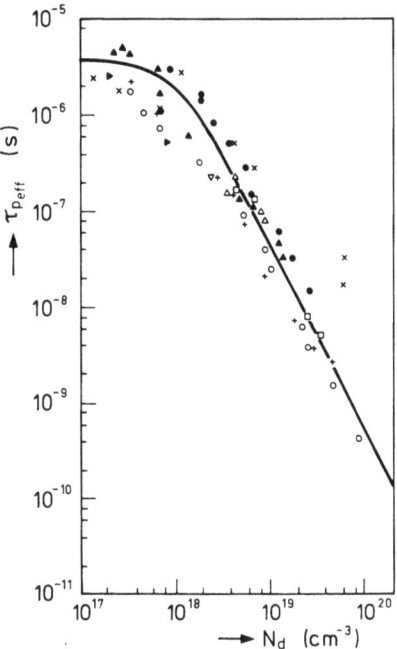

Fig. 2.9 Effective hole lifetime ($\tau_{p_{\text{eff}}}$) in heavily doped n-type silicon, as a function of the dope concentration

The effective hole lifetime in N^+ regions therefore becomes

$$\tau_{p_{\text{eff}}} \approx \frac{\tau_{p_0}}{1 + C_n \tau_{p_0} n^2} \,. \tag{2.74}$$

Eq. (2.74) is plotted in Fig. 2.9 against $n = N_d$ for $C_n = 1.5 * 10^{-31}$ cm^6 s^{-1} and $\tau_{p_0} = 4 * 10^{-6}$ s, and compared with experimental values [2.24].
The recombination at interfaces and surfaces is often characterized by a recombination velocity (s):

$$\left. \begin{array}{l} J_p = qs(p - p_0) \\ \\ J_n = qs(n - n_0). \end{array} \right\} \tag{2.75}$$

or

Here p_0 and n_0 are the equilibrium concentrations.

2.7 Avalanche Multiplication

The counterpart of recombination is generation of electron-hole pairs. The thermal generation is already described by the expression for the net recombination rate (Eq. (2.72)) when $pn < n_i^2$. It may also occur by means of light irradiation or electron beam bombardment, but the most important phenomenon in most silicon devices is the electron-hole pair generation by means of impact ionization, which, physically speaking, is the inverse of Auger recombination [2.25]. Electrons and/or holes gain so much energy in a high electric field that they can create extra electron-hole pairs by exciting electrons from the valence band into the conduction band. In this way an avalanche of free carriers may arise and the initial flux of cariers is multiplied, until complete breakdown occurs.

The generation rate G_{av} for this avalanche multiplication is given by

$$G_{av} = \alpha_n n v_n + \alpha_p p v_p$$

$$= \alpha_n \frac{J_n}{q} + \alpha_p \frac{J_p}{q}. \tag{2.76}$$

The ionization coefficients α_n and α_p are strongly dependent on the magnitude of the electric field F.

The simplest expression for this dependence is the semi-empirical formula of Chynoweth [2.26]:

$$\alpha_{n,p} = A_{n,p} \exp\left(-\frac{b_{n,p}}{|F|}\right). \tag{2.77}$$

Different sets of values for the parameters A and b can be found in the literature. We will use the values as given in [2.27]:

$A_n = 7.03 * 10^5$ cm^{-1} and $b_n = 1.23 * 10^6$ V cm^{-1},

$A_p = 1.58 * 10^6$ cm^{-1} and $b_p = 2.04 * 10^6$ V cm^{-1}.

The values are obtained from measurements in bulk regions. Recently measurements have also been performed in surface layers, which resulted in different values [2.28]:

$A_n = 2.45 * 10^6$ cm^{-1} and $b_n = 1.92 * 10^6$ V cm^{-1}.

In order to calculate the total generated avalanche current, I_{av}, let us assume that an initial current, $I_n(0)$, of electrons enters at $x = 0$ in Fig. 2.10 a region with high electric fields and avalanche multiplication.

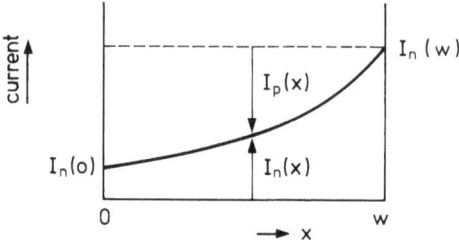

Fig. 2.10 Avalanche multiplication of an initial electron current $I_n(0)$ to the value $I_n(w)$

2.7 Avalanche Multiplication

This current leaves that region at $x = w$ and has then grown to $I_n(w) = I = M_n I_n(0)$, where M_n is called the multiplication factor for electrons. A hole current has also been generated, in such a way that everywhere $I = I_n(x) + I_p(x) = $ constant.

According to (2.76) we have

$$dI_n = \alpha_n I_n \, dx + \alpha_p I_p \, dx = (\alpha_n - \alpha_p) I_n \, dx + \alpha_p I \, dx$$

or

$$\frac{dI_n}{dx} = (\alpha_n - \alpha_p) I_n + \alpha_p I. \qquad (2.78)$$

The avalanche current $I_{av} = I - I_n(0) = [1 - (1/M_n)]I$. Solving the differential equation (2.78) leads to

$$\left(1 - \frac{1}{M_n}\right) = \int_0^w \alpha_n \exp\left\{-\int_0^x (\alpha_n - \alpha_p) \, dx'\right\} dx. \qquad (2.79)$$

We will now limit ourselves to the case of weak avalanche, that is

$$I_{av} = I - I_n(0) = I\left(1 - \frac{1}{M_n}\right) \approx I_n(0)\left(1 - \frac{1}{M_n}\right).$$

The quantity $[1 - (1/M_n)]$ is given by Eq. (2.79). For $\alpha_n = \alpha_p$, as is the case in GaAs, we then get

$$I_{av} = I_n(0) \int_0^w \alpha_n \, dx. \qquad (2.80)$$

In silicon we have $\alpha_p \approx \gamma \alpha_n$ with $\gamma \approx 0.1$, so

$$1 - \frac{1}{M_n} = \int_0^w \alpha_n \exp\left\{-(1 - \gamma) \int_0^x \alpha_n \, dx'\right\} dx$$

$$= \frac{1}{1 - \gamma}\left[1 - \exp\left\{-(1 - \gamma) \int_0^w \alpha_n \, dx\right\}\right]$$

$$\approx \int_0^w \alpha_n \, dx,$$

if

$$\int_0^w \alpha_n \, dx \ll 1.$$

(The weak avalanche condition again!) We thus get the same result as in Eq. (2.80).

If the initial current is a hole current $I_p(0)$, we must integrate α_p over the avalanche region.

In complete breakdown situations $M_{n,p}$ is infinite and the ionization integral in Eq. (2.79) equals one. Such a situation cannot be described by a weak avalanche formula as Eq. (2.80).

2.8 Noise Sources

Since transistors show spontaneous fluctuations in the terminal currents, we discuss here several mechanisms underlying these fluctuations.

2.8.1 Shot Noise

Let a stationary random variable $x(t)$ be the sum of a large number of independent events $F_i(t)$ occurring at random at an average rate n, so that

$$x(t) = \sum_i F(t - t_i),$$

where $F(t - t_i) = 0$ for $t < t_i$ and F represents the event starting at $t = t_i$. Since $\langle x(t) \rangle = 0$, usually the variance

$$\langle \Delta x^2(t) \rangle = \langle (x(t) - \langle x \rangle)^2 \rangle$$

is taken as a measure for the noise.
Making use of spectral analysis, where

$$\langle \Delta x^2(t) \rangle = \int_0^\infty S_x(f) \, df,$$

it can be proved that the spectral density $S_x(f)$ is given by [2.29]

$$S_x(f) = 2n |\varphi(f)|^2,$$

where

$$\varphi(f) = \int_{-\infty}^{+\infty} F(t) \exp(-j\omega t) \, dt$$

is the Fourier transform of $F(t)$. This result is known as Carson's theorem. As an example we consider the fluctuations in the number n of electrons emitted per second by a thermionic process. Since the electrons are emitted independently and at random, n follows a Poisson distribution. Since $F(t - t_i)$ can be considered as a delta function, $|\varphi(f)| = 1$ and the spectral density of the current $I = -qn(t)$ becomes

$$S_i(f) = 2q \langle I \rangle. \tag{2.81}$$

This is known as Schottky's theorem [2.30]. It holds for any current consisting of independent events all carrying the charge $-q$. It therefore not only holds for saturated thermionic diodes, but also for carriers crossing potential barriers, such as in $p - n$ junction diodes or bipolar transistors. Due to transit time effects however, the noise increases rapidly at high frequencies.

2.8 Noise Sources

2.8.2 Diffusion Noise and Thermal Noise

In any conductor at zero bias spontaneous voltage fluctations are observed at the terminals. This is caused by the random motion of the carriers due to collisions with the lattice vibrations. Assuming the conductor to be divided up into rectangular boxes $\Delta x\, \Delta y\, \Delta z$, the electrons can make random jumps from one box to an adjacent one with individual jumps being treated as independent. Considering two adjacent boxes (k, l, m) and $(k + 1, l, m)$ the particle current from (k, l, m) to $(k + 1, l, m)$ is given by [2.31]

$$\Phi_{k,k+1} = a n(k, l, m)\, \Delta x\, \Delta y\, \Delta z,$$

where a is a constant, $n(k, l, m)$ is the electron concentration and in the reverse direction

$$\Phi_{k+1,k} = a\left[n(k, l, m) + \frac{\partial n}{\partial x}\bigg|_{k,l,m} \Delta x \right] \Delta x\, \Delta y\, \Delta z.$$

However, since the net particle current

$$\Phi = \Phi_{k,k+1} - \Phi_{k+1,k} = -a \frac{\partial n}{\partial x}\bigg|_{k,l,m} \Delta x^2\, \Delta y\, \Delta z$$

must be independent of the way in which the semiconductor is divided, $a \cdot \Delta x^2$ must be a constant. Calling this constant the electron diffusion constant D_n, we have

$$\Phi = -D_n \frac{\partial n}{\partial x}\bigg|_{k,l,m} \Delta y\, \Delta z.$$

Hence

$$\Phi_{k,k+1} = D_n n(k, l, m) \frac{\Delta y\, \Delta z}{\Delta x}.$$

Since the currents $\Phi_{k,k+1}$ and $\Phi_{k+1,k}$ are made up of independent jumps occurring at random, both currents should show full shot noise. Therefore the spectral current density is given by

$$S_i(f) = q^2(2\Phi_{k,k+1} + 2\Phi_{k+1,k}) = 4q^2 D_n n(k, l, m)\frac{\Delta y\, \Delta z}{\Delta x}.$$

In this form the relation is known as diffusion noise.
When the conductor is in thermal equilibrium the diffusion noise reduces to thermal noise and the above relation may be rewritten, using Einstein's relation (Eq. (2.41)):

$$S_i(f) = 4k_B T q\mu n \frac{\Delta y\, \Delta z}{\Delta x} = \frac{4k_B T}{\Delta R}, \qquad (2.82)$$

where ΔR is the resistance of the box $(\Delta x\, \Delta y\, \Delta z)$.

In the latter form the relation is known as Nyquist's formula [2.32]. Since the noise does not depend on the frequency up to $f \approx 10^{12}$ Hz, an alternative name is white noise.

The Nyquist theorem holds for any conductor at thermal equilibrium and can therefore be applied to the channel of field-effect transistors.

It also holds for most conductors through which d.c. current is flowing, as long as the carrier density in the sample does not fluctuate. In the latter case excess noise is observed (often of the generation-recombination type). When the field effect device is subjected to high fields, so that hot electron effects occur, Einstein's relation no longer holds and the noise must be treated using the more general expression for diffusion noise.

2.8.3 Flicker Noise

Although there are several mechanisms causing resistance fluctuations, which are observed under biased conditions, in most practical devices only flicker noise appears to be of interest. Unfortunately the exact mechanism is as yet unknown. In some cases the resistance fluctuations have been explained by assuming trapping of charge carriers at the semiconductor interface [2.33], in other cases the observed noise could be better interpreted in terms of mobility fluctuations [2.34]. However the frequency dependence of the spectral density, which is usually proportional to $1/f$, requires additional detailed mechanisms to be explained correctly.

In most cases the noise spectrum of a current flowing through a segment Δx can be expressed as follows:

$$S_i(x,f) = \frac{\alpha_H}{fn(x)\,\Delta x} I^2(x), \tag{2.83}$$

where $n(x)$ is the carrier density and α_H is called the Hooge constant. Owing to the lack of an exact physical mechanism this constant should be considered as an empirical constant [2.35].

In addition to the above type of noise, planar transistors show a type of l.f. noise known as burst noise. Strictly this phenomenon is not noise in the original sense, since it is most probably caused by a random flow of several thousands of electrons via a local breakdown of the emitter-base potential barrier. Generally the associated spectrum differs from the $1/f$ type.

References

[2.1] M. H. B. Stiddard: The Elementary Language of Solid State Physics. Academic Press, London (1975).
[2.2] S. M. Sze: Physics of Semiconductor Devices, 2nd edn. John Wiley & Sons, New York (1981).
[2.3] J. S. Blakemore: Solid State Physics, 2nd edn. Cambridge University Press, Cambridge (1985).

References

[2.4] L. A. Pipes: Applied Mathematics for Engineers and Physicists. McGraw-Hill, New York (1958).
[2.5] R. A. Smith: Semiconductors. Cambridge University Press, Cambridge (1961).
[2.6] J. Lindmayer, Ch. Y. Wrigley: Fundamentals of Semiconductor Devices. Van Nostrand, Princeton (1965).
[2.7] S. S. Li, W. R. Thurber: The Dopant Density and Temperature Dependence of Electron Mobility and Resistivity in n-Type Silicon. Solid-State Electronics 20, 609–616 (1977); S. S. Li: The Dopant Density and Temperature Dependence of Hole Mobility and Resistivity in Boron Doped Silicon. Solid-State Electronics 21, 1109–1117 (1987).
[2.8] B. R. Nag: Theory of Electrical Transport in Semiconductors. Pergamon Press, Oxford (1972).
[2.9] A. H. Marshak, C. M. van Vliet: Electrical Current and Carrier Density in Degenerate Materials with Non-Uniform Band Structure. Proc. IEEE 72, 148–164 (1984).
[2.10] G. Baccarani, M. Rudan, R. Guerrieri, P. Ciampolini: Physical Models for Numerical Device Simulation. In: Process and Device Modeling, (W.L. Engl., ed.). North-Holland, Amsterdam (1986).
[2.11] R. Stratton: Diffusion of Hot and Cold Electrons in Semiconductor Barriers. Phys. Rev. 126, 2002–2014 (1962).
[2.12] G. Baccarani, M. R. Wordeman: An Investigation of Steady-State Velocity Overshoot in Silicon. Solid-State Electronics 28, 407–416 (1985).
[2.13] E. H. Putley, W. H. Mitchell: The Electrical Conductivity and Hall Effect of Silicon. Proc. Phys. Soc. London $A72$, 193–200 (1958).
[2.14] J. W. Slotboom, H. C. de Graaff: Measurement of Bandgap Narrowing in Si Bipolar Transistors. Solid-State Electronics 19, 857–862 (1976).
[2.15] J. del Alamo, S. Swirhun, R. M. Swanson: Simultaneous Measurement of Hole Lifetime, Hole Mobility and Bandgap Narrowing in Heavily Doped n-Type Silicon. IEDM Techn. Digest 290–293 (1985).
[2.16] M. S. Bennett: Improved Concepts for Predicting the Electrical Behavior of Bipolar Structures in Silicon. IEEE Trans. Electr. Dev. ED-30, 920–927 (1983).
[2.17] W. Shockley: Electrons and Holes in Semiconductors. Van Nostrand, Princeton (1950).
[2.18] T. Ando, A. B. Fowler, F. Stern: Electronic Properties of 2D Systems. Rev. Modern Phys. 54, 437–671 (1982).
[2.19] D. M. Caughey, R. E. Thomas: Carrier Mobilities in Silicon Empirically Related to Doping and Field. Proc. IEEE 52, 2192–2193 (1967).
[2.20] W. E. Beadle, J. C. C. Tsai, R. D. Plummer: Quick Reference Manual for Silicon Integrated Circuit Technology. John Wiley & Sons, New York (1985).
[2.21] A. J. Walker, P. H. Woerlee: Mobility Model for Silicon Inversion Layers, Proc. ESSDERC 1987. North-Holland, Amsterdam (1987), pp. 667–670.
[2.22] W. Shockley, W. T. Read: Statistics of the Recombination of Holes and Electrons. Phys. Rev. 87, 835–842 (1952); R. N. Hall: Electron-Hole Recombination in Germanium. Phys. Rev. 87, 387 (1952).
[2.23] J. Dziewior, W. Schmid: Auger Coefficient for Highly Doped and Highly Excited Silicon. Appl. Phys. Lett. 31, 346–348 (1977).
[2.24] J. del Alamo: Minority Carrier Transport in Heavily Doped n-Type Silicon. Ph. D. Thesis, Stanford (1985).
[2.25] G. A. Baraff: Distribution Function and Ionization Rates for Hot Electrons in Semiconductors. Phys. Rev. 128, 2507–2517 (1962).
[2.26] A. G. Chynoweth: Ionization Rates for Electrons and Holes in Silicon. Phys. Rev. 109, 1537–1540 (1958).
[2.27] R. J. van Overstraeten, H. J. de Man: Measurement of the Ionization Rates in Diffused Silicon $p-n$ Junctions. Solid-State Electronics 13, 583–608 (1970).
[2.28] J. W. Slotboom, G. Streutker, G. J. T. Davids, P. B. Hartog: Surface Impact Ionization in Silicon Devices. IEDM Techn. Digest 494 (1987).

[2.29] A. v. d. Ziel: Noise; Sources, Characterization, Measurements. Prentice Hall, Englewood Cliffs (1970).
[2.30] W. Schottky: Über spontane Stromschwankungen in Elektricitätsleitern. Ann. Physik 57, 541 (1918).
[2.31] A. G. Th. Becking: For details see [2.29].
[2.32] H. Nyquist: Thermal Agitation of Electric Charge in Conductors. Phys. Rev. 32, 110 (1928).
[2.33] A. L. McWhorter: $1/f$ Noise and Related Surface Effects. Ph.D. Thesis, MIT (1955).
[2.34] T. G. M. Kleinpenning: $1/f$ Noise in Thermo emf of Semiconductors. Physica 77, 78 (1974).
[2.35] F. N. Hooge: $1/f$ Noise is no Surface Effect. Phys. Lett. 29A, 139 (1969).

Modelling of Bipolar Device Phenomena 3

In this chapter we will first discuss some general problems in bipolar device modelling, namely the choice between injection and transport models and the validity of the charge control principle. After that we will show how the various device phenomena like main currents, Early effect, depletion capacitance etc., can be described by means of compact, explicit and analytical mathematical expressions. Unless stated otherwise, the device structure considered here is that of a vertical *npn* transistor. In most cases the vertical *pnp* transistor only needs a change of sign in its model formulas. The lateral *pnp* transistor, which is quite different, will be treated in a separate chapter.

3.1 Injection and Transport Models

3.1.1 Solution of the Continuity Equations

The continuity equation for electrons is given in chapter 2 as (Eq. (2.48))

$$\frac{\partial n}{\partial t} - \frac{1}{q} \nabla_r \cdot \mathbf{J}_n = G - R$$

and the current density $\mathbf{J}_n$ as (Eq. (2.43))

$$\mathbf{J}_n = q\mu_n n \mathbf{F} + qD_n \nabla_r n.$$

We will apply these equations to the neutral base region of a bipolar device under the following simplifying assumptions:

— current flow is one-dimensional and dominated by the diffusion component;
— the depletion layer edges of the e-b and b-c junctions are abrupt and situated at $x = 0$ and $x = W_b$, respectively;
— the region in between ($0 < x < W_b$) is (quasi-)neutral and $p(x) \approx N_a + n(x)$;
— the dope concentration is constant;
— the right-hand side of the continuity equation is determined by recombination and can be written as $G - R = -(n - n_0)/\tau_{n_0}$. This follows

from Eq. (2.72): under low injection conditions pn is equal to $N_a n$ and $\tau_{no}(p + p_1) \approx \tau_{no} N_a \gg \tau_{po}(n + n_1)$.

With all these assumptions Eqs. (2.43) and (2.48) lead to the following differential equation for the excess minority carrier concentration $(n - n_0)$ in the static case:

$$D_n \frac{d^2(n - n_0)}{dx^2} = \frac{n - n_0}{\tau_{no}}. \tag{3.1}$$

The general solution of Eq. (3.1) is

$$n(x) - n_0 = A \exp\left(-\frac{x}{L_n}\right) + B \exp\left(\frac{x}{L_n}\right),$$

where $L_n = \sqrt{D_n \tau_{no}}$ is called the diffusion-recombination length. The integration constants A and B are determined by the boundary conditions:

at $x = 0$:
$$n(0)\{N_a + n(0)\} = n_i^2 \exp(V_{be}/U_T) \tag{3.2}$$

and

at $x = W_b$:
$$n(W_b) \approx 0. \tag{3.3}$$

Under low injection conditions $n(0) \ll N_a$ and

$$n(0) \approx \frac{n_i^2}{N_a} \exp(V_{be}/U_T) \quad \text{or} \quad n(0) - n_0 = \frac{n_i^2}{N_a} \{\exp(V_{be}/U_T) - 1\}.$$

The voltage $U_T = k_B T/q$, where T denotes the temperature of the device in Kelvin. The boundary condition (3.3) implies that the electrons at the b-c depletion edge have infinite velocity. The maximum velocity, however, is the saturated drift velocity v_{sat} ($= 10^7$ cm/s in silicon), as follows from Eq. (2.67). So the condition (3.3) is not exactly true; if we take $n(W_b) - n_0 = 0$ we are not far from reality. With these modified boundary conditions we get from the general solution

$$n(x) - n_0 = \{n(0) - n_0\} \frac{\sinh\left(\dfrac{W_b - x}{L_n}\right)}{\sinh\left(\dfrac{W_b}{L_n}\right)} \tag{3.4}$$

and

$$J_n = qD_n \frac{dn}{dx}\bigg|_{x=0}$$

or

$$J_n(x) = -qD_n \frac{n_i^2}{N_a L_n} \frac{\cosh\left(\dfrac{W_b - x}{L_n}\right)}{\sinh\left(\dfrac{W_b}{L_n}\right)} \{\exp(V_{be}/U_T) - 1\}. \tag{3.5}$$

J_n flows in the negative x direction.

3.1 Injection and Transport Models

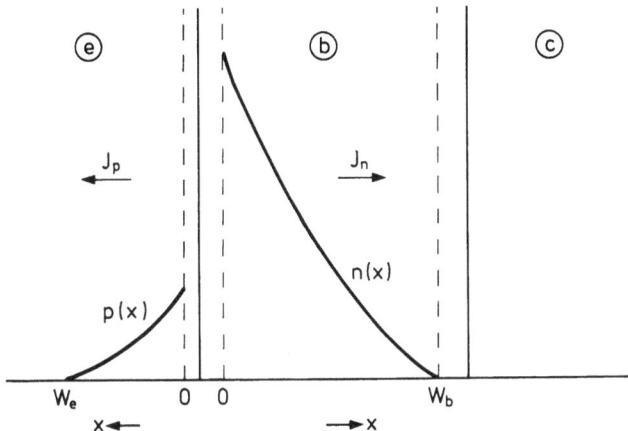

Fig. 3.1 Minority carrier distributions in the emitter and base of an *npn* transistor with homogeneous doping profiles

For the hole current in the emitter we can derive a similar expression, with the same simplifying assumptions, including an infinite recombination velocity (see Eq. (2.75)) at the contact:

$$J_p(x) = qD_p \frac{n_i^2}{N_d L_p} \frac{\cosh\left(\frac{W_e - x}{L_p}\right)}{\sinh\left(\frac{W_e}{L_p}\right)} \{\exp(V_{be}/U_T) - 1\}. \tag{3.6}$$

The situation is sketched in Fig. 3.1.

Because the doping levels in emitter and base are quite different, the quantity n_i^2 in Eqs. (3.5) and (3.6) is not the same, due to the effect of bandgap narrowing (see section 2.3). We can introduce different Gummel numbers [3.1, 3.2] for base and emitter:

$$G_b = \frac{n_{i0}^2}{n_{ib}^2} \frac{N_a}{D_n} L_n \tanh\left(\frac{W_b}{L_n}\right)$$

and

$$G_e = \frac{n_{i0}^2}{n_{ie}^2} \frac{N_d}{D_p} L_p \tanh\left(\frac{W_e}{L_p}\right).$$

These Gummel numbers indicate a total effective impurity charge, divided by a diffusion constant.

Multiplying the current densities by the emitter area gives for the total d.c. currents (as positive quantities)

$$I_e = \{-J_n(0) + J_p(0)\} A_{em} = I_0\{\exp(V_{be}/U_T) - 1\}$$

with

$$I_0 = qn_{i0}^2 \left(\frac{1}{G_e} + \frac{1}{G_b}\right) A_{em}, \tag{3.7}$$

$$I_c = -J_n(W_b) A_{em} = -\alpha_f J_n(0) A_{em} = qn_{i0}^2 \frac{\alpha_f}{G_b} A_{em}\{\exp(V_{be}/U_T) - 1\} \tag{3.8}$$

with

$$\alpha_f = \frac{1}{\cosh\left(\dfrac{W_b}{L_n}\right)}$$

and

$$I_b = \{J_p(0) - J_n(0) + J_n(W_b)\} A_{em}$$

$$= qn_{i0}^2 \left\{\frac{1}{G_e} + (1 - \alpha_f)\frac{1}{G_b}\right\} A_{em}\{\exp(V_{be}/U_T) - 1\}. \tag{3.9}$$

In Eq. (3.9) the term with $J_p(0)$ indicates the hole current that recombines in the emitter, whereas $J_n(0) - J_n(W_b)$ represents recombination in the base. Usually $G_e \gg G_b$, so from Eq. (3.7) it follows that

$$I_e \approx \frac{qn_{i0}^2}{G_b} A_{em}\{\exp(V_{be}/U_T) - 1\}.$$

3.1.2 Injection Model

If recombination in the base dominates I_b, that is if $(1 - \alpha_f)/G_b \gg 1/G_e$, Eqs. (3.8) and (3.9) become

$$\left.\begin{array}{l} I_c = \alpha_f I_e \\[4pt] I_b = (1 - \alpha_f) I_e. \end{array}\right\} \tag{3.10}$$

and

Eq. (3.10) represents the injection type of model: only part of the injected emitter current arrives at the collector, the rest equals the base current. The current gain in the common emitter configuration is

$$h_{FE} = \frac{\alpha_f}{1 - \alpha_f}.$$

3.1.3 Transport Model

The foregoing is an adequate description for transistors with a rather thick base region, $W_b \geqslant 0.15 L_n$. However, in most modern types of transistors L_n

3.1 Injection and Transport Models

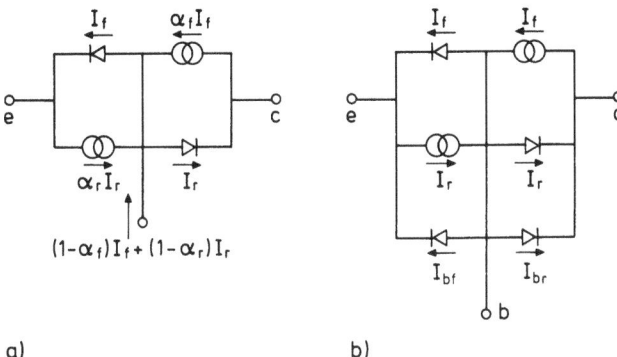

Fig. 3.2 Simple equivalent circuit diagrams for a. the injection model, b. the transport model

is about 20 to 50 μm and we easily obtain $W_b < 0.1\, L_n$. In that case we have $\cosh(W_b/L_n) \approx 1$ and $\alpha_f = 1$, so

and
$$\left.\begin{array}{l} I_c \approx \dfrac{qn_{i0}^2}{G_b} A_{em}\{\exp(V_{be}/U_T) - 1\} \\[2ex] I_b \approx \dfrac{qn_{i0}^2}{G_e} A_{em}\{\exp(V_{be}/U_T) - 1\}. \end{array}\right\} \quad (3.11)$$

Eq. (3.11) is the formulation for the transport type of model. Its base current is given by the hole current injected into the emitter, recombination in the base region is neglected. The current gain h_{FE} equals G_e/G_b.

Fig. 3.2.a gives the circuit diagram for the injection model not only for the normal forward mode of operation (α_f and I_f) as was discussed here, but also for the reverse mode (α_r and I_r) when the b-c junction is forward biased and the e-b junction is reverse biased. Fig. 3.2.b shows the same for the transport model; the base currents (I_{bf} and I_{br}) are now given by equations like Eq. (3.11).

As the transport models are more appropriate to modern bipolar transistors, we will restrict the discussion, throughout the remainder of the bipolar part of this book, to models of that type.

In transport models we have $W_b \ll L_n$, so

$$G_b \approx \left(\frac{n_{i0}}{n_{ib}}\right)^2 \frac{N_a W_b}{D_n}.$$

When the doping profile is not constant n_{ib}, N_a and D_n are position-dependent (cf. Eqs. (2.41), (2.55) and (2.63)) and the more general expression for G_b becomes

$$G_b = \int_0^{W_b} \left(\frac{n_{i0}}{n_{ib}}\right)^2 \frac{N_a(x)}{D_n(x)}\, dx. \quad (3.12)$$

The general expression for the emitter Gummel number is more complex. If the emitters are not completely transparent for holes ($W_e \sim L_p$), $\tanh(W_e/L_p) \neq W_e/L_p$ and has to be replaced by a weighting function $g(x) = J_p(x)/J_p(0)$ [3.2].

Also the boundary condition $p(W_e) = 0$ is not necessarily true: s (Eq. (2.75)) can be relatively low. Taking all this into account we get for G_e [3.2]:

$$G_e = \frac{1}{s}\left[\left(\frac{n_{i0}}{n_{ie}}\right)^2 g N_d\right]_{x=W_e} + \int_0^{W_e}\left(\frac{n_{i0}}{n_{ie}}\right)^2 \frac{N_d(x)}{D_p(x)} g(x)\, dx. \qquad (3.13)$$

For completely transparent emitters $g(x) \equiv 1$.

3.2 The Quasi-Static Approximation and the Charge Control Principle

Expressions for the currents I_e, I_c and I_b as functions of the applied voltage V_{be}, like Eqs. (3.7) to (3.11), were derived for the static, d.c. situation. If V_{be} varies with time, all the non-equilibrium quantities like currents and carrier concentrations will also vary with time and it is no longer true that $\partial n/\partial t = 0$, as was assumed in Eq. (3.1). So the solution of the differential equation (3.1) is no longer given by Eq. (3.4) if $V_{be} = V_{be}(t)$ and the same holds for the other Eqs. (3.7) to (3.11) which are based on that solution.

The problem is now that the exact solutions of the differential equation (3.1), including the $\partial n/\partial t$ term, become quite complicated, in particular the transient solutions in the time domain, but also the small-signal a.c. solutions in the frequency domain. To simplify things it is common use in compact device modelling for circuit analysis purposes to assume that the given solution of Eq. (3.1) is still valid for $V_{be} = V_{be}(t)$.

The same is then true for the current equations (3.7) to (3.11); this is called the "quasi-static approximation". In its general form it states that all static, d.c. relations between currents, charges and applied voltages can, as an approximation, also be used in dynamic, time-dependent cases. As a consequence the small-signal a.c. relations are obtained by linearizing the d.c. ones.

In addition to this quasi-static approximation the charge control principle is also used in compact modelling of bipolar devices: It prescribes how currents and charges are related. The starting point is the continuity equation for holes (in an *npn* transistor; in a *pnp* we take the continuity equation for electrons), see Eq. (2.50):

$$\frac{\partial p}{\partial t} + \frac{1}{q}\nabla_r \cdot \mathbf{J}_p = G - R.$$

Multiplying both sides by q and integrating over the volume of the device we have:

3.2 The Quasi-Static Approximation and the Charge Control Principle

$$\frac{\partial}{\partial t} q \int p \, dx \, dy \, dz + \int \nabla_r \cdot \mathbf{J}_p \, dx \, dy \, dz = q \int (G - R) \, dx \, dy \, dz.$$

The first term on the left-hand side is the rate of change of the total hole charge Q_p within the device. The second term is the net hole current flow $I_{ep} - I_b$, where I_{ep} is the hole current at the emitter contact. The term on the right-hand side represents the recombination current I_{rec}. So we have

$$I_b(t) = I_{ep}(t) + I_{rec}(t) + \frac{dQ_p(t)}{dt}. \tag{3.14}$$

I_{ep} represents the recombination of holes at the emitter contact. With the help of Eq. (3.6) we get

$$I_{ep}(t) = \frac{qn_{i0}^2}{G_e} \frac{A_{em}}{\cosh\left(\frac{W_e}{L_p}\right)} [\exp\{V_{be}(t)/U_T\} - 1]. \tag{3.15}$$

If the emitters are fully transparent to holes ($W_e/L_p \ll 1$) then $\cosh(W_e/L_p) \approx 1$. In the opposite case ($W_e/L_p \gg 1$) I_{ep} becomes very small. The bulk recombination current I_{rec} comprises both base and emitter recombination. If the latter dominates, $I_{ep}(t) + I_{rec}(t)$ is approximately given by Eq. (3.11) and the general, time-dependent base current expression becomes

$$I_b(t) \approx I_{b0}[\exp\{V_{be}(t)/U_T\} - 1] + \frac{dQ_p(t)}{dt}. \tag{3.16}$$

We may say that the total base current consists of a recombination component, formulated by means of the quasi-static approximation, plus a charging current component dQ_p/dt.

The collector current I_c can also be related to an injected charge. If we consider the neutral base region (see Fig. 3.1) we can write for the carrier transit time

$$\tau_b = \int_0^{W_b} \frac{dx}{v_{dr}} = -\int_0^{W_b} \frac{qn}{J_n} dx = \frac{-1}{J_n} \int_0^{W_b} qn \, dx$$

or

$$|J_n| = \frac{Q_n}{\tau_b}. \tag{3.17}$$

In the derivation of Eq. (3.17) use has been made of $J_n = -qnv_{dr}$, where v_{dr} is the carrier drift velocity.

Eq. (3.17) is a concise formulation of the charge control principle, which states that the current and its pertaining charge are linearly proportional to each other and that the proportionality constant is given by the transit time of the carriers in the region considered [3.4, 3.5]. We will also apply the quasi-static approximation to Eq. (3.17), so $J_n(t) = Q_n(t)/\tau_b$.

Many compact bipolar models are based on equations like (3.16) and (3.17). We then use d.c. relations for the charges as functions of the applied voltages, assume that they are valid for the time-dependent case as well and then use Eqs. (3.16) and (3.17) to calculate I_b and I_c. The errors made in this way arise from the time delays and phase shifts, on the one hand between the currents and their related charges and on the other hand between the charges and the applied voltages. A more detailed discussion is given in section 3.10 on time and frequency dependent behaviour.

3.3 Collector Currents and Stored Charges

We will derive in the first place a general relation between the collector current and the charges in a bipolar transistor. The result will then be applied to a special formula, the integral charge control relation, also using the charge control principle. This section ends with an alternative approach that avoids the charge control principle.

3.3.1 General Relation Between Collector Current and Charges

Let us consider a one-dimensional *npn* structure and let us apply Eq. (2.36) to the current density J_n in that structure. Again using Eq. (2.59) we then get

$$J_n = -q\mu_n n \frac{d\varphi_n}{dx} = -q\mu_n n_i \exp(\psi/U_T) \exp(-\varphi_n/U_T) \frac{d\varphi_n}{dx}.$$

Integration from emitter to collector gives

$$\int_e^c J_n \frac{\exp(-\psi/U_T)}{q\mu_n n_i} dx = -\int_e^c \exp(-\varphi_n/U_T) \, d\varphi_n$$

$$= -U_T \{\exp(-\varphi_{ne}/U_T) - \exp(-\varphi_{nc}/U_T)\}. \tag{3.18}$$

In Fig. 3.3 a typical distribution of φ_p, φ_n and ψ is sketched for the forward mode of operation.

The difference in quasi-Fermi levels on each side of a *pn* junction equals the applied voltage. We again neglect the recombination in the base region (transport model), so we put $J_p = 0$. According to Eq. (2.52) we then have $\nabla_r \varphi_p = 0$ in the base region, or $\varphi_p(x) = $ constant, where the constant might just as well be taken zero. As a result, the forward voltage applied to the *b-e* junction is

$$-V_{be} = \varphi_{ne} - \varphi_{pb} = \varphi_{ne}$$

and the reverse voltage of the *b-c* junction is

$$V_{cb} = \varphi_{nc} - \varphi_{pb} = \varphi_{nc}.$$

3.3 Collector Currents and Stored Charges

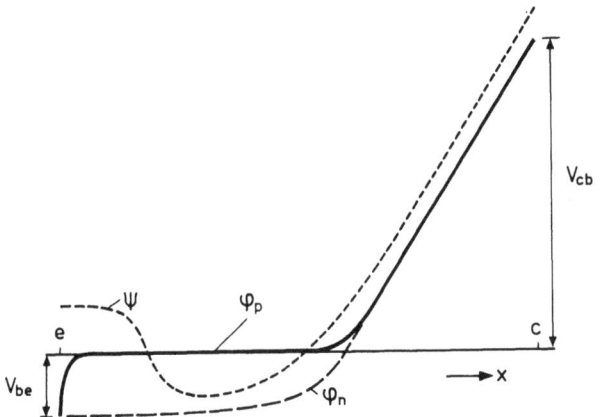

Fig. 3.3 One-dimensional distribution functions for the quasi Fermi-levels $\varphi_p(x)$ and $\varphi_n(x)$ and for the electrostatic potential $\psi(x)$ in an npn transistor in the forward mode of operation $(V_{be} > 0, V_{cb} > 0)$

Neglect of generation and recombination also makes J_n constant, so Eq. (3.18) can be rewritten as

$$J_n \int_e^c \frac{\exp(-\psi/U_T)}{qD_n n_i} dx = -\{\exp(V_{be}/U_T) - \exp(-V_{cb}/U_T)\}.$$

Although the integration on the left-hand side of this equation goes from emitter to collector, only the base region makes a significant contribution, but here $\varphi_p = 0$, so we can easily add φ_p to the exponent of the integrand:

$$\int_e^c \frac{\exp(-\psi/U_T)}{qD_n n_i} dx = \int_e^c \frac{\exp\{-\psi + \varphi_p)/U_T\}}{qD_n n_i} dx$$

$$= \int_e^c \frac{p\, dx}{qD_n n_i^2}.$$

Finally,

$$|J_n| = \frac{\exp(V_{be}/U_T) - \exp(-V_{cb}/U_T)}{\int_e^c \frac{p}{qD_n n_i^2} dx}. \tag{3.19}$$

The denominator can be rewritten:

$$\int_e^c \frac{p\, dx}{qD_n n_i^2} = \frac{1}{qn_{i0}^2} \int_{\text{base}} \frac{p(x)}{D_n(x)} \left(\frac{n_{i0}}{n_{ib}}\right)^2 dx = \frac{G_b}{qn_{i0}^2}.$$

The base Gummel number G_b is a generalized form of Eq. (3.12). For low injection levels we may put $p \approx N_a^-$.

If we multiply $|J_n|$ from Eq. (3.19) by the emitter area A_{em} we get the total collector current:

$$I_c = I_s\{\exp(V_{be}/U_T) - \exp(-V_{cb}/U_T)\}, \tag{3.20}$$

where

$$I_s = \frac{qn_{i0}^2}{G_b} A_{em}.$$

Splitting I_c into a forward current I_f and a reverse current I_r leads to

$$\left.\begin{aligned} I_f &= I_s\{\exp(V_{be}/U_T) - 1\} \\ \text{and} \\ I_r &= I_s\{\exp(-V_{cb}/U_T) - 1\}. \end{aligned}\right\} \tag{3.21}$$

These expressions are the basis of the Ebers-Moll model [3.6].
The denominator in Eq. (3.19) can be reshaped as follows:

$$\int_e^c \frac{p\,dx}{qD_n n_i^2} = \frac{1}{q^2 \overline{(D_n n_i^2)}_b A_{em}} \int_e^c qA_{em} p\,dx,$$

where

$$\overline{(D_n n_i^2)}_b = \frac{\int_e^c p\,dx}{\int_e^c \frac{p}{D_n n_i^2}\,dx}$$

is a mean value, averaged over mainly the neutral base region.
The integral in the numerator denotes the total charge of one type (holes or electrons) in the active part of the transistor and is called simply the base charge Q_b:

$$Q_b = \int_e^c qA_{em} p\,dx. \tag{3.22}$$

Its value at zero bias is Q_{b0}; this is a characteristic quantity of the device, closely related to the base Gummel number G_b and the sheet resistance of the active base under the emitter (see Eq. (2.69)):

$$Q_{b0}/A_{em} = \frac{1}{\bar{\mu}_p \rho_\square} = \frac{qn_{i0}^2 G_b}{\overline{(D_n n_i^2)}_b}. \tag{3.23}$$

The saturation value of the collector current in Eq. (3.20) can also be expressed by means of Q_{b0}:

$$I_s = \frac{q^2 \overline{(D_n n_i^2)}_b A_{em}^2}{Q_{b0}}. \tag{3.24}$$

Here $\overline{(D_n n_i^2)}_b$ is a function of the maximum base dope concentration, shown in Fig. 3.4. See also sections 2.4 and 2.5.
We have replaced p by N_a^- in the integral in the denominator of Eq. (3.19). This is correct in the neutral base region at low injection levels, provided

3.3 Collector Currents and Stored Charges

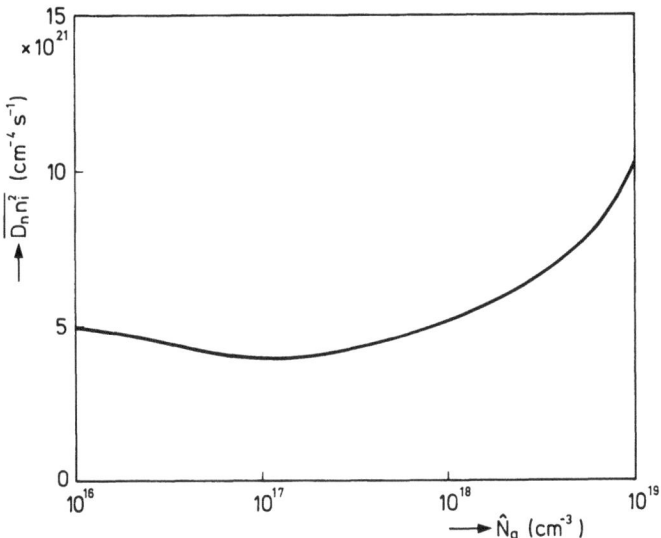

Fig. 3.4 The product $\overline{D_n n_i^2}$ as a function of the top dope concentration ($\hat{N}_a$) in the base

the base region is not too thin. In reality, depletion region edges are not abrupt, but do have tails. For shallow base regions (~ 0.1 μm) these tails of the e-b and b-c junctions can mix and there is no such thing as a neutral base region. Under these circumstances $p < N_a^-$, but Q_{b0} still denotes the hole charge at zero bias.

At high injection levels we get $p > N_a^-$ and $Q_b > Q_{b0}$. The general expression for the collector current becomes

$$I_c = I_s Q_{b0} \frac{\exp(V_{be}/U_T) - \exp(-V_{cb}/U_T)}{Q_b}. \qquad (3.25)$$

This Eq. (3.25) is known as the Moll-Ross relation [3.7]. The total base charge can be divided into the following parts:

$$Q_b = Q_{b0} + Q_{Te} + Q_{Tc} + Q_{be} + Q_{bc}. \qquad (3.26)$$

Q_{b0} is the zero bias base charge, as discussed before, Q_{Te} and Q_{Tc} are the depletion charges of the e-b and b-c junctions, respectively, and Q_{be} and Q_{bc} are the stored charges of the minority carriers, injected respectively from the emitter and collector.

For the moment we will disregard the depletion charges ($Q_{Te} = Q_{Tc} = 0$); these quantities will be discussed in section 3.5 and 3.6.

3.3.2 The Integral Charge Control Relation

The stored charges Q_{be} and Q_{bc} can be related to the forward and reverse currents I_f and I_r of Eq. (3.21) by means of the charge-control principle of

section 3.2,
$$Q_{be} = \tau_f I_f \quad \text{and} \quad Q_{bc} = \tau_r I_r \tag{3.27}$$

where the constants τ_f and τ_r are called the forward and reverse base transit times. Combining Eqs. (3.21), (3.25), (3.26) and (3.27) we get

$$Q_b = Q_{b0} + \tau_f I_s Q_{b0} \frac{\exp(V_{be}/U_T) - 1}{Q_b} + \tau_r I_s Q_{b0} \frac{\exp(-V_{cb}/U_T) - 1}{Q_b}$$

or

$$\frac{Q_b}{Q_{b0}} = \frac{1}{2} + \frac{1}{2}$$

$$\times \sqrt{1 + 4\frac{\tau_f I_s}{Q_{b0}} \left\{ \exp\left(\frac{V_{be}}{U_T}\right) - 1 \right\} + 4\frac{\tau_r I_s}{Q_{b0}} \left\{ \exp\left(\frac{-V_{cb}}{U_T}\right) - 1 \right\}}. \tag{3.28}$$

Substitution of Eq. (3.28) into Eq. (3.25) and splitting up into forward and reverse currents gives

$$I_f = I_s \frac{\exp(V_{be}/U_T) - 1}{Q_b/Q_{b0}}$$

and $\qquad\qquad\qquad\qquad\qquad\qquad\qquad\qquad\qquad\qquad\qquad\qquad\qquad$ (3.29)

$$I_r = I_s \frac{\exp(-V_{cb}/U_T) - 1}{Q_b/Q_{b0}}.$$

In the normal forward mode ($V_{cb} > 0$) we then have

$$I_f = I_s \frac{\exp(V_{be}/U_T) - 1}{\frac{1}{2} + \frac{1}{2}\sqrt{1 + 4I_s/I_{kf}\{\exp(V_{be}/U_T) - 1\}}}, \tag{3.30}$$

where Q_{b0}/τ_f has been replaced by I_{kf}. I_{kf} can be considered as the current value at which high injection in the base starts.

From Eq. (3.30) it follows that for low values of V_{be} (low injection)

$$I_f \approx I_s \{\exp(V_{be}/U_T) - 1\},$$

whereas for high values of V_{be} (high injection)

$$I_f \approx \sqrt{I_s I_{kf}} \exp(V_{be}/2U_T).$$

The low and high injection asymptotes intersect at a current value I_{kf}, see Fig. 3.5.

For the reverse mode of operation ($V_{cb} < 0$, $V_{be} < 0$) we get likewise

$$I_e = I_r = I_s \frac{\exp(-V_{cb}/U_T) - 1}{\frac{1}{2} + \frac{1}{2}\sqrt{1 + 4I_s/I_{kr}\{\exp(-V_{cb}/U_T) - 1\}}}, \tag{3.31}$$

where $I_{kr} = Q_{b0}/\tau_r$.

3.3 Collector Currents and Stored Charges

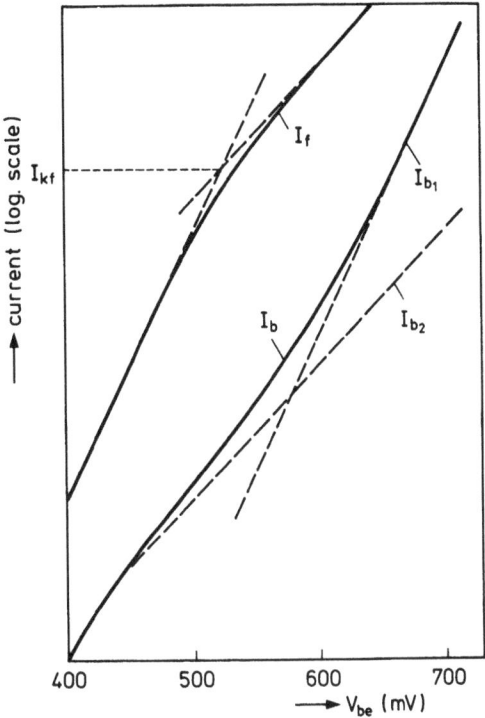

Fig. 3.5 The Gummel plot $I_f(V_{be})$ and $I_b(V_{be})$ with the various components indicated by dashed lines

Equations like (3.29), (3.30) and (3.31) are called Integral Charge Control (ICC) relations [3.8] and form the core of the Gummel-Poon model [3.9]. The main parameters are here I_s, Q_{b0}, τ_f and τ_r. In reality the base transit times τ_f and τ_r are current-dependent, e.g. $\tau_f = \tau_{f0}\{1 + f(I_f)\}$ with $f(I_f)$ as a fit formula, depending on the built-in electric field in the base [3.10].

The stored charges Q_{be} and Q_{bc} are given by the charge control relations (3.27). The storage or diffusion capacitances are

$$C_{be} = \frac{dQ_{be}}{dV_{be}} = \tau_f \frac{dI_f}{dV_{be}} = \tau_f r_0^{-1}$$

and

$$C_{bc} = -\frac{dQ_{bc}}{dV_{cb}} = -\tau_r \frac{dI_r}{dV_{cb}} = \tau_r r_0'^{-1}.$$

(3.32)

Here r_0 and r_0' are the differential resistances of the junctions.
If we calculate for example Q_{be} by integrating Eq. (3.4) we find, using Eq. (3.5):

$$Q_{be} = qA_{em} \int_0^{W_b} \{n(x) - n_0\} \, dx$$

$$= qA_{em} \frac{L_n}{\sinh\left(\frac{W_b}{L_n}\right)} \left\{\cosh\left(\frac{W_b}{L_n}\right) - 1\right\} \frac{n_i^2}{N_a + n(0)} \{\exp(V_{be}/U_T) - 1\}$$

$$= \frac{L_n^2}{D_n} \left\{\cosh\left(\frac{W_b}{L_n}\right) - 1\right\} I_n(0)$$

$$= \tau_f I_n(0),$$

so

$$\tau_f = \frac{L_n^2}{D_n} \left\{\cosh\left(\frac{W_b}{L_n}\right) - 1\right\}.$$

If we can neglect recombination in the neutral base we have $W_b \ll L_n$ and $\tau_f = W_b^2/2D_n$. The same thing can be derived for τ_r.
We have also charge storage in the emitter, for which a similar derivation leads to

$$Q_e = qA_{em} \int_0^{W_e} \{p(x) - p_0\} \, dx = \frac{L_p^2}{D_p} \left\{\cosh\left(\frac{W_e}{L_p}\right) - 1\right\} I_p(0).$$

For $W_e \ll L_p$ and with $I_p(0) = I_f/h_{FE}$ we get

$$Q_e \approx \tau_e I_f, \qquad \tau_e \approx \frac{W_e^2}{2D_p h_{FE}}$$

(h_{FE} is the d.c. current gain). Q_e can usually be neglected against Q_b.

3.3.3 Current, Charges and Minority Carrier Concentrations

An alternative approach to the description of currents and charges is to relate them to the injected minority carrier concentrations in the base; these carrier concentrations, in turn, depend on the applied junction voltages [3.11]. In this way we avoid using the charge control relations of Eq. (3.27). Again we consider a one-dimensional structure with a doping profile as sketched in Fig. 3.6. In the neutral base region ($0 < x < W_b$) we have charge neutrality ($p \approx n + N_a$) and no recombination ($J_p = 0$ and $J_n =$ constant). The doping profile is approximated by

$$N_a(x) = \hat{N}_a \exp(-\eta x/W_b). \tag{3.33}$$

Under low injection conditions ($p \approx N_a$) this non-constant doping profile causes a built-in electric field, which is given by

3.3 Collector Currents and Stored Charges

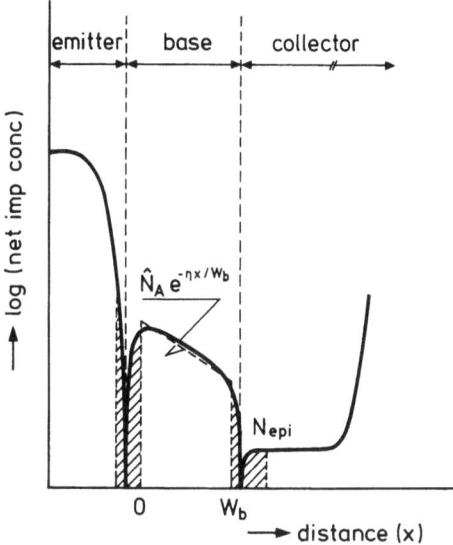

Fig. 3.6 Typical doping profile in an *npn* transistor. Hatched areas denote depletion regions. The dashed line gives the doping profile approximation of Eq. (3.33)

$$F = \frac{D_p}{\mu_p} \frac{1}{p} \frac{dp}{dx} \approx \frac{k_B T}{q} \frac{1}{N_a} \frac{dN_a}{dx}$$

$$= -U_T \frac{\eta}{W_b}. \tag{3.34}$$

Use has been made here of the Einstein relation (Eq. (2.41)) and the assumption of no recombination ($J_p = 0$), see Eq. (2.52).
The parameter η is thus a measure of the strength of the built-in field in the base; the value $\eta = 0$ represents a constant base dope concentration.
The electron current density in the region $0 < x < W_b$ (see Fig. 3.6) is given by Eq. (2.52):

$$J_n = qD_n \frac{dn}{dx} + q\mu_n nF$$

$$= qD_n \frac{dn}{dx} + q\mu_n \frac{D_p}{\mu_p} \frac{n}{p} \frac{dp}{dx}$$

$$= qD_n \left(\frac{dn}{dx} + \frac{n}{p} \frac{dp}{dx} \right)$$

$$= qD_n \left(\frac{2n + N_a}{n + N_a} \frac{dn}{dx} + \frac{n}{n + N_a} \frac{dN_a}{dx} \right). \tag{3.35}$$

Eq. (3.35) constitutes a differential equation for $n(x)$ in the quasi-neutral base region; J_n has a negative value and is independent of the position x. In general Eq. (3.35) has no analytic solution; only in two asymptotic cases (low injection and high injection) an analytic solution is possible.

3.3.3.1 The Low-Injection Case: $n(x) \ll N_a(x)$

Eq. (3.35) simplifies to

$$J_n = qD_n \left(\frac{dn}{dx} + \frac{n}{N_a} \frac{dN_a}{dx} \right) = \frac{qD_n}{N_a} \frac{d(nN_a)}{dx}.$$

After integration we get

$$J_n \int_0^x N_a \, dx = qD_n \int_0^x d(nN_a)$$

or, using Eq. (3.33),

$$n(x) = n(0) \exp(\eta x/W_b) + \frac{|J_n| W_b}{qD_n \eta} \{\exp(\eta x/W_b) - 1\}.$$

We introduce here the built-in field factor

$$f_b = \exp(\eta). \tag{3.36}$$

At $x = W_b$ we now have

$$n(W_b) = n(0) \cdot f_b - \frac{|J_n|}{qD_n} \frac{W_b}{\eta} (f_b - 1)$$

or

$$|J_n| = \frac{qD_n}{W_b} \eta \frac{f_b}{f_b - 1} \left\{ n(0) - \frac{1}{f_b} n(W_b) \right\}.$$

Substituting this in the expression for $n(x)$ finally results in:

$$n(x) = \frac{f_b - \exp(\eta x/W_b)}{f_b - 1} n(0) + \frac{\exp(\eta x/W_b) - 1}{f_b - 1} n(W_b). \tag{3.37}$$

Fig. 3.7 shows two examples of the minority carrier distribution according to Eq. (3.37).
For $\eta = 0$ we have triangular distributions; with a strong built-in field ($\eta = 8$) the distributions shift towards the collector side.
Multiplying $|J_n|$ by the emitter area gives for the electron (or collector) current

$$I_c = I_n = \frac{qD_n A_{em}}{W_b} \left\{ \eta \frac{f_b}{f_b - 1} n(0) - \eta \frac{1}{f_b - 1} n(W_b) \right\}. \tag{3.38}$$

Here too we can distinguish a forward current (I_f), depending on $n(0)$, and

3.3 Collector Currents and Stored Charges

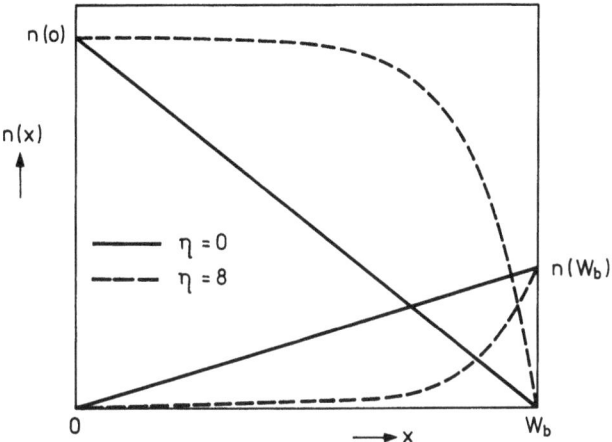

Fig. 3.7 Electron distributions in a p-type base for different values of the built-in field

a reverse current (I_r) depending on $n(W_b)$:

$$\left.\begin{aligned} I_f &= \frac{qD_n A_{em}}{W_b} \eta \frac{f_b}{f_b - 1} n(0), \\ I_r &= \frac{qD_n A_{em}}{W_b} \eta \frac{1}{f_b - 1} n(W_b). \end{aligned}\right\} \quad (3.39)$$

The total stored base charge is

$$Q_b = qA_{em} \int_0^{W_b} n(x)\, dx$$

$$= qA_{em} W_b \left(\frac{f_b}{f_b - 1} - \frac{1}{\eta}\right) n(0) + qA_{em} W_b \left(\frac{1}{\eta} - \frac{1}{f_b - 1}\right) n(W_b).$$

The zero bias base charge is

$$Q_{b0} = qA_{em} \int_0^{W_b} N_a(x)\, dx = q\hat{N}_a A_{em} W_b \frac{1}{\eta} \frac{f_b - 1}{f_b}. \quad (3.40)$$

Again, Q_b can be divided into a forward (Q_{be}) and a reverse base charge (Q_{bc}):

$$\left.\begin{aligned} Q_{be} &= Q_{b0} \frac{(\eta - 1)f_b^2 + f_b}{(f_b - 1)} \frac{n(0)}{\hat{N}_a}, \\ Q_{bc} &= Q_{b0} \frac{f_b(f_b - \eta - 1)}{(f_b - 1)^2} \frac{n(W_b)}{\hat{N}_a}. \end{aligned}\right\} \quad (3.41)$$

For $\eta = 0$ we get a familiar result: $Q_{b0} = qN_a A_{em} W_b$, $Q_{be} = \frac{1}{2}qA_{em} W_b n(0)$ and $Q_{bc} = \frac{1}{2}qA_{em} W_b n(W_b)$, see also Fig. 3.7. Here the base transit times are not parameters of a charge control relation as in Eq. (3.27), but they follow in a natural way from their definition $\tau = Q/I$. Eqs. (3.39) and (3.41) give the low injection limits:

$$\left.\begin{array}{l}\tau_f = \dfrac{Q_{be}}{I_f} = \dfrac{W_b^2}{D_n}\dfrac{\eta - 1 + (1/f_b)}{\eta^2}, \\[2ex] \tau_r = \dfrac{Q_{bc}}{I_r} = \dfrac{W_b^2}{D_n}\dfrac{f_b - \eta - 1}{\eta^2}.\end{array}\right\} \qquad (3.42)$$

With $\eta = 0$ they become $\tau_f = \tau_r = W_b^2/2D_n$.

3.3.3.2 The High-Injection Case: $n(x) \gg N_a(x)$

The differential equation (3.35) now becomes:

$$J_n = 2qD_n \frac{dn}{dx}$$

and, after integration,

$$|J_n| = \frac{2qD_n}{W_b}\{n(0) - n(W_b)\}.$$

The forward and reverse currents are then

$$\left.\begin{array}{l} I_f = \dfrac{2qD_n A_{em}}{W_b} n(0) \\[3ex] I_r = \dfrac{2qD_n A_{em}}{W_b} n(W_b). \end{array}\right\} \qquad (3.43)$$

and

The solution for $n(x)$ is

$$n(x) = n(0) - \{n(0) - n(W_b)\}\frac{x}{W_b},$$

and therefore the stored charges are:

$$\left.\begin{array}{l} Q_{be} = Q_{b0}\dfrac{\eta}{2}\dfrac{f_b}{f_b - 1}\dfrac{n(0)}{\hat{N}_a} \\[3ex] Q_{bc} = Q_{b0}\dfrac{\eta}{2}\dfrac{f_b}{f_b - 1}\dfrac{n(W_b)}{\hat{N}_a}. \end{array}\right\} \qquad (3.44)$$

and

The transit times are $\tau_f = \tau_r = W_b^2/4D_n$.

3.3 Collector Currents and Stored Charges

So far we have established relations between the currents and charges on the one hand and injected minority carrier concentrations on the other. We will now relate these concentrations to the applied junction voltages by means of the *pn* products at $x = 0$ and $x = W_b$:

$$\{n(0) + \hat{N}_a\} n(0) = n_i^2 \exp(V_{be}/U_T),$$

$$\left\{n(W_b) + \frac{\hat{N}_a}{f_b}\right\} n(W_b) = n_i^2 \exp(-V_{cb}/U_T). \quad (3.45)$$

If an epitaxial layer is present in the collector region (see Fig. 3.6) and if this epitaxial layer contains a stored hole charge, which occurs when the c-b junction is no longer reverse biased (in the reverse mode of operation or in hard or quasi-saturation), the *pn* products on both sides of the junction are equal. The boundary condition for $n(W_b)$ changes into

$$\left\{n(W_b) + \frac{\hat{N}_a}{f_b}\right\} n(W_b) = \{p(0) + N_{\text{epi}}\} p(0), \quad (3.46)$$

where $p(0)$ is the hole concentration at the base side of the collector epilayer. We can combine the boundary conditions of Eq. (3.45) with the current and charge formulas of Eqs. (3.39), (3.41), (3.43) and (3.44) and get the voltage dependence of currents and charges. For e.g. the forward current we find in the case of low injection

$$I_f = I_s\{\exp(V_{be}/U_T) - 1\},$$

and for high injection $\quad (3.47)$

$$I_f = \sqrt{I_s I_k}\{\exp(V_{be}/2U_T) - 1\}.$$

Here

$$I_s = qA_{em} \frac{D_n n_i^2}{\hat{N}_a W_b} \eta \frac{f_b}{f_b - 1}$$

$$= \frac{q^2 A_{em}^2 D_n n_i^2}{Q_{b0}}.$$

If we realize that $D_n n_i^2$ is in fact the mean value over the neutral base region, this is exactly the same result as reached before (cf. Eq. (3.24)). The intersection of the low and high injection asymptotes is at a current

$$I_k = \frac{4qD_n A_{em} \hat{N}_a}{W_b} \frac{1}{\eta} \frac{f_b - 1}{f_b} = \frac{4D_n}{W_b^2} Q_{b0}.$$

This I_k is two times higher than the I_{kf} in Eq. (3.30), so the high injection asymptote lies $\sqrt{2}$ times higher than in Fig. 3.5! The advantage of this is that at a given Q_{b0}, I_k does not depend on τ_f, as I_{kf} does. From the viewpoint of device physics this is more realistic, because the starting point for high injection depends primarily on Q_{b0} and not on τ_f.

3.3.3.3 The General Case

The differential equation (3.35) has no general analytic solution, but we can obtain generally applicable fit formulas by smoothly connecting the low and high injection solutions [3.11]. To this end we first define the following three quantities, determined by the built-in electric field:

$$a_1 = \frac{2}{\eta} \frac{f_b - 1}{f_b},$$

$$a_2 = \frac{(\eta - 1)f_b^2 + f_b}{(f_b - 1)^2}, \qquad (3.48)$$

$$a_3 = \frac{f_b(f_b - \eta - 1)}{(f_b - 1)^2}.$$

The forward current, as given by Eqs. (3.39) and (3.43) then becomes:

for low injection: $\quad I_f = \dfrac{2qD_n A_{em}}{W_b} \dfrac{1}{a_1} n(0)$

and

for high injection: $\quad I_f = \dfrac{2qD_n A_{em}}{W_b} n(0).$

For the total current range we now write

$$I_f = \frac{2qD_n A_{em}}{W_b} g_{I_f} n(0) \qquad (3.49)$$

with

$$g_{I_f} = \frac{2 + a_1 + 4n(0)/\hat{N}_a}{(2 + a_1)a_1 + 4n(0)/\hat{N}_a}.$$

g_{I_f} is a fit function of η and $n(0)/\hat{N}_a$ that behaves monotonically with respect to these variables.

For the reverse current we have in a similar way at low injection

$$I_r = \frac{2qD_n A_{em}}{W_b} \frac{1}{f_b} \frac{1}{a_1} n(W_b)$$

and at high injection

$$I_r = \frac{2qD_n A_{em}}{W_b} n(W_b).$$

A general fit formula is here

$$I_r = \frac{2qD_n A_{em}}{W_b} g_{I_r} n(W_b) \qquad (3.50)$$

3.3 Collector Currents and Stored Charges

with

$$g_{I_r} = \frac{1/f_b + n(W_b)/\hat{N}_a}{a_1 + n(W_b)/\hat{N}_a}.$$

Finally, the stored base charges can be formulated as:

low injection, $\quad Q_{be} = Q_{b0} \cdot a_2 \cdot \dfrac{n(0)}{\hat{N}_a}$,

high injection, $\quad Q_{be} = Q_{b0} \dfrac{1}{a_1} \dfrac{n(0)}{\hat{N}_a}$,

general, $\quad Q_{be} = g_{Q_f} Q_{b0} \dfrac{n(0)}{\hat{N}_a}$, \hfill (3.51)

with

$$g_{Q_f} = \frac{2 + a_1 + n(0)/\hat{N}_a}{2 + a_1 + a_1 a_2 n(0)/\hat{N}_a} a_2,$$

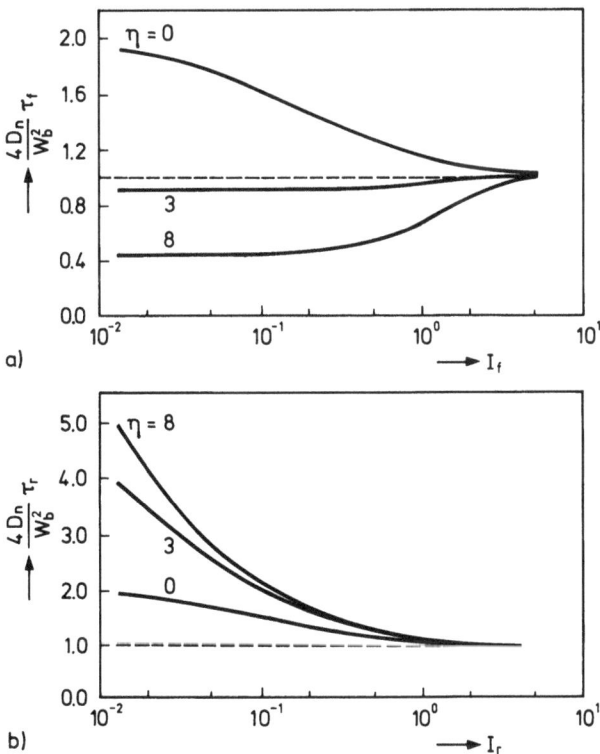

Fig. 3.8 Current dependence of the forward (τ_f) and reverse (τ_r) base transit times. The Early effect has been neglected

and

low injection,
$$Q_{bc} = Q_{b0} a_3 \frac{n(W_b)}{\hat{N}_a},$$

high injection,
$$Q_{bc} = Q_{b0} \frac{1}{a_1} \frac{n(W_b)}{\hat{N}_a},$$

general,
$$Q_{bc} = Q_{b0} g_{Q_r} \frac{n(W_b)}{\hat{N}_a}, \qquad (3.52)$$

with
$$g_{Q_r} = \frac{a_3 + f_b n(W_b)/\hat{N}_a}{1 + a_1 f_b n(W_b)/\hat{N}_a}.$$

The fit formulas of Eqs. (3.49), (3.50), (3.51) and (3.52), together with the boundary conditions of Eq. (3.45) give a complete description of the main currents and stored charges. We need four parameters: I_s, I_r, Q_{b0} and η. This is the same number as for the Integral Charge Control approach (I_s, Q_{b0}, τ_f and τ_r). The advantages now are that we have explicitly incorporated the effects of the built-in field in the base and that τ_f and τ_r are no longer constants, but automatically bias-dependent.

Fig. 3.8 gives $\tau_f(I_f)$ and $\tau_r(I_r)$ for several values of η. At low currents they follow Eq. (3.42), at high currents they have $W_b^2/4D_n$ as the limit.

3.4 Base Currents

For the transport models we had already found (Eq. (3.11)) that the base current injected into the emitter was determined by the emitter Gummel number (see Eq. (3.13)). A similar method is often used for the c-b junction. Introducing the current gain parameters β_f and β_r leads then to a base current expression for both normal and reverse bias:

$$I_{b_1} = \frac{I_s}{\beta_f}\{\exp(V_{be}/U_T) - 1\} + \frac{I_s}{\beta_r}\{\exp(-V_{cb}/U_T) - 1\}. \qquad (3.53)$$

Apart from these "ideal" components we may also encounter "non-ideal" components,

$$I_{b_2} = I_{bf}\{\exp(V_{be}/m_f U_T) - 1\} + I_{br}\{\exp(-V_{cb}/m_r U_T) - 1\}, \qquad (3.54)$$

where m_f and m_r are parameters and called non-ideality factors. They usually have values between one and two. Physically speaking these non-ideal components arise from recombination in the depletion layers of forward biased junctions. A more physical alternative to Eq. (3.54) can be derived as follows. We start with the SRH recombination formula (2.72):

3.4 Base Currents

$$R = \frac{pn - n_i^2}{\tau_{n_0}(p + p_1) + \tau_{p_0}(n + n_1)}.$$

For the sake of simplicity we put $\tau_{n_0} = \tau_{p_0} = \tau = (\sigma v_{th} N_t)^{-1}$. The maximum recombination takes place where $p(x) = n(x) = n_i \exp(V_{be}/2U_T)$, so

$$R_{\max} = \frac{1}{\tau} \frac{n_i^2\{\exp(V_{be}/U_T) - 1\}}{2n_i \exp(V_{be}/2U_T) + 2n_i \cosh(\Delta E_t/k_B T)}.$$

Here ΔE_t stands for $\Delta E_t = E_t - E_i - k_B T \ln g_t$ and it is mainly determined by the energy level E_t of the traps involved.

The total recombination current I_{rec} is approximately given by $I_{rec} = qA_{em}R_{\max}$; so we get

$$I_{b_2} = \frac{qA_{em}n_i}{2\tau} \frac{\exp(V_{be}/U_T) - 1}{\exp(V_{be}/2U_T) + \cosh(\Delta E_t/k_B T)}$$

or

$$I_{b_2} = I_{bf} \frac{\exp(V_{be}/U_T) - 1}{\exp(V_{be}/2U_T) + \exp(V_{LF}/2U_T)}. \quad (3.55)$$

I_{bf} and V_{LF} are the two parameters; for $V_{be} \gg V_{LF}$, I_{b_2} is proportional to $\exp(V_{be}/2V_T)$ and it then becomes rapidly smaller than I_{b_1}; see Fig. 3.5. From this figure it is clear that the d.c. current gain $h_{FE} = I_f/I_b$ as a function of I_f has the typical shape sketched in Fig. 3.9. At low values of I_f the gain increases with I_f, it reaches a maximum ($\sim \beta_f$) and then falls off due to high injection in the base or to quasi-saturation; see section 3.7.

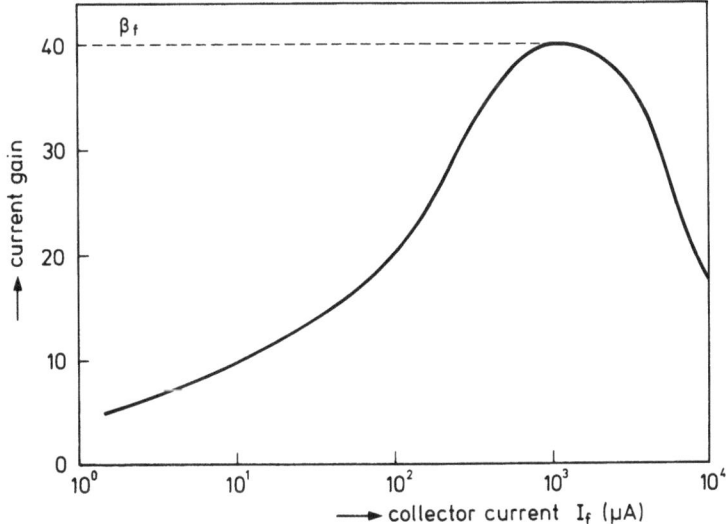

Fig. 3.9 Typical shape of the current gain $h_{FE}(I_f)$ without Early effect

3.5 Depletion Charges and Capacitances

On both sides of a *pn* junction we usually find space charge layers, on the *p*-side with a negative space charge due to ionized acceptors, on the *n*-side a positive one, due to ionized donors [3.12]. Both space charges have the same magnitude. Depletion charge (Q_T) is defined here as the total space charge on one side of the junction, with the convention that at zero bias $Q_T = 0$ and that Q_T is positive at forward bias and negative at reverse bias. The depletion capacitance C_T is then defined as $C_T = dQ_T/dV_j$, where V_j is the junction voltage. We also have

$$Q_T = \int_0^{V_j} C_T(V_j)\, dV_j. \tag{3.56}$$

The depletion capacitance C_T depends on V_j and can be modelled as

$$C_T = \frac{C_0}{(1 - V_j/V_d)^p}. \tag{3.57}$$

Here C_0 is the zero bias value of C_T, V_d is the built-in junction voltage (also sometimes called diffusion voltage) and p is the grading coefficient. So we need three parameters.

For abrupt junctions $p = \frac{1}{2}$, which can be proved as follows: Let us consider an abrupt junction with sharp edges at $-x_p$ and $+x_n$ (Fig. 3.10).
Both space charges have equal magnitudes, so $x_p N_a^- = x_n N_d^+$. Applying Poisson's equation (2.51) we arrive at

$$\frac{dF}{dx} = \frac{q}{\varepsilon} N_d^+ = \frac{-q}{\varepsilon} N_a^-$$

and

$$F_{\max} = \frac{qN_a^-}{\varepsilon} x_p = \frac{qN_d^+}{\varepsilon} x_n.$$

The total potential difference across the junction is

$$-V_j + V_d = \int F\, dx = \frac{q}{2\varepsilon}(N_a^- x_p^2 + N_d^+ x_n^2) = \frac{qN_d^+}{2\varepsilon}\left(1 + \frac{N_d^+}{N_a^-}\right) x_n^2.$$

So we have

$$x_n = \sqrt{\frac{2\varepsilon}{q}(-V_j + V_d)\frac{N_a^-/N_d^+}{N_a^- + N_d^+}},$$

$$x_p = \sqrt{\frac{2\varepsilon}{q}(-V_j + V_d)\frac{N_d^+/N_a^-}{N_a^- + N_d^+}}.$$

If the junction area is denoted by A_j, the capacitance C_T follows from

3.5 Depletion Charges and Capacitances

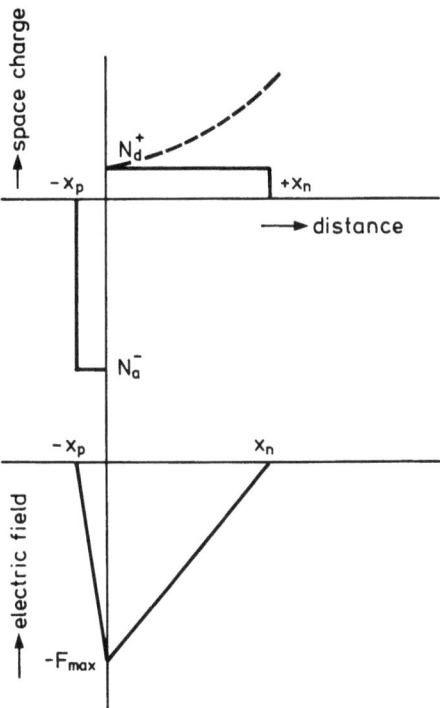

Fig. 3.10 Space charge and field distribution of a reverse biased junction under the abrupt depletion edge assumption. The dashed line represents a non-constant doping profile in the n region

$$C_T = \frac{\varepsilon A_j}{x_p + x_n} = A_j \sqrt{\frac{q\varepsilon(N_a^- + N_d^+)}{2V_d\left(\frac{N_a^-}{N_d^+} + \frac{N_d^+}{N_a^-}\right)}} \cdot \frac{1}{(1 - V_j/V_d)},$$

which proves Eq. (3.57) with $p = \frac{1}{2}$.
For a linear junction ($N(x) = ax$) we can prove that $p = \frac{1}{3}$ [3.12]. In practice p-values are mostly between $\frac{1}{2}$ and $\frac{1}{3}$, although other values might occur: A doping profile, indicated in Fig. 3.10 by the dashed line, makes $p < \frac{1}{3}$.
From Eqs. (3.56) and (3.57) it follows that

$$Q_T = \frac{1}{1-p} C_0 V_d \{1 - (1 - V_j/V_d)^{1-p}\}. \tag{3.58}$$

The expressions (3.57) and (3.58) have a singularity at $V_j = V_d$, which is unacceptable for numerical evaluation. It can be remedied in several ways:
1. Replace [3.13] $(1 - V_j/V_d)$ by

$$\tfrac{1}{2}\{(1 - V_j/V_d) + \sqrt{(1 - V_j/V_d)^2 + K}\}.$$

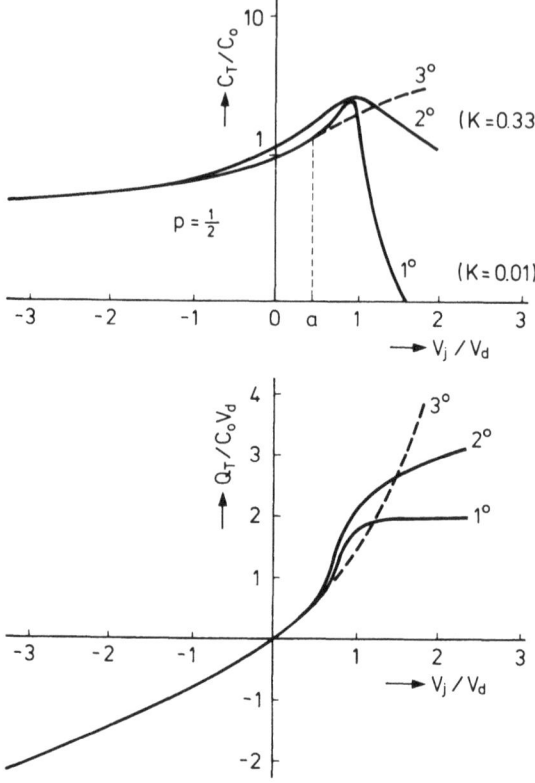

Fig. 3.11 Depletion capacitance C_T and depletion charge Q_T as a function of junction voltage V_j, for three different models.

2. Replace [3.14] $(1 - V_j/V_d)^{1-p}$ by

$$\frac{1 - V_j/V_d}{\{(1 - V_j/V_d)^2 + K\}^{p/2}}.$$

3. At $V_j = aV_d$, with $0 < a < 1$, we continue the function $C_T(V_j)$ by its tangent.

All three methods require an extra parameter (K or a); methods 1 and 2 keep C_T finite at $V_j = V_d$ and let C_T go to zero for $V_j \gg V_d$ and are therefore preferable from a physical viewpoint. Fig. 3.11 shows the behaviour of C_T and Q_T in the three cases.

3.5.1 Influence of Current on Q_{T_c}

A complication arises with the c-b junction when current flows: In the normal forward mode of transistor operation this current is determined by

3.5 Depletion Charges and Capacitances

V_{be} and not by V_{cb}. So the state of the c-b junction is governed by two independent variables (V_{cb} and I_n). At high electric fields the mobility decreases and the drift velocity saturates at a value v_s; see Eq. (2.68). When now current flows, this will change the initial space charge by an amount $-J_n/v_s$. We consider again an abrupt junction, as in Fig. 3.10, and assume furthermore that the n-side is an epitaxial collector layer with $N_d^+ = N_{epi}$ and thickness W_{epi}.

The depletion charge Q_{T_c} is

$$Q_{T_c} = \left(qN_{epi} - \frac{J_n}{v_s}\right)x_n = qN_{epi}\left(1 - \frac{J_n}{J_{hc}}\right)x_n,$$

where [3.15]

$$J_{hc} = qN_{epi}v_s. \tag{3.59}$$

So the positive charge of the ionized donors is counteracted by the negative charge of the electrons of the current flow.

A second modification is necessary because of the ohmic voltage drop in the collector epilayer. At $x = x_n$ in Fig. 3.10 the electric field

$$F(x_n) = \rho J_n = \frac{J_n}{q\mu_n N_{epi}}$$

according to Eq. (2.68). This increases $\int F\,dx$ by an amount

$$\rho W_{epi} J_n = \frac{\rho W_{epi}}{A_j} I_n = R_{epi} I_n,$$

where R_{epi} is the ohmic resistance of the total epilayer. To account for this ohmic voltage drop we must replace $V_{cb} + V_d$ by $(V_{cb} + V_d - I_n R_{epi})$. Together with the charge modulation we then get:

$$x_n = \left\{\frac{2\varepsilon N_a}{q(N_{epi} + N_a)N_{epi}} \frac{1 + \frac{N_{epi}}{N_a}I_n/I_{hc}}{1 - I_n/I_{hc}}(V_{cb} + V_d - I_n R_{epi})\right\}^{1/2}$$

and

$$x_p = \left\{\frac{2\varepsilon N_{epi}}{q(N_{epi} + N_a)N_a} \frac{1 - I_n/I_{hc}}{1 + \frac{N_{epi}}{N_a}I_n/I_{hc}}(V_{cb} + V_d - I_n R_{epi})\right\}^{1/2}.$$

$$\tag{3.60}$$

Note that voltage drop and space charge modulation counteract each other at the n-side, but reinforce each other at the p-side. So, when I_n increases, x_p decreases and the neutral base widens [3.15]; x_n also decreases if $V_{cb} + V_d < I_{hc}R_{epi}$, otherwise it increases. The electrical junction, i.e. the point where the space charge changes sign, does not move in an abrupt junction. In a linear junction, in contrast with the foregoing, the electrical junction moves to the right in Fig. 3.10 when current increases, but the thickness of the depletion layer ($x_p + x_n$) does not change and the space charge modulation terms in Eq. (3.60) are absent (I_{hc} is infinite).

Practical cases are somewhere in between the abrupt and the linear junction, but collector depletion charge is mostly formulated without the space charge modulation, for the practical reason that the determination of the required model parameter is difficult.

3.6 Early Effect

In this section we return to Eq. (3.26) and we will see what happens if we no longer neglect the depletion charges Q_{T_e} and Q_{T_c}. These depletion charges depend on their junction voltages and will modulate the fixed zero bias base charge Q_{b0}. This effect is called the Early effect [3.16]. We can take it into account by putting $Q_{b0} + Q_{T_e} + Q_{T_c} = Q_{b0}(1 + q_1)$ instead of Q_{b0}; $1 + q_1$ is called the Early factor and

$$q_1 = \frac{Q_{T_e} + Q_{T_c}}{Q_{b0}}. \tag{3.61}$$

Q_{T_e} and Q_{T_c} are defined as functions of V_{be} and V_{cb}, respectively:

$$Q_{T_e}(V_{be}) = \int_0^{V_{be}} C_{T_e}(V_j) \, dV_j,$$

$$Q_{T_c}(V_{cb}) = \int_0^{-V_{cb}} C_{T_c}(V_j) \, dV_j, \tag{3.62}$$

In the normal mode of operation we have $Q_{T_e} > 0$ and $Q_{T_c} < 0$.
Often the expressions (3.62) are approximated by using the mean values over the range of interest for the capacitances ($\bar{C}_T = 1/V \int_0^V C_T \, dV_j$)

$$Q_{T_e} \approx \bar{C}_{T_e} V_{be}$$

and (3.63)

$$Q_{T_c} \approx -\bar{C}_{T_c} V_{cb},$$

or by using so-called Early voltages ($V_{ea} = Q_{b0}/\bar{C}_T$):

$$Q_{T_e} \approx \frac{Q_{b0}}{V_{ear}} V_{bc}$$

and (3.64)

$$Q_{T_c} \approx \frac{Q_{b0}}{V_{eaf}} V_{cb}.$$

In the latter case the Early factor becomes

3.6 Early Effect

$$1 + q_1 = 1 + \frac{V_{be}}{V_{ear}} - \frac{V_{cb}}{V_{eaf}}.$$

Note that the forward Early voltage (V_{eaf}) belongs to V_{cb} and the reverse (V_{ear}) to V_{be}.

The Early effect influences the main currents I_f and I_r, the stored base charges Q_{be} and Q_{bc}, the knee currents I_k, I_{kf} and I_{kr} and also the transit times τ_f and τ_r. It usually does not affect the base current because this current mainly arises from recombination in the emitter. The I_s value of the main current I_n is inversely proportional to Q_{b0} (Eq. (3.24)), so with Early effect I_n becomes

$$I_n = \frac{I_f - I_r}{1 + \dfrac{Q_{T_e} + Q_{T_c}}{Q_{b0}}} = \frac{I_f - I_r}{1 + q_1}. \tag{3.65}$$

From Eq. (3.41) it follows that Q_{be} and Q_{bc} have to be multiplied by $(1 + q_1)$ and Eq. (3.42) shows that τ_f and τ_r must increase by a factor $(1 + q_1)^2$. It is also clear that the parameters I_{kf} and I_{kr} in Eqs. (3.30) and (3.31) should be divided by $(1 + q_1)$ as should the quantity I_k in Eq. (3.47); however, the ratios I_s/I_{kf} and I_s/I_{kr} remain unaltered.

As an example of the influence of the Early effect on the device characteristics, Fig. 3.12 shows the forward characteristics $I_c(V_{ce})$ for I_b = constant. This figure reveals that the formulation with Early voltage (V_{eaf}) is a rough approximation, for this voltage is not a constant parameter but a bias-dependent quantity.

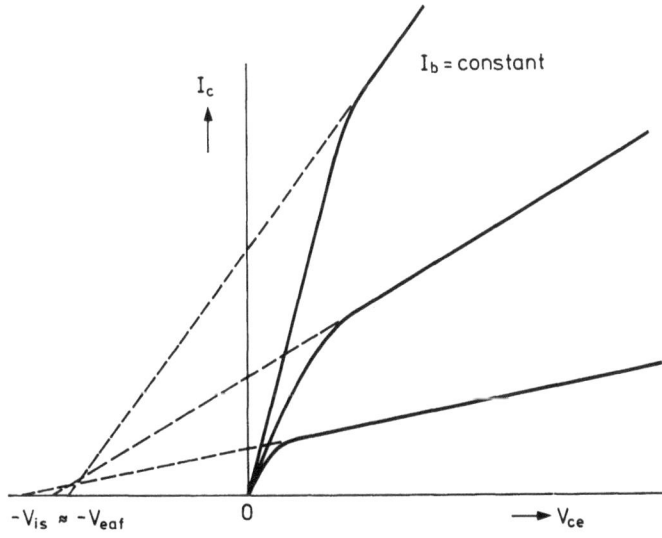

Fig. 3.12 $I_c(V_{ce})$ characteristics with indication of the Early voltage V_{eaf}

Thus, we have

$$\frac{dI_c}{dV_{cb}} = -\frac{I_f}{(1+q_1)^2}\frac{dq_1}{dV_{cb}}$$

or

$$\frac{I_c}{V_{is}+V_{ce}} = -\frac{I_c}{1+q_1}\frac{dq_1}{dV_{cb}}.$$

This follows from Eq. (3.65) with $I_r = 0$. The intersection of the tangent with the V_{ce} axis is at V_{is} and this V_{is} follows from

$$V_{is} + V_{ce} = -\frac{1+q_1}{dq_1/dV_{cb}} = \frac{(1+q_1)Q_{b0}}{C_{T_c}} \tag{3.66}$$

or, using Eq. (3.64),

$$V_{is} + V_{ce} = (1+q_1)V_{eaf},$$

which makes it clear that $V_{is} \approx V_{eaf}$. The current dependence of V_{is} also follows from Eq. (3.66). First we have a slight increase with current because Q_{T_e} in q_1 increases. At higher currents V_{is} falls, because C_{T_c} increases, due to the ohmic voltage drop in the collector epilayer (cf. Eq. (3.60)).

Fig. 3.13 shows the result of a computer simulation [3.17] that confirms Eq. (3.66).

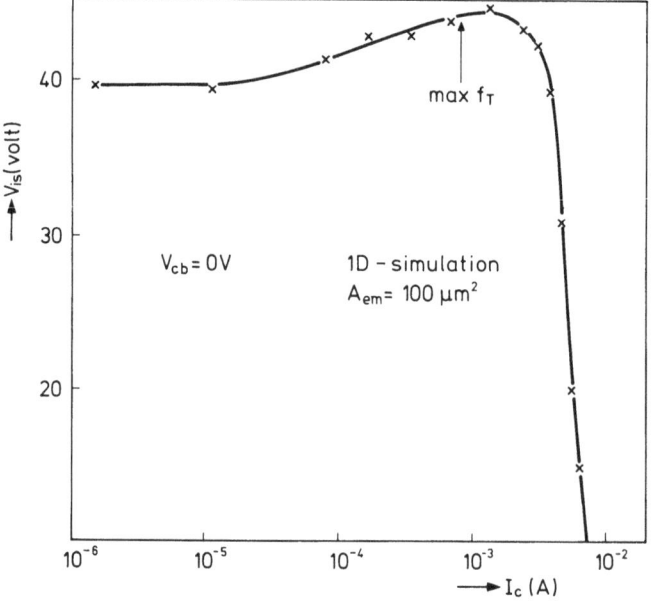

Fig. 3.13 The current dependence of the intercept voltage V_{is} of Fig. 3.12. 1-D computer simulation results

A further consequence of the Early effect is that the $I_c(V_{be})$ characteristic is modified because of the V_{be} dependence of Q_{T_e}; it makes the differential resistance dV_{be}/dI_c larger, even at low injection levels. This effect becomes more pronounced when Q_{b0} becomes relatively small, as is the case in very high frequency transistors. The Early effect lowers the current gain: $h_{FE} = \beta_f/(1 + q_1)$ instead of β_f, see Fig. 3.9.

The punch-through effect occurs when Q_{T_c} has become so negative that it compensates for $Q_{b0} + Q_{T_e}$: The base region is completely depleted by the reverse bias voltage of the collector. The denominator of Eq. (3.65) is then zero and I_n goes to infinity.

3.7 Quasi-Saturation, Base Widening and Kirk Effect

Due to an internal voltage drop in the collector region the base-collector junction might become forward biased at high currents although the external base-collector voltage remains in the reverse bias. This internal voltage drop arises in the active part of the transistor, under the emitter. We will call this effect quasi-saturation [3.18–3.22]. The c-b junction being locally in forward bias results in an increase of the reverse current I_r and in a charge storage (Q_{epi}) in the collector epilayer, which must be added to the already existing Q_b, according to the definition in Eq. (3.22). This enlargement of Q_b is sometimes called base widening or base push-out.

An increase of I_r and Q_b reduces the collector current and also the gain h_{FE}. Recombination of the stored charge Q_{epi} increases the base current and causes a further reduction of the h_{FE}. The total transit time will be enlarged by an amount of Q_{epi}/I_c, so the cut-off frequency will decrease.

Quasi-saturation is different from the phenomenon of hard saturation, when the externally applied base-collector voltage (V_{cb}) makes the total c-b junction (not only the active part) forward biased.

In a space charge region with high electric field the carriers move with the saturated drift velocity (v_s) and their density modulates the space charge. This is sometimes called the Kirk effect; it is present in the epilayer for current densities higher than $J_{hc} = qN_{epi}v_s$. See also section 3.5 for the resemblance with the collector depletion charge.

3.7.1 The Charge Storage in the Epilayer

The above-mentioned phenomena can be calculated by assuming the following [3.22]:

- The collector epilayer, located between the metallurgical junction ($x = 0$) and the buried layer ($x = W_c$), has a constant dope concentration N_{epi}; see Fig. 3.6.
- In that part where the injected charge is stored ($0 < x < x_i$), the so-called injection region, we have quasi-neutrality $p + N_{epi} \approx n$.

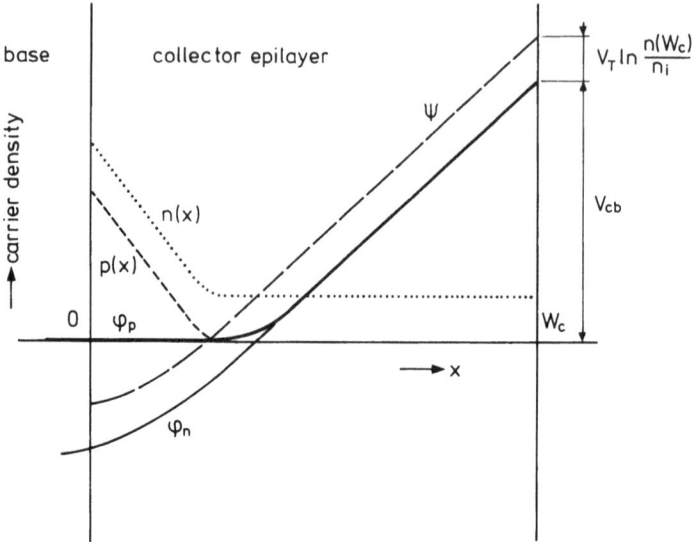

Fig. 3.14 Distributions of $\varphi_n(x)$, $\varphi_p(x)$, $\psi(x)$, $n(x)$ and $p(x)$ in an n-type collector epilayer under quasi-saturation conditions

— Recombination in the injection region is negligible, so $J_p = 0$ and the quasi-Fermi level $\varphi_p = \text{constant} (= 0)$.

The typical situation of quasi-saturation is sketched in Fig. 3.14 for the potential ψ and the quasi-Fermi levels φ_n and φ_p.
In a similar derivation as for Eq. (3.35) we now get

$$J_n = qD_n\left(1 + \frac{n}{p}\right)\frac{dn}{dx}$$

$$= qD_n\left(2 + \frac{N_{\text{epi}}}{p}\right)\frac{dp}{dx}. \tag{3.67}$$

For an npn transistor as discussed here, J_n flows in the negative direction; $p(x)$ in Eq. (3.67) is the minority carrier concentration in the collector epilayer, injected from the base. We solve Eq. (3.67) with the boundary condition $p(x_i) = n_i$ and get

$$x_i = \frac{2qD_n}{|J_n|}\left\{p(0) - n_i + \frac{1}{2}N_{\text{epi}}\ln\frac{p(0)}{n_i}\right\}. \tag{3.68}$$

Eq. (3.68) relates the thickness of the injection region x_i to the injected carrier concentration $p(0)$ at the junction; see Fig. 3.14. Both $p(0)$ and x_i depend on the applied V_{cb}, and the current J_n: V_{cb} and J_n are independent variables as far as the c-b junction behaviour is concerned.
Using the Einstein relation (Eq. (2.41)) and introducing the resistivity $\rho_{\text{epi}} = (q\mu N_{\text{epi}})^{-1}$ gives us from Eq. (3.68)

3.7 Quasi-Saturation, Base Widening and Kirk Effect

$$2\frac{p(0)}{N_{epi}} + \ln\frac{p(0)}{N_{epi}} = \frac{1}{U_T}\left(\rho_{epi}|J_n|x_i - \frac{1}{2}V_{dc}\right), \tag{3.69}$$

where

$$V_{dc} = 2U_T\left(\ln\frac{N_{epi}}{n_i} - \frac{n_i}{N_{epi}}\right) \approx 2U_T \ln\frac{N_{epi}}{n_i}$$

denotes the built-in or diffusion voltage [3.22] of the c-b junction. We can now define an internal collector voltage V_{ci} by putting $\rho_{epi}|J_n|x_i - \frac{1}{2}V_{dc} = V_{ci}$; Eq. (3.69) then simply becomes

$$2\frac{p(0)}{N_{epi}} + \ln\frac{p(0)}{N_{epi}} = \frac{V_{ci}}{U_T}. \tag{3.70}$$

The total charge storage in the collector epilayer is given by

$$Q_{epi} = qA_{em}\int_0^{x_i} p\,dx = qA_{em}\int_0^{x_i}\frac{p\,dp}{dp/dx}$$

$$= qA_{em}\int_{n_i}^{p(0)}\frac{qD_n}{|J_n|}(2p + N_{epi})\,dp$$

$$= \frac{q^2 D_n A_{em}^2}{I_c}\int_{n_i}^{p(0)}(2p + N_{epi})\,dp.$$

Use has been made of Eq. (3.67) and of $I_c = |J_n|A_{em}$. Evaluation of the integral gives

$$Q_{epi} = \frac{q^2 D_n A_{em}^2}{I_c}\{p^2(0) + p(0)N_{epi} - n_i^2 - n_i N_{epi}\},$$

or, neglecting n_i,

$$Q_{epi} * I_c \approx D_n(qN_{epi}A_{em})^2\left\{\left(\frac{p(0)}{N_{epi}}\right)^2 + \frac{p(0)}{N_{epi}}\right\}. \tag{3.71}$$

Because $p(0)$ is a function of the internal collector voltage V_{ci}, see Eq. (3.70), we can write for low injection in the epilayer ($p(0) \ll N_{epi}$),

$$Q_{epi} * I_c \approx D_n(qN_{epi}A_{em})^2 \exp(V_{ci}/U_T),$$

and for high injection ($p(0) \gg N_{epi}$)

$$Q_{epi} * I_c \approx D_n(qN_{epi}A_{em})^2\left(\frac{V_{ci}}{2U_T}\right)^2.$$

3.7.2 Influence of I_c: Ohmic and Hot Carrier Behaviour (Kirk Effect)

Eqs. (3.70) and (3.71) define Q_{epi} as a function of the internal collector voltage V_{ci}. The question how this V_{ci} depends on V_{cb} and I_c must be answered next. To this end we first observe that, using Eq. (2.59), we have at $x = W_c$:

$$\psi(W_c) - \psi(x_i) = \psi(W_c) - 0 = V_{cb} + U_T \ln \frac{n(W_c)}{n_i}. \qquad (3.72)$$

Secondly, we will distinguish between two cases for the situation in the region $x_i < x < W_c$.

1. This region is ohmic; the field strength $-d\psi/dx = \rho_{epi} J_c$, so $\psi(W_c) - \psi(x_i) = \rho_{epi}(W_c - x_i)J_c$ or

$$\rho_{epi} J_c x_i = \rho_{epi} W_c J_c - V_{cb} - U_T \ln \frac{n(W_c)}{n_i}.$$

Furthermore $n(W_c) = N_{epi}$ and therefore

$$\rho_{epi} J_c x_i = \rho_{epi} W_c J_c - V_{cb} - \tfrac{1}{2} V_{dc}.$$

The internal voltage in Eq. (3.70) becomes

$$V_{ci} = \rho_{epi} J_c x_i - \tfrac{1}{2} V_{dc} = \rho_{epi} W_c J_c - V_{cb} - V_{dc}.$$

If we replace $\rho_{epi} W_c J_c$ by $(\rho_{epi} W_c/A_{em}) J_c A_{em} = R_{epi} I_c$ we get

$$V_{ci} = I_c R_{epi} - (V_{cb} + V_{dc}).$$

2. This region is a space charge region with high electric field. The carrier density J_c/qv_s modulates the space charge (Kirk effect).
The field strength follows here from

$$dF/dx = -\frac{qN_{epi}}{\varepsilon}\left(\frac{J_c}{J_{hc}} - 1\right),$$

so

$$F(x) = F(x_i) + \frac{qN_{epi}}{\varepsilon}\left(\frac{J_c}{J_{hc}} - 1\right)(x - x_i)$$

and

$$\psi(W_c) = -\int_0^{W_c} F\, dx = (W_c - x_i) F(x_i) + \frac{qN_{epi}}{2\varepsilon}\left(\frac{J_c}{J_{hc}} - 1\right)(W_c - x_i)^2. \qquad (3.73)$$

Because $J_c > J_{hc}$ we must take $n(W_c) = (J_c/J_{hc})N_{epi}$ in Eq. (3.72). Combining Eqs. (3.72) and (3.73) leads to

$$\left.\begin{array}{l} A(W_c - x_i)^2 + B(W_c - x_i) = V \\[4pt] \text{and} \\[4pt] V = V_{cb} + U_T \ln\left(\dfrac{I_c}{I_{hc}}\dfrac{N_{epi}}{n_i}\right), \\[4pt] \text{where} \\[4pt] A = \dfrac{qN_{epi}}{2\varepsilon}\left(\dfrac{I_c}{I_{hc}} - 1\right), \quad B = F(x_i) = F_{hc}, \end{array}\right\} \qquad (3.74)$$

3.7 Quasi-Saturation, Base Widening and Kirk Effect

the field strength at which the drift velocity completely saturates. The solution of the quadratic equation in (3.74) is

$$(W_c - x_i) = \frac{-B + \sqrt{B^2 + 4AV}}{2A} = \frac{2V}{B + \sqrt{B^2 + 4AV}},$$

therefore

$$\rho_{epi} J_c x_i = \rho_{epi} J_c W_c - \rho J_c \frac{2V}{B + \sqrt{B^2 + 4AV}}.$$

With this the internal voltage V_{ci} now becomes

$$V_{ci} = I_c R_{epi} - I_c R_{epi} \frac{2V}{BW_c + \sqrt{(BW_c)^2 + 4AW_c^2 V}} - \frac{1}{2} V_{dc}. \quad (3.75)$$

Eq. (3.75) is the saturated drift velocity or hot carrier version of the formula for the internal voltage. If we redefine A, B and V as follows, Eq. (3.75) may be used over the entire current range:

$$A \equiv 0 \qquad \text{for} \quad I_c < I_{hc}$$

and

$$\begin{aligned}
A &= \frac{qN_{epi}}{2\varepsilon} \left(\frac{I_c}{I_{hc}} - 1 \right) & \text{for} \quad I_c \geqslant I_{hc}, \\
BW_c &= I_c R_{epi} & \text{for} \quad I_c < I_{hc}, \\
BW_c &= \frac{v_s}{\mu_{epi}} W_c = F_{hc} W_c & \text{for} \quad I_c \geqslant I_{hc}
\end{aligned} \quad (3.76)$$

and

$$\begin{aligned}
V &= V_{cb} + \tfrac{1}{2} V_{dc} & \text{for} \quad I_c < I_{hc}, \\
V &= V_{cb} + \frac{1}{2} V_{dc} + U_T \ln \frac{I_c}{I_{hc}} & \text{for} \quad I_c \geqslant I_{hc}.
\end{aligned}$$

Note that the function $A(I_c)$, $B(I_c)$ and $V(V_{cb}, I_c)$ have a singularity at I_{hc}, which must be removed for numerical evaluation. An often used trick is to replace the term

$$(I_c/I_{hc} - 1) \text{ by } \tfrac{1}{2}\{I_c/I_{hc} - 1 + \sqrt{(I_c/I_{hc} - 1)^2 + \delta}\}.$$

As already mentioned, the Kirk effect of modulating the space charge by means of the carriers moving with the saturated drift velocity, may also be important for the behaviour of the depletion charge in the collector. So injection and depletion together cover the whole range of phenomena in the collector epilayer. Fig. 3.15 shows the possible field and depletion charge distributions. For $J_c < J_{hc}$ we have the ohmic depletion and injection cases, for $J_c \geqslant J_{hc}$ the saturated drift velocity cases.

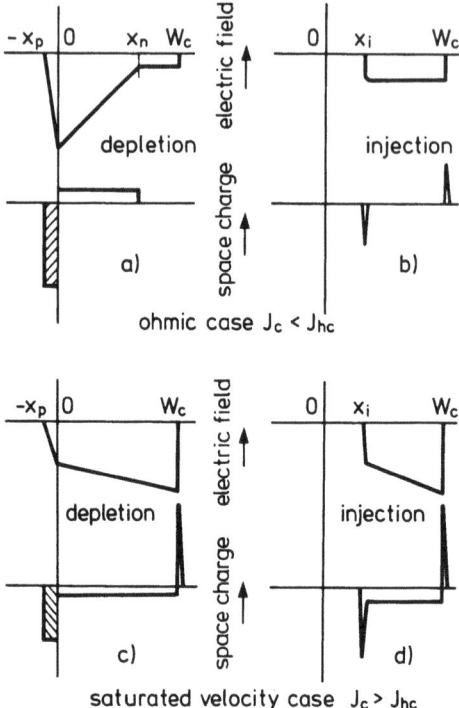

Fig. 3.15 The field and space charge distributions in the collector epilayer (n-type) for the various depletion and injection conditions

The boundary between depletion and injection is given by the condition $V_{ci} = 0$ in Eq. (3.75). For $I_c < I_{hc}$ this leads to

$$V_{cb} + V_{dc} = I_c R_{epi}. \tag{3.77}$$

The same condition makes $x_n + x_p$ in Eq. (3.60) zero as well, but for $I_c > I_{hc}$ there exists a transition gap between $V_{ci} = 0$ and $x_n + x_p = 0$.
The injection region ($0 < x < x_i$) can also be considered as an extension of the neutral base region: The quasi-saturation is then described as base widening or base push-out. It modulates the collector epilayer resistance [3.23]

$$R_{cv} = \frac{R_{epi}}{1 + \exp\left\{-2\left(1 - 2\frac{x_i}{W_c}\right)\right\}}, \tag{3.78}$$

where x_i depends on V_{cb} and I_c according to Eq. (3.74). Another result of the base push-out is an increase of the forward base transit time.
Eq. (3.27) is rewritten as [3.24]

3.7 Quasi-Saturation, Base Widening and Kirk Effect

$$Q_{be} + Q_{epi} = B_f I_f = \left(1 + \frac{x_i}{W_b}\right)^2 \tau_f I_f; \qquad (3.79)$$

the base transit time is proportional to the square of the base width.

3.7.3 Inverse Mode of Operation

The afore-mentioned calculation of charge injection into the collector epilayer can also be used in the inverse mode of operation. Then $V_{cb} < 0$ and $x_i = W_c$. The boundary condition $p(x_i) = n_i$ is replaced by $p(W_c)$ as given by

$$p(W_c)\{p(W_c) + N_{epi}\} = n_i^2 \exp(-V_{cb}/U_T)$$

or

$$\frac{p(W_c)}{N_{epi}} = -\frac{1}{2} + \frac{1}{2}\sqrt{1 + 4\frac{n_i^2}{N_{epi}^2}\exp\left(-\frac{V_{cb}}{U_T}\right)}$$

$$= -\tfrac{1}{2} + \tfrac{1}{2}\sqrt{1 + 4\exp\{-(V_{cb} + V_{dc})/U_T\}}. \qquad (3.80)$$

Eq. (3.68) is transformed into

$$W_c = \frac{2qD_n}{|J_n|}\left\{p(0) - p(W_c) + \frac{1}{2}N_{epi}\ln\frac{p(0)}{p(W_c)}\right\}$$

or, similar to Eq. (3.69),

$$2\frac{p(0)}{N_{epi}} + \ln\frac{p(0)}{N_{epi}} = \frac{I_c R_{epi}}{U_T} + \frac{2p(W_c)}{N_{epi}} + \ln\frac{p(W_c)}{N_{epi}}.$$

This is again in the form of Eq. (3.70), but now

$$V_{ci} = I_c R_{epi} + U_T\left\{2\frac{p(W_c)}{N_{epi}} + \ln\frac{p(W_c)}{N_{epi}}\right\}. \qquad (3.81)$$

Thus Eqs. (3.70), (3.80) and (3.81) completely define the injection in the reverse mode.
However, Eq. (3.80) is also useful in the forward mode with $V_{cb} > 0$. The exponential term is very small and $p(W_c) \approx N_{epi}\exp\{-(V_{cb} + V_{dc})/U_T\}$. For V_{ci} we now have $V_{ci} = I_c R_{epi} - V_{cb} - V_{dc}$, exactly as in the ohmic case of Eq. (3.75). By means of integration we can define a charge Q_0 depending on $p(0)$ and therefore on V_{ci}, and a charge Q_{W_c}, depending on $p(W_c)$ and V_{cb} [3.25]; $Q_0 + Q_{W_c} = Q_{epi}$.
Finally we can point out that the boundary condition (3.46) for $n(W_b)$ is now completed with Eqs. (3.70), (3.80) and (3.81). As an example we take the forward mode

$$\frac{p(W_c)}{N_{epi}} \approx \exp\{-(V_{cb} + V_{dc})/U_T\} \ll 1$$

and
$$V_{ci} \approx (I_c R_{epi} - V_{cb} - V_{dc})$$
$$\approx -(V_{cb} + V_{dc}).$$

Then
$$p(0) \approx N_{epi} \exp\left(-\frac{V_{cb} + V_{dc}}{U_T}\right)$$

and
$$n(W_b)\frac{\hat{N}_a}{f_b} \approx N_{epi}^2 \exp\left(-\frac{V_{cb} + V_{dc}}{U_T}\right)$$
$$= n_i^2 \exp\left(-\frac{V_{cb}}{U_T}\right).$$

The concentration $n(W_b)$ behaves as if V_{cb} was directly applied to the junction.

3.8 Avalanche Multiplication

In section 2.6 we have derived that under weak avalanche conditions the generated avalanche current is given by Eq. (2.80):
$$I_{av} = I_n(0) \int_0^W \alpha_n \, dx.$$

The total current is $I_n(0) + I_{av} = M_n I_n(0)$ with M_n as the multiplication factor (for electrons).
In a bipolar transistor the multiplication usually takes place around the b-c junction, because there the electric field has the highest values. For the initial current $I_n(0)$ we must take here the (electron) collector current $I_c = I_n$ from Eq. (3.65). The avalanche multiplication is incorporated in the bipolar circuit model by adding a current source I_{av} parallel to the internal b-c juction (b'c' in Fig. 3.16): this increases the total collector current, but decreases the total base current. The latter may even change sign! We will discuss now two methods for the evaluation of I_{av}.

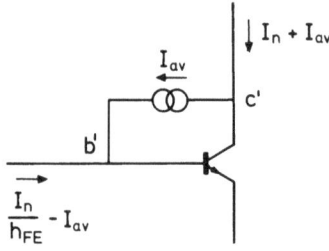

Fig. 3.16 Incorporation of the avalanche current source (I_{av})

3.8 Avalanche Multiplication

The first method uses Miller's [3.26] empirical formula for the multiplication factor M:

$$M = \frac{1}{1 - \left(\dfrac{V_{cb}}{V_{cb0}}\right)^{n_1}}. \tag{3.82}$$

The voltage V_{cb} is the (internal) base-collector voltage, V_{cb0} is the breakdown voltage ($M \to \infty$) and n_1 is a fit parameter whose value is usually between 4 and 7 for silicon. The I_{av} is now given by $I_{av} = (M - 1)I_n$, but Dutton [3.27] introduces an extra fit parameter k_1 and writes

$$I_{av} = k_1(M - 1)I_n. \tag{3.83}$$

Equations (3.82) and (3.83) together give a complete description of the generated avalanche current as a function of I_n and V_{cb}; three parameters are required (n_1, k_1, V_{cb0}), but k_1 has no physical meaning, and is only added for a better fit. According to Eq. (3.83), I_{av} increases linearly with the initial collector current I_n, which is not always found in practice. The above-mentioned description can be improved by making k_1 and n_1 current-dependent [3.28]:

$$\left. \begin{array}{l} k_1' = k_1 I_n^{k_2} \exp(k_3 I_n), \\ n_1' = n_1 I_n^{n_2} \exp(n_3 I_n). \end{array} \right\} \tag{3.84}$$

We have now seven parameters: $k_1 - k_2 - k_3$, $n_1 - n_2 - n_3$ and V_{cb0}. Given a proper choice of these parameters, I_{av} first increases and then decreases with I_n.

The second method uses Chynoweth's [2.26] formula for α_n (see Eq. (2.77)) and tries to evaluate $\int \alpha_n \, dx$ [3.29]:

$$I_{av} = I_n \int_0^{x_n} A_n \exp\left(-\frac{b_n}{|F|}\right) dx. \tag{3.85}$$

With an electric field distribution as sketched in Figs. 3.10 or 3.15 a the series expansion for $|F|$ becomes

$$\frac{1}{|F|} = \frac{1}{F_{max}} \left(1 + \frac{x}{x_n} + \cdots \right).$$

Substituting this in Eq. (3.85) we get

$$I_{av} \approx I_n A_n \int_0^{x_n} \exp\left\{-\frac{b_n}{F_{max}}\left(1 + \frac{x}{x_n}\right)\right\} dx$$

$$= I_n A_n \exp\left(-\frac{b_n}{F_{max}}\right) x_n \int_0^1 \exp\left(-\frac{b_n}{F_{max}} \frac{x}{x_n}\right) d\left(\frac{x}{x_n}\right)$$

$$= I_n A_n x_n \exp\left(-\frac{b_n}{F_{max}}\right)\left[-\frac{F_{max}}{b_n}\left\{\exp\left(-\frac{b_n}{F_{max}}\right) - 1\right\}\right].$$

Because $\exp(-b_n/F_{max}) \ll 1$, we finally have

$$I_{av}/I_n \approx +\frac{A_n}{b_n} x_n F_{max} \exp\left(-\frac{b_n}{F_{max}}\right). \tag{3.86}$$

For A_n and b_n we use the values mentioned in section 2.6; x_n is given by Eq. (3.60) and from the fact that

$$dF/dx = \frac{qN_{epi}}{\varepsilon}\left(1 - \frac{I_n}{I_{hc}}\right)$$

F_{max} follows:

$$F_{max} = \frac{qN_{epi}}{\varepsilon}\left(1 - \frac{I_n}{I_{hc}}\right)x_n. \tag{3.87}$$

The model for the avalanche current is completely described by Eqs. (3.60), (3.86) and (3.87) and several effects are included: the internal voltage drop and the space charge modulation due to saturated drift velocity (Kirk effect). Neglected, however, is the voltage drop on the p-side of the b-c junction. If we take this into account as well, F_{max} in Eq. (3.87) must be lowered by an amount ΔF_{max} [3.29], which is a complicated functon of I_n and V_{cb} that requires many parameters for a good fit. An advantage of the formulation of Eqs. (3.60), (3.86) and (3.87) is its sound physical basis. At low currents I_{av} increases with I_n, but for high I_n values the internal voltage drop and the Kirk effect lower F_{max} and therefore I_{av}.

We can also use Eq. (3.86) in such a manner that in principle no extra parameters are required, i.e. using the already existing description of the collector depletion charge and capacitance, given in section 3.5. The depletion layer width x_d must replace x_n in Eq. (3.86),

$$x_d = x_n + x_p = \frac{\varepsilon A_{em}}{C_{T_c}}, \tag{3.88}$$

while

$$F_{max} = \frac{Q_{T_c} + C_{0c}V_{dc}}{\varepsilon A_{em}}. \tag{3.89}$$

Q_{T_c} and C_{T_c} are related to the active part of the transistor. At equilibrium $Q_{T_c} = 0$ in Eq. (3.89), but $F_{max} \neq 0$, hence the additional term $C_{0c}V_{dc}$. C_{T_c} and Q_{T_c} follow from Eqs. (3.57) and (3.58), so we get

$$x_d F_{max} = \frac{1}{1-p_c}(V_{dc} - V_j)$$

and

$$b_n/F_{max} = \frac{b_n \varepsilon A_{em}}{Q_{T_c} + C_{0c}V_{dc}} = \frac{b_n \varepsilon A_{em}(1-p_c)}{C_{0c}V_{dc}^{p_c}}(V_{dc} - V_j)^{p_c - 1}.$$

Substitution of these into Eq. (3.86) gives

3.9 Series Resistances

$$I_{av}/I_n = AV_1(V_{dc} - V_j)\exp\{-AV_2(V_{dc} - V_j)^{p_c-1}\}$$

with (3.90)

$$AV_1 = \frac{A_n}{b_n}\frac{1}{1-p_c} \quad \text{and} \quad AV_2 = \frac{\varepsilon b_n(1-p_c)A_{em}}{C_{0c}V_{dc}^{p_c}}.$$

A_n, b_n and ε are known quantities for silicon, while p_c, C_{0c} and V_{dc} are existing depletion parameters, so no extra parameters seem to be needed. However, I_{av} is very sensitive to the value of AV_2, so it is wise to make AV_2 a separate parameter.

Eq. (3.90) gives I_{av}/I_n as a function of the internal junction voltage V_j, so collector voltage drops are taken into account.

A drawback of using Eq. (2.80) as a starting point is its limitation to weak avalanche ($I_{av} \ll I_n$). Equations like (3.86) and (3.90) cannot be used near or at the junction breakdown voltage. As a practical limit we can use the sustaining voltage BV_{ce0} [3.30] for base current drive.

Another limitation is that field distributions like the ones sketched in Figs. 3.15 c and d are not represented.

Eq. (3.82) has no such limitations, but a good fit to the experimental data requires a number of fit parameters without clear physical significance.

3.9 Series Resistances

In the devices internal voltage drops will arise as soon as currents start to flow. Because of these voltage drops the internal junction voltage differs from the externally applied voltage; see Fig. 3.17. We can distinguish in Fig. 3.17 the emitter, base and collector series resistances.

3.9.1 Emitter Series Resistance

The emitter series resistance consists of contact resistance plus bulk resistance and is usually modelled as a lumped linear resistor. Its value

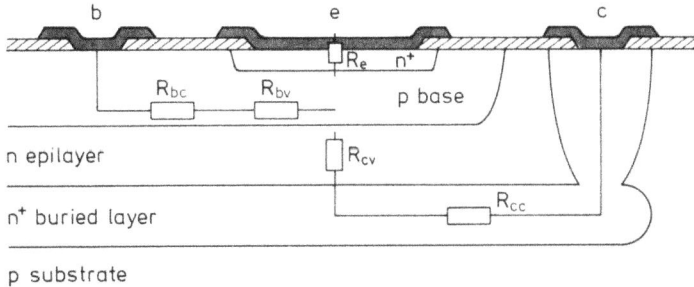

Fig. 3.17 Cross-section of a typical *npn* transistor with indication of the series resistances

depends on the type of contact and the contact and emitter areas. For silicide contacts we can reckon with 10 Ω for a square micron of contact area; aluminium contacts have 50–100 Ωμm² and for polysilicon we may use 100–200 Ωμm².

3.9.2 Base Resistance

The base resistance has a constant part R_{bc} for the inactive region outside the emitter and a variable part R_{bv} for the base region under the emitter (pinched base).
In Fig. 3.18 a the base current I_b enters the pinched base region at $x = 0$ and diminishes gradually in the positive x-direction because an increasing part disappears into the emitter.
We have [3.31]:

$$\frac{dI(x)}{dx} = \frac{J_e(x)}{h_{FE}} L_e. \tag{3.91}$$

$J_e(x)/h_{FE}$ is the density of the current that disappears into the emitter. The lateral current $I(x)$ causes a lateral voltage drop in the pinched base, the base voltage $V(x)$ becomes position dependent and so does the emitter current,

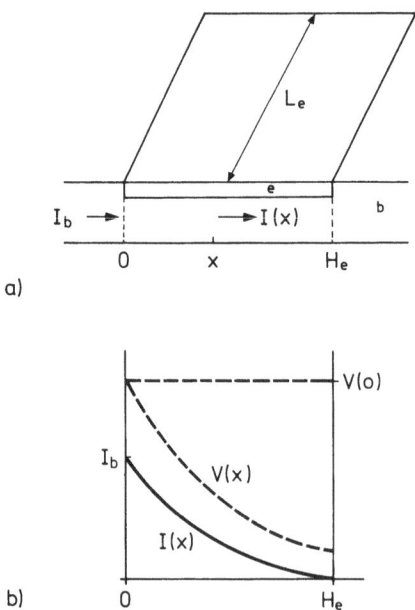

Fig. 3.18 *a.* geometry of the pinched base region; *b.* current and potential distribution in the pinched base

3.9 Series Resistances

$$J_e(x) = J_0 \exp\{V(x)/U_T\}. \tag{3.92}$$

This effect is called emitter-base current crowding. When the lateral voltage drop reaches the value of several $k_B T$, the emitter current injection has become very inhomogeneous: The emitter current is pushed to the edges. The lateral voltage drop is given by

$$\frac{dV(x)}{dx} = -\frac{\rho_\square}{L_e} I(x), \tag{3.93}$$

where $\rho_\square$ is the sheet resistance of the pinched base region. Combining Eqs. (3.91), (3.92) and (3.93) results in [3.31]

$$\frac{d^2 I}{dx^2} = -\frac{1}{U_T} \frac{\rho_\square}{L_e} I \frac{dI}{dx} \tag{3.94}$$

with the boundary conditions $I(0) = I_b$ and $I(H_e) = 0$, see Fig. 3.18 b. We get a rather complicated solution:

$$\left. \begin{array}{l} I(x) = I_b \dfrac{\tan Z(1 - x/H_e)}{\tan Z}, \\[6pt] V(x) = V(0) - 2U_T \ln \dfrac{\cos Z(1 - x/H_e)}{\cos Z}, \\[6pt] Z \tan Z = \dfrac{1}{2} \dfrac{H_e}{L_e} \rho_\square \dfrac{I_b}{U_T}. \end{array} \right\} \tag{3.95}$$

The resistance R_{bv} can be defined in various ways [3.32], e.g. by means of dissipated power

$$P_{\text{diss.}} = I_b^2 R_{bv} = -\int_0^{H_e} I \, dV = -\int_0^{H_e} I \frac{dV}{dx} dx = \frac{\rho_\square}{L_e} \int_0^{H_e} I^2 \, dx.$$

Thus R_{bv} is defined as

$$R_{bv} = \frac{\rho_\square}{L_e} \frac{1}{I_b^2} \int_0^{H_e} I^2(x) \, dx. \tag{3.96}$$

The results of Eqs. (3.95) and (3.96) can be simplified by neglecting the current crowding: $J_e(x)$ is constant, $dI/dx = -I_b/H_e$. We then get

$$\left. \begin{array}{l} I(x) = I_b(1 - x/H_e), \\[4pt] V(x) = V(0) - \dfrac{\rho_\square}{L_e} I_b(1 - \tfrac{1}{2} x/H_e) x, \\[6pt] R_{bv} = \dfrac{1}{3} \rho_\square \dfrac{H_e}{L_e}. \end{array} \right\} \tag{3.97}$$

This value of R_{bv} may be regarded as the low current limit (crowding negligible). It only depends on the $\rho_\square$ and the geometry.

For two base contacts on either side of the emitter R_{bv} becomes $\frac{1}{12}\rho_\square(H_e/L_e)$, that is two resistors in parallel, each with the value $\frac{1}{3}\rho_\square(\frac{1}{2}H_e/L_e)$.
With equal contacts on all four sides we get

$$R_{bv} = \frac{1}{6}\rho_\square \frac{H_e L_e}{(H_e + L_e)^2}.$$

A square emitter thus gives $\rho_\square/24$, whereas for a circle R_{bv} amounts to $\rho_\square/8\pi$. The general solution of Eq. (3.95) can be modelled with good accuracy [3.33, 3.34] by means of a diode in parallel with a resistor r_{b_2},

$$I_b = \frac{2U_T}{r_{b_2}}\left\{\exp\left(\frac{V_{bb'}}{U_T}\right) - 1\right\} + \frac{V_{bb'}}{r_{b_2}}, \qquad (3.98)$$

Here

$$r_{b_2} = \rho_\square \frac{H_e}{L_e} \qquad \text{for 1 base contact},$$

$$= \frac{1}{4}\rho_\square \frac{H_e}{L_e} \qquad \text{for 2 base contacts, etc.}$$

Fig. 3.19 shows how R_{bv} decreases with current, and what differences the various definitions [3.32] ($V_{bb'}/I_b$, $dV_{bb'}/dI_b$ and Eq. (3.96)) give.
If the conductivity of the pinched base is modulated by minority carrier injection (stored charges Q_{be} and Q_{bc}) or by depletion layer widths (depletion charges Q_{T_e} and Q_{T_c}), the sheet resistance $\rho_\square$ in the foregoing formulas must

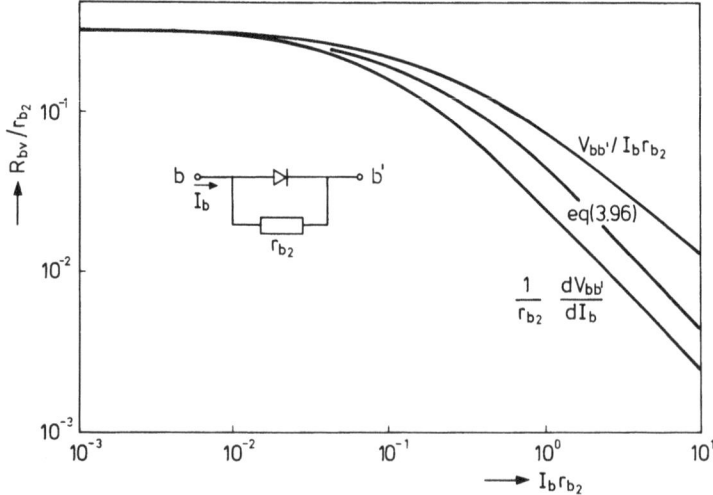

Fig. 3.19 Current dependence of the pinched base resistance R_{bv} as d.c. and a.c. resistance and according to the dissipation criterion

3.9 Series Resistances

be replaced by

$$\frac{\rho_{\square}}{1 + \frac{Q_{T_e} + Q_{T_c} + Q_{be} + Q_{bc}}{Q_{b0}}}.$$

3.9.3 Collector Series Resistance

The collector series resistance is formed by the contributions of the collector contact diffusion plus the lateral resistance of the buried layer (R_{cc} in Fig. 3.17) and the collector epilayer (R_{cv}). This epilayer resistance can be modulated by e.g. base widening, see Eq. (3.78), but the unmodulated value is $R_{epi} = \rho_{epi} \cdot W_c/A_{em}$ with $A_{em} = H_e \cdot L_e$. If the emitter dimensions are small with respect to the thickness of the collector epilayer, we have current spreading in the epilayer and R_{epi} becomes

$$R_{epi} = \rho_{epi} \frac{W_c}{A_{eff}},$$

where $A_{eff} > A_{em} = H_e L_e$. A good approximation [3.35] is

$$A_{eff} = \frac{A_{em}}{1 - 0.84 \frac{W_c}{\sqrt{A_{em}}}}, \qquad (3.99)$$

provided that $W_c < \frac{1}{2}\sqrt{A_{em}}$.

For multi-emitter structures the calculation of the collector spreading resistance is rather complicated; we can use the power dissipation criterion again [3.36]:

$$I_c^2 R_{epi} = \iint VJ \, dA = \sum_{j=1}^{N} I_j \bar{V}_j$$

and

$$\bar{V}_j = \sum_{k=1}^{N} I_k r_{jk},$$

so (3.100)

$$R_{epi} = \sum_{j=1}^{N} \sum_{k=1}^{N} \frac{I_j I_k}{I_c^2} r_{jk}.$$

Here $\bar{V}_j$ is the mean potential value of the j-th emitter stripe, as caused by currents I_k flowing through all the N emitter stripes. The resistance coefficients r_{jk} have the property that $r_{jk} = r_{kj}$ and that for equal emitter stripes $r_{11} = r_{22} = r_{33}$ etc.

Applying Eq. (3.100) on the assumption that all emitter stripes carry the same current $[I_j = I_k = (1/N)I_c]$ gives for

1 stripe $\quad R_{epi} = R_1 = r_{11}$,

2 stripes $\quad R_{epi} = R_2 = \frac{1}{2}r_{11} + \frac{1}{2}r_{12}$,

3 stripes $\quad R_{epi} = R_3 = \frac{1}{3}r_{11} + \frac{4}{9}r_{12} + \frac{2}{9}r_{13}$, etc.

The self-resistance coefficients (r_{jj}) may use Eq. (3.99), the mutual resistance coefficients (r_{jk}) are approximated [3.36] by

$$r_{jk} = \frac{\rho_{epi}}{2\pi}\left(\frac{1}{d_{jk}} - \frac{1}{\sqrt{d_{jk}^2 + 2W_c^2}}\right). \tag{3.101}$$

Here d_{jk} is the distance between the centres of the j-th and k-th emitter stripes.

3.10 Time- and Frequency-Dependent Behaviour

3.10.1 Charge Control and Quasi-Static Approach

In this section we will first use the quasi-static approach and the charge control formulation in a.c. and transient cases. Each variable $I(t)$ can be split into the d.c. value I and a time-dependent value $\tilde{i}$. The harmonic time dependences in particular will be written as

$$\tilde{i}\exp(j\omega t),$$

so

$$I(t) = I + \tilde{i}\exp(j\omega t).$$

We will only consider the normal forward mode of operation. Linearizing Eq. (3.47) we find for the small-signal a.c. current $\tilde{i}_f$

$$\tilde{i}_f = \frac{I_f}{U_T}\tilde{v}_{be} = g_m\tilde{v}_{be}, \tag{3.102}$$

where g_m is called the transconductance.
The charge control relation (3.27) gives

$$q_{be} = \tau_f\tilde{i}_f = \tau_f g_m\tilde{v}_{be}. \tag{3.103}$$

The a.c. base current follows from Eqs. (3.16) and (3.53):

$$\tilde{i}_b = \frac{I_f}{\beta_f U_T}\tilde{v}_{be} + j\omega\tilde{q}_{be}$$

$$= g_m\left(\frac{1}{\beta_f} + j\omega\tau_f\right)\tilde{v}_{be}. \tag{3.104}$$

In common-emitter configuration the input admittance is

$$Y_{ie} = \left.\frac{\tilde{i}_b}{\tilde{v}_{be}}\right|_{\tilde{v}_{ce}=0} = \frac{g_m}{\beta_f} + j\omega\tau_f g_m$$

and the transfer admittance is

3.10 Time- and Frequency-Dependent Behaviour

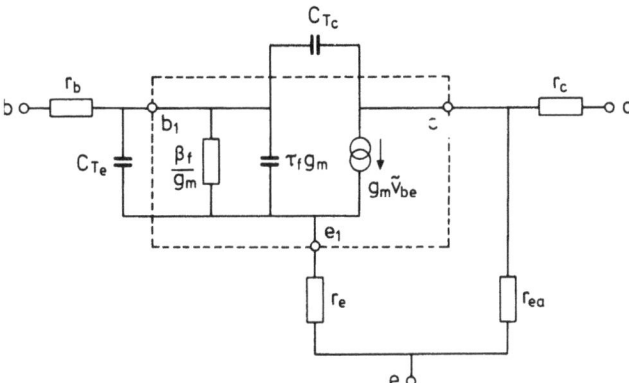

Fig. 3.20 Linear a.c. equivalent circuit. Outside the dotted line area the parasitic elements have been added

$$Y_{fe} = \frac{\tilde{i}_f}{\tilde{v}_{be}}\bigg|_{\tilde{v}_{ce}=0} = g_m.$$

In Fig. 3.20 the equivalent circuit diagram is shown within the dotted lines. The diagram is completed by adding (shown outside the dotted lines) series resistances, depletion capacitances and Early effect (parallel resistor r_{ea}). The a.c. current gain

$$h_{fe} = \frac{Y_{fe}}{Y_{ie}} = \frac{\beta_f}{1 + j\omega\beta_f\tau_f};$$

$|h_{fe}| = 1$ at $\omega = \omega_T = \tau_f^{-1}$, the cut-off frequency, provided that $\beta_f \gg 1$.
For the common base configuraton we have

$$Y_{ib} = (1 + 1/\beta_f)g_m + j\omega\tau_f g_m \quad \text{and} \quad Y_{fb} = -g_m,$$

so the current gain is now

$$h_{fb} = \frac{Y_{fb}}{Y_{ib}} = -\frac{\alpha_f}{1 + j\omega\alpha_f\tau_f} \quad \text{with} \quad \alpha_f = \frac{\beta_f}{1 + \beta_f}.$$

At the alpha cut-off frequency,

$$\omega_{ca} = (\alpha_f\tau_f)^{-1} = \frac{1 + \beta_f}{\beta_f}\omega_T,$$

$|h_{fb}|$ is a factor of $\sqrt{2}$ lower than at low frequencies.

3.10.2 Exact One-Dimensional Solution

In order to check the accuracy of the charge control and quasi-static approximations we shall compare the foregoing results with the "exact" solutions of the one-dimensional differential equation for the minority carrier concentration in the neutral base region (see e.g. Fig. 3.6). This differential equation is obtained from the electron continuity equation (2.48) without generation and recombination ($G - R = 0$) and the flux equation (2.43) [3.37],

$$q\frac{\partial n}{\partial t} = \frac{\partial J_n}{\partial x},$$

$$J_n = q\mu_n n F + qD_n\frac{\partial n}{\partial x}.$$

If we take for F the value for the built-in electric field as given by Eq. (3.34),

$$F = -U_T \eta/W_b,$$

we get

$$q\frac{\partial n}{\partial t} = qD_n\frac{\partial^2 n}{\partial x^2} - \mu_n k_B T \frac{\eta}{W_b}\frac{\partial n}{\partial x}.$$

Separating the d.c. and a.c. parts ($\bar{n}$ and $\tilde{n}$, respectively) and using Einstein's relation we finally have

$$j\omega\tilde{n} = D_n\frac{d^2\tilde{n}}{dx^2} - D_n\frac{\eta}{W_b}\frac{d\tilde{n}}{dx}. \tag{3.105}$$

The solution of Eq. (3.105) is [3.36]

$$\tilde{n}(x) = \tilde{n}(0)\exp\left(\frac{\eta x}{2W_b}\right)\frac{\sinh\left\{\left(1 - \frac{x}{W_b}\right)\sqrt{\left(\frac{\eta}{2}\right)^2 + j\Omega}\right\}}{\sinh\sqrt{\left(\frac{\eta}{2}\right)^2 + j\Omega}}, \tag{3.106}$$

where $\tilde{n}(0) = \bar{n}(0)(\tilde{v}_{be}/U_T)$, as follows from Eq. (3.45) for low injection conditions. The normalized frequency is

$$\Omega = \frac{W_b^2}{D_n}\omega = \frac{\eta^2}{\eta - 1 + 1/f_b}\tau_f\omega,$$

according to Eq. (3.42). The electron current $\tilde{i}_n(x)$ follows from

$$\tilde{i}_n(x) = qD_n A_{em}\left(\frac{d\tilde{n}}{dx} - \frac{\eta}{W_b}\tilde{n}\right).$$

The a.c. emitter current equals $\tilde{i}_n(0)$,

$$\tilde{i}_e = qD_n A_{em}\left\{\frac{d\tilde{n}}{dx}\bigg|_{x=0} - \frac{\eta}{W_b}\tilde{n}(0)\right\}$$

and the collector current $\tilde{i}_c$ equals $\tilde{i}_n(W_b)$,

3.10 Time- and Frequency-Dependent Behaviour

$$\tilde{i}_c = \tilde{i}_f = qD_n A_{em} \frac{d\tilde{n}}{dx}\bigg|_{x=W_b}.$$

With the solution (3.106) we get, after extensive mathematical manipulation, for the admittances in common base [3.36]

$$Y_{ib} = \left(1 + \frac{1}{\beta_f}\right) \frac{g_m}{\eta} \frac{f_b - 1}{f_b} \left\{\frac{\eta}{2} + \sqrt{\left(\frac{\eta}{2}\right)^2 + j\Omega} \coth \sqrt{\left(\frac{\eta}{2}\right)^2 + j\Omega}\right\}$$
(3.107)

and

$$Y_{fb} = -g_m \frac{\sinh\left(\frac{\eta}{2}\right)}{\eta/2} \frac{\sqrt{\left(\frac{\eta}{2}\right)^2 + j\Omega}}{\sinh \sqrt{\left(\frac{\eta}{2}\right)^2 + j\Omega}}.$$
(3.108)

To simplify things we will now assume a constant base doping, so $\eta = 0$. The admittances then become

$$Y_{ib} = g_m \left(1 + \frac{1}{\beta_f}\right) \sqrt{j\Omega} \coth \sqrt{j\Omega}$$

and

$$Y_{fb} = -g_m \frac{\sqrt{j\Omega}}{\sinh \sqrt{j\Omega}}.$$

For $\eta = 0$, $\Omega = 2\omega\tau_f$.
By making a series expansion we get in the low-frequency limit

$$Y_{ib} \approx g_m \left(1 + \frac{1}{\beta_f}\right) \sqrt{j\Omega} \left(\frac{1}{\sqrt{j\Omega}} + \frac{1}{3}\sqrt{j\Omega} + \cdots\right)$$

$$\approx g_m \left(1 + \frac{1}{\beta_f}\right) + \frac{1}{3} g_m \left(1 + \frac{1}{\beta_f}\right) j\Omega$$

$$= g_m \left(1 + \frac{1}{\beta_f}\right) + j\omega \frac{2}{3} g_m \tau_f$$

and

$$Y_{fb} \approx -g_m \frac{\sqrt{j\Omega}}{\sqrt{j\Omega} + \frac{1}{6} j\Omega \sqrt{j\Omega} + \cdots}$$

$$\approx \frac{-g_m}{1 + \frac{1}{3} j\omega \tau_f}.$$

In common emitter configuration the admittances become

$$Y_{ie} = Y_{ib} + Y_{fb} \approx g_m\left(1 + \frac{1}{\beta_f}\right) + \frac{2}{3}j\omega\tau_f g_m - g_m + \frac{1}{3}j\omega\tau_f g_m$$

$$= \frac{g_m}{\beta_f} + j\omega\tau_f g_m$$

and

$$Y_{fe} = -Y_{fb} = \frac{g_m}{1 + \frac{1}{3}j\omega\tau_f}.$$

In comparison with the quasi-static results we notice the following important differences:

— in common base the input capacitance is $\frac{2}{3}g_m\tau_f$ instead of $g_m\tau_f$,
— both transadmittances Y_{fb} and Y_{fe} are no longer real, but become complex: they show an extra phase shift.

These essential differences remain, even when $\eta \neq 0$; in the latter case the numerical factors of $\frac{1}{3}$ and $\frac{2}{3}$ become functions of η. Thus the "exact" solution for the frequency behaviour has shown a few essential limitations of the quasi-static approach. There are various ways to remedy this.

3.10.3 Time Delays

The first method is to introduce time delays between I_f and Q_{be} and between Q_{be} and V_{be} [3.38],

$$I_f(t) = \frac{1}{\tau_f} Q_{be}(t - \tau_1)$$

and

$$Q_{be}(t) = Q_{be}\{V_{be}(t - \tau_2)\}.$$

It can be shown [3.38] that the delay times τ_1 and τ_2 depend on the doping profile in the base and therefore on η,

$$\left.\begin{array}{l} \tau_1 = (\frac{1}{6} + \frac{2}{45}\eta + \cdots)\tau_f, \\ \tau_2 = (\frac{1}{6} + \frac{1}{15}\eta + \cdots)\tau_f. \end{array}\right\} \tag{3.109}$$

For $\eta = 0$ the admittances become

$$Y_{fe} = -Y_{fb} = g_m \exp\{-j\omega(\tau_1 + \tau_2)\} = g_m \exp(-\frac{1}{3}j\omega\tau_f)$$

$$\approx \frac{g_m}{1 + \frac{1}{3}j\omega\tau_f}$$

and

3.10 Time- and Frequency-Dependent Behaviour

$$Y_{ib} = g_m \exp\left(-\frac{1}{3}j\omega\tau_f\right) + g_m\left(\frac{1}{\beta_f} + j\omega\tau_f\right)\exp\left(-\frac{1}{6}j\omega\tau_f\right)$$

$$\approx g_m\left(1 + \frac{1}{\beta_f}\right) + \frac{2}{3}j\omega\tau_f g_m.$$

This indeed equals the low-frequency results of the "exact" solution.

3.10.4 Base Charge Partitioning

The second method starts with the electron continuity equation (2.50) in one-dimensional form [3.39],

$$\frac{\partial n}{\partial t} - \frac{1}{q}\frac{\partial J_n}{\partial x} = G - R = 0$$

(generation-recombination neglected).
Multiplying by the emitter area A_{em} and integrating from 0 to x in the base region gives

$$\frac{\partial}{\partial t}\int_0^x qA_{em}n(x', t)\,dx' = I_n(x, t) - I_n(0, t)$$

$$= I_n(x, t) - I_e(t).$$

We now integrate this from 0 to W_b and get

$$I_e(t) = \frac{-1}{W_b}\int_0^{W_b} I_n(x, t)\,dx + \frac{d}{dt}qA_{em}\int_0^{W_b}\left(1 - \frac{x}{W_b}\right)n(x, t)\,dx. \tag{3.110}$$

Integrating inversely, from W_b to 0, gives

$$I_c(t) = I_f(t) = \frac{-1}{W_b}\int_0^{W_b} I_n(x, t)\,dx - \frac{d}{dt}qA_{em}\int_0^{W_b}\frac{x}{W_b}n(x, t)\,dx. \tag{3.111}$$

Under low injection conditions and with $\eta = 0$ we then have

$$Q_{be}(t) = \int_0^{W_b} qA_{em}n(x, t)\,dx,$$

$$\frac{2}{3}Q_{be}(t) = \int_0^{W_b} qA_{em}\left(1 - \frac{x}{W_b}\right)n(x, t)\,dx,$$

$$\frac{1}{3}Q_{be}(t) = \int_0^{W_b} qA_{em}\frac{x}{W_b}n(x, t)\,dx.$$

We further assume that the "old" quasi-static collector current

$$I_f^{\text{q.s.}} = \frac{-1}{W_b} \int_0^{W_b} I_n(x, t)\, dx.$$

So

$$I_e(t) = I_f^{\text{q.s.}}(t) + \frac{2}{3} \frac{dQ_{be}}{dt},$$

$$I_c(t) = I_f^{\text{q.s.}}(t) - \frac{1}{3} \frac{dQ_{be}}{dt},$$

$$I_b(t) = \frac{dQ_{be}}{dt}.$$

There is no d.c. base current here, because recombination has been assumed to be zero, but we can easily add a term $I_f^{\text{q.s.}}(t)/\beta_f$. The linearized a.c. equations then again give the "exact" low-frequency solutions in the case $\eta = 0$. It is also possible to start with a time-dependent integral charge-control equation like Eq. (3.19). A similar derivation [3.40] then leads to a more elaborate, but essentially the same result.

3.10.5 Second-Order Differential Operators

The third method makes use of second order differential operators; we use the results of the quasi-static approach and apply the differential operators afterwards,

$$I_f(s) = \frac{I_f^{\text{q.s.}}}{1 + as\tau_f + b(s\tau_f)^2}. \tag{3.112}$$

In the frequency domain we have $s = j\omega$, and Eq. (3.112) becomes

$$\tilde{i}_f(j\omega) = \frac{\tilde{i}_f^{\text{q.s.}}}{1 - b(\omega\tau_f)^2 + j\omega a\tau_f},$$

where $\tilde{i}_f^{\text{q.s.}}$ is the quasi-static a.c. solution. In the time domain $I_f(t)$ is found by solving Eq. (3.112) for $s = d/dt$,

$$b\tau_f^2 \frac{d^2 I_f}{dt^2} + a\tau_f \frac{dI_f}{dt} + I_f = I_f^{\text{q.s.}}(t).$$

$I_f^{\text{q.s.}}(t)$ is the quasi-static solution in the time domain.
The differential operator in Eq. (3.112) can be a Bessel polynomial [3.41], with $b = \frac{1}{3}a^2$. We may also go back to Eq. (3.108) and use a series expansion in $j\Omega$. After some algebraic manipulation we find

$$a = \frac{f_b}{(\eta - 1)f_b + 1} \left\{ \frac{\eta}{\tanh(\eta/2)} - 2 \right\},$$

$$b = \frac{f_b^2}{\{(\eta - 1)f_b + 1\}^2} \left\{ 6 + \frac{1}{2}\eta^2 - \frac{3\eta}{\tanh(\eta/2)} \right\}.$$

For $\eta = 0$ we have $a = \frac{1}{3}$ and $b = \frac{1}{30}$, which we already knew from the approximations for Y_{fe} and Y_{fb}.

A similar differential operator can be applied to Q_{be}, the forward stored charge in the base

$$Q_{be}(s) = \frac{Q_{be}^{q.s.}}{1 + cs\tau_f + d(s\tau_f)^2}. \tag{3.113}$$

When we use Eq. (3.106) for $\tilde{n}(x)$, we can evaluate

$$q_{be} = qA_{em} \int_0^{W_b} \tilde{n}\, dx$$

and make an expansion in terms of $j\Omega$. This will give us the coefficients c and d in Eq. (3.113). The outcome of this procedure is

$$c \approx \frac{1}{2}a \quad \text{and} \quad d \approx \frac{-1}{180} + \frac{1}{30}\eta.$$

The foregoing results relate to low-injection conditions; at high injection the built-in field disappears and a, b, c and d are no longer functions of η, but they reach fixed values:

$$a = \frac{1}{3}, \quad b = \frac{1}{30}, \quad c = \frac{1}{6} \quad \text{and} \quad d = \frac{-1}{180}.$$

The corrections to the quasi-static and charge-control procedures have been applied to the quantities I_f and Q_{be}. They are, in a modified way, also applicable to I_r, Q_{bc}, Q_{epi} and Q_e but we then get a rather complex compact model, that requires a lot of CPU time for evaluation. A second reason to limit the application to I_f and Q_{be} is that often the improvements are overshadowed by phase shifts and delay times brought about by the RC products of series resistances and depletion capacitances (see Fig. 3.20). The delay times associated with the depletion charges Q_{T_e} and Q_{T_c} are determined by the dielectric relaxation of the space charge regions. They are so short that we can assume an instantaneous response of Q_T to a variation in junction voltage.

3.11 Transit Time and Cut-Off Frequency f_T

The a.c. current gain h_{fe} in common emitter follows from

$$h_{fe} = \frac{Y_{fe}}{Y_{ie}} \approx \frac{g_m}{\left(1 + \frac{1}{3}j\omega\tau_f\right)\left(\frac{g_m}{\beta_f} + j\omega\tau_f g_m\right)} \tag{3.114}$$

$$= \frac{\beta_f}{(1 + \frac{1}{3}j\omega\tau_f)(1 + j\omega\tau_f\beta_f)},$$

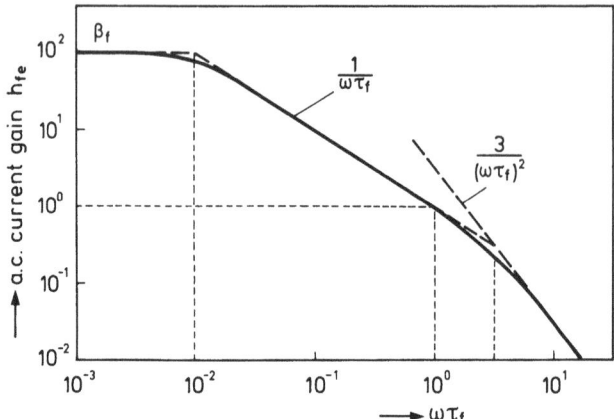

Fig. 3.21 Sketch of the a.c. small-signal current gain (h_{fe}) as a function of frequency. The cut-off frequency f_T is reached at $\omega\tau_f = 1$

see also section 3.10. The h_{fe} as a function of $\omega\tau_f$ is given in Fig. 3.21. Provided that $\beta_f \gg 1$, a certain frequency interval exists, where $|h_{fe}| \approx 1/\omega\tau_f$, with a fall-off of 6 dB per octave. If in this interval the $|h_{fe}|$ reaches the value one, the cut-off frequency f_T is defined as the frequency at which this happens, so $\omega_T = 2\pi f_T = 1/\tau_f$. Using the charge-control relation (3.42) we have $2\pi f_T = I_f/Q_{be}$. All this is only valid for that part of the equivalent circuit in Fig. 3.20 that lies within the dotted lines. The complete equivalent circuit, with the series resistances and depletion capacitances, has more complicated admittances Y_{fe} and Y_{ie}, and $h_{fe}(\omega)$ becomes very cumbersome.
Fortunately the f_T can be calculated by using the charge-control principle, which is a good approximation in this case. So we use

$$\tau_{tot} = 1/2\pi f_T = \frac{\sum Q}{I_c},$$

or even more appropriate, in a.c. cases

$$\tau_{tot} = \sum \frac{dQ}{dI_c}\bigg|_{\tilde{v}_{ce}=0}. \tag{3.115}$$

The equivalent circuit of Fig. 3.20 can be refined still further by splitting up the r_b, r_c and C_{T_c}, as shown in Fig. 3.22.
In Fig. 3.22 r_{bv}, r_{cv} and C_{T_c} are associated with the active transistor part, r_{bc}, r_{cc} and $C_{T_{ex}}$ with the inactive part. For $I_c > I_{hc}$, r_{cv} is current-dependent. Application of Eq. (3.115) gives for the various contributions:

— emitter transit time

$$\tau_e = \frac{dQ_e}{dI_c};$$

it is current dependent [3.42],

3.11 Transit Time and Cut-Off Frequency f_T

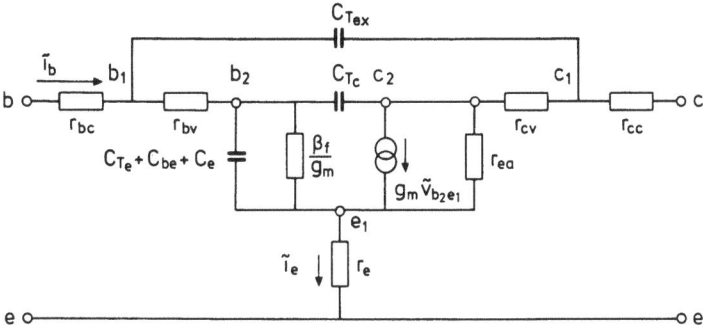

Fig. 3.22 Refinement of the equivalent circuit of Fig. 3.20 by splitting up the base and collector series resistances and the collector depletion capacitance

— base transit time

$$\tau_f = \frac{dQ_{be}}{dI_c};$$

see Fig. 3.8,
— emitter depletion charge

$$\frac{dQ_{T_e}}{dI_c} = \frac{dQ_{T_e}}{dV_{b_2e_1}} \frac{dV_{b_2e_1}}{dI_c} = C_{T_e} \frac{U_T}{I_c} = \frac{C_{T_e}}{g_m} = C_{T_e} r_0,$$

— collector depletion charge in the active region. If the charge modulation due to mobile carriers [Eq. (3.60)] is incorporated, we have for $N_a \gg N_{epi}$

$$x_n + x_p = \sqrt{\frac{2\varepsilon}{qN_{epi}} \frac{V_{c_2b_2} + V_{dc}}{(1 - I_c/I_{hc})}}$$

and

$$Q_{T_c}(V_{c_2b_2}, I_c) = -qN_{epi} A_{em}(1 - I_c/I_{hc})(x_n + x_p);$$

so

$$\left.\frac{dQ_{T_c}}{dI_c}\right|_{\tilde{v}_{ce}=0} = \left.\frac{\partial Q_{T_c}}{\partial V_{c_2b_2}}\right|_{I_c=\text{const.}} \times \frac{\tilde{v}_{c_2b_2}}{\tilde{i}_c} + \left.\frac{\partial Q_{T_c}}{\partial I_c}\right|_{V_{c_2b_2}=\text{const.}}$$

For the abrupt junction we are considering

$$\frac{\partial Q_{T_c}}{\partial I_c} = \frac{x_n + x_p}{2v_s}$$

that is half the carrier transit time in the collector depletion layer [3.42]. Furthermore

$$\tilde{v}_{ce} = \tilde{i}_c(r_{cc} + r_{cv}) + \tilde{v}_{c_2b_2} + \tilde{i}_c r_0 + \left(1 + \frac{1}{h_{fe}}\right) r_e \tilde{i}_c = 0,$$

$$\frac{\tilde{v}_{c_2b_2}}{\tilde{i}_c} = -\left\{\left(1 + \frac{1}{h_{fe}}\right) r_e + r_0 + r_{cc} + r_{cv}\right\}.$$

This makes the collector contribution equal to

$$\left\{\left(1 + \frac{1}{h_{fe}}\right) r_e + r_0 + r_{cc} + r_{cv}\right\} C_{T_c} + \frac{x_n + x_p}{2v_s}.$$

— collector depletion charge in the inactive part of the c-b junction:

$$\left.\frac{dQ_{T_{ex}}}{dI_c}\right|_{\tilde{v}_{ce}=0} = \frac{dQ_{T_{ex}}}{dV_{c_1b_1}} \frac{\tilde{v}_{c_1b_1}}{\tilde{i}_c}.$$

From

$$\tilde{v}_{ce} = \tilde{i}_c r_{cc} + \tilde{v}_{c_1b_1} + \frac{\tilde{i}_c}{h_{fe}} r_{bv} + \tilde{v}_{b_2e_1} + \left(1 + \frac{1}{h_{fe}}\right) \tilde{i}_c r_e = 0$$

it follows that

$$\frac{dQ_{T_{ex}}}{dV_{c_1b_1}} \frac{\tilde{v}_{c_1b_1}}{\tilde{i}_c} = C_{T_{ex}} \left\{ r_{cc} + \frac{r_{bv}}{h_{fe}} + r_0 + \left(1 + \frac{1}{h_{fe}}\right) r_e \right\}.$$

All these contributions together constitute the total delay time between collector and emitter,

$$\tau_{tot} \approx \tau_e + \tau_f + (r_e + r_{cc})(C_{T_c} + C_{T_{ex}}) + r_{cv} C_{T_c}$$

$$+ \frac{r_{bv}}{h_{fe}} C_{T_{ex}} + \frac{x_n + x_p}{2v_s} + \frac{U_T}{I_c}(C_{..} + C_{T_c} + C_{T_{ex}}). \quad (3.116)$$

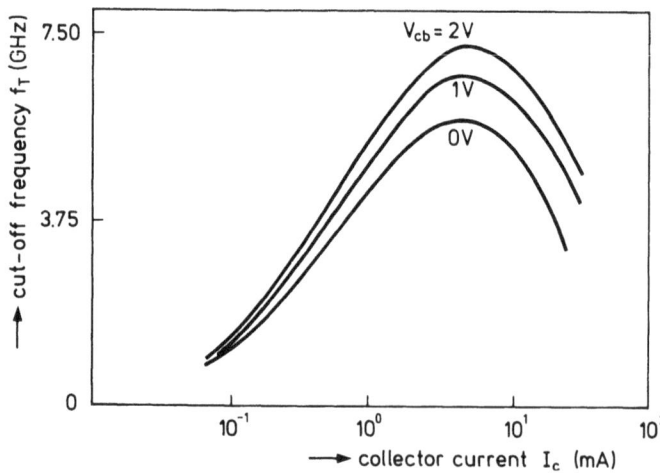

Fig. 3.23 Typical example of the bias dependence of the cut-off frequency f_T

3.12 Noise Behaviour

At low currents the last term in Eq. (3.116) is dominant. At high currents f_T falls off, due to high injection effects in the base that increase τ_e and τ_f, and/or to quasi-saturation. The latter gives rise to delay time contributions like dQ_{epi}/dI_c and dQ_{bc}/dI_c.

The V_{cb}-dependence arises from the direct influence on C_{T_c} and $C_{T_{ex}}$ and the indirect influence via the Early effect on Q_{be} and I_c. Fig. 3.23 gives a typical example of $f_T(V_{cb}, I_c)$.

3.12 Noise Behaviour

There are several noise sources in a bipolar transistor. We can distinguish between "white" noise sources and frequency dependent noise sources (see also section 2.7). The main white noise sources are the thermal noise sources associated with the series resistances and the shot noise sources arising from currents that flow across junctions.

From Eq. (2.82) it follows that we can write for the thermal noise voltage of a resistance R

$$\langle e^2 \rangle = 4k_B T R \, \Delta f. \tag{3.117}$$

Here Δf is the frequency range and T the temperature of the resistance in Kelvin. As equivalent circuit we can use an a.c. resistance (R) in series with a voltage source $\sqrt{4k_B T R \, \Delta f}$ or in parallel with a current source $\sqrt{4k_B T \Delta f / R}$. Thermal noise sources are thus added to the series resistances in the circuit diagram of Fig. 3.22, i.e. so to r_e, r_{cc} and r_{bc}. The base resistance r_{bv} is of a special kind; it is related to the phenomenon of current crowding (see section 3.9), which changes the thermal noise [3.44]. The temperature T in Eq. (3.117) must be replaced by $t_{eff} \cdot T$ with $t_{eff} = 1$ for negligible current crowding and $t_{eff} \approx 0.5$ for heavy current crowding. The a.c. resistance r_{bv} follows from Eq. (3.98),

$$r_{bv} = \frac{dV_{bb'}}{dI_b} = \frac{r_{b_2}}{1 + 2 \exp(V_{bb'}/U_T)}$$

or

$$r_{bv}(V_{bb'})/r_{bv}(0) = \frac{3}{1 + 2 \exp(V_{bb'}/U_T)}.$$

The quantity t_{eff} can be calculated as a function of $r_{bb'}(V_{bb'})/r_{bv}(0)$ [3.44]. A good fit to that calculation is the following,

$$t_{eff} \approx \left\{ \frac{r_{bv}(V_{bb'})}{r_{bv}(0)} \right\}^{1/4} = \left\{ \frac{3}{1 + 2 \exp(V_{bb'}/U_T)} \right\}^{1/4}. \tag{3.118}$$

For the noise current source parallel to r_{bv} we find with the aid of Eqs. (3.117) and (3.118)

$$\langle i_{bv}^2 \rangle = \frac{4k_B T \Delta f}{r_{bv}} t_{\text{eff}}$$

$$= 1.316 \{1 + 2 \exp(V_{bb'}/U_T)\}^{3/4} \frac{4k_B T \Delta f}{r_{b_2}}. \quad (3.119)$$

From Eq. (2.81) it follows that the shot noise for currents across junctions is given by

$$\langle i^2 \rangle = 2qI \Delta f, \quad (3.120)$$

where I is the d.c. current. Such shot noise sources are added to the collector current I_c and to the base current components as given by Eqs. (3.53), (3.54) and (3.55).

In addition to these white noise sources we must also mention the flicker noise or $1/f$ noise which, in bipolar transistors, is mainly attributed to the base current [3.45].

In accordance with Eq. (2.83) we may write

$$\langle i_b^2 \rangle = K_f I_b^{a_f} \frac{\Delta f}{f}. \quad (3.121)$$

The parameters are K_f and a_f; the latter usually has a value between one and two.

Fig. 3.24 shows the equivalent circuit diagram of Fig. 3.22, with all the uncorrelated noise sources added. Although these sources are frequency-independent (with the exception of the flicker noise) the overall noise behaviour of the equivalent circuit does depend on frequency. The noise figure usually increases at high frequencies.

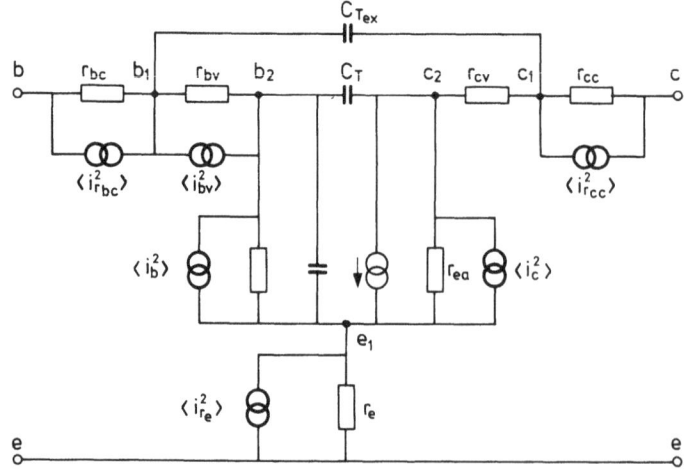

Fig. 3.24 Addition of the relevant white noise current sources to the small-signal equivalent circuit (common emitter configuration)

3.13 Temperature Dependences

Bipolar device characteristics vary with temperature for three reasons: The intrinsic concentration n_i is very sensitive to temperature changes, mobility and diffusion constant depend on temperature, and the impurity charge in the base may not be fully ionized and is then a function of temperature.
The intrinsic concentration n_i varies with temperature according to Eq. (2.53) as

$$n_i^2 = 9.61 * 10^{32} \, T^3 \, \exp(-qV_{g0}/k_P T) \quad (\text{cm}^{-6}).$$

V_{g0} is the bandgap voltage and depends on the dope concentration of the region involved (see also section 2.3).

The diffusion voltage of the depletion capacitance is influenced by n_i. For strong asymmetric junctions we have for the diffusion voltage (see section 3.7)

$$V_d = 2\frac{k_B T}{q} \ln \frac{N_{\text{dope}}}{n_i}. \tag{3.122}$$

Together with Eq. (2.53) for n_i we get

$$V_d(T) = V_{g0} + \frac{k_B T}{q} \ln\left(\frac{N_{\text{dope}}^2}{9.61 * 10^{32} \, T^3}\right). \tag{3.123}$$

Sometimes it is easier to express such functions by means of a reference value V_r at the reference temperature T_r,

$$V_d(T) = \frac{T}{T_r} V_r + \left(1 - \frac{T}{T_r}\right) V_{g0} - 3\frac{k_B T}{q} \ln\left(\frac{T}{T_r}\right).$$

The depletion capacitance depends on V_d and thus on T. The zero volt value for abrupt junctions is (section 3.5)

$$C_0 = \sqrt{\frac{q\varepsilon N}{2V_d(T)}}$$

or, more generally

$$C_0 = \text{const.} \{V_d(T)\}^{-p}. \tag{3.124}$$

The fixed base charge at zero bias (Q_{b0}) depends on temperature because not all the impurities are ionized and the charges in the depletion layers at the emitter and collector sides change with temperature. So we get

$$Q_{b0} = qA_{em} \int_{x_{je}}^{x_{jc}} N_a^-(x) \, dx - \int_0^{V_{de}} C_{T_e} \, dV_j - \int_0^{V_{dc}} C_{T_c} \, dV_j,$$

$$Q_{b0}(T) = g_i(T)qA_{em} \int_{x_{je}}^{x_{jc}} N_a(x) \, dx - \frac{C_{0e}V_{de}}{1-p_e} - \frac{C_{0c}V_{dc}}{1-p_c}. \tag{3.125}$$

C_0 and V_d are given by Eqs. (3.123) and (3.124). The ionization factor g_i is

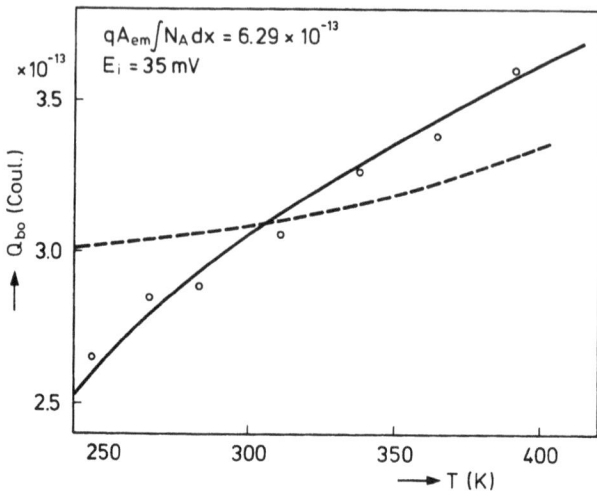

Fig. 3.25 The zero bias base charge as a function of temperature. Dashed line: only the charges in the depletion regions are taken into account. Fully drawn line: incomplete ionization of the dope is also considered. Dots: experimental results

taken for the maximum base dope concentration and follows from Eq. (2.22):

$$g_i(T) \approx \frac{\hat{N}_a^-(T)}{\hat{N}_a} = \frac{2r}{r + \sqrt{r^2 + 8r}} \tag{3.126}$$

with

$$r = \frac{N_v}{\hat{N}_a} \exp\left(-\frac{\Delta E_a}{k_B T}\right).$$

ΔE_a is the acceptor ionization energy, N_v the effective density of states for the valence band (Eq. 2.13): $N_v \propto T^{3/2}$.

Fig. 3.25 gives an example of $Q_{bo}(T)$. The dashed line represents the case when only the temperature dependence of the depletion layer charge is taken into account.

At high doping levels ΔE_a diminishes; for $N_{\text{dope}} > 10^{19}$ cm^{-3} we hardly see freeze-out of impurities, not even at liquid nitrogen temperatures [3.46].

The (majority carrier) mobilities vary with temperature as given by Eq. (2.65),

$$\mu_{n,p} \propto T^{-s_{n,p}}. \tag{3.127}$$

This is an approximation for the range 250–450 K. The exponents $s_{n,p}$ are functions of the dope concentration, see Fig. 2.6. Using the Einstein relation we get for the diffusion constants

$$D_{n,p} \propto T^{1-s_{n,p}}. \tag{3.128}$$

The sheet resistance of the pinched base under the emitter is (see Eq. (3.23))

3.13 Temperature Dependences

$$\rho_\square = \frac{A_{em}}{\bar{\mu}_p Q_{bo}} \propto T^{a_b}. \tag{3.129}$$

For a maximum base dope of $\hat{N}_a \approx 4*10^{17}$ cm^{-3} we have $a_b \approx 0.7$. That is definitely smaller than s_p, because of the temperature dependence of Q_{bo}. The saturation value of the collector current follows from Eq. (3.24);

$$I_s \propto \frac{D_n n_{ib}^2}{Q_{bo}} \propto D_n n_{ib}^2 \bar{\mu}_p \rho_\square,$$

so we can write

$$I_s = \text{const. } T^{m_b} \exp(-qV_{gb}/k_B T) \tag{3.130}$$

in the range 250–450 K.
The value of m_b follows from Eqs. (3.127), (3.128) and (3.129): $m_b = 4 - s_n - s_p + a_b$. Typical values are $m \approx 2.7$. V_{gb} in Eq. (3.130) is the bandgap voltage in the base [3.47].
For the ideal component of the forward base current we have an expression similar to Eq. (3.130); from Eq. (3.53) it follows that

$$I_s/\beta_f = \text{const. } T^{m_e} \exp(-qV_{ge}/k_B T). \tag{3.131}$$

V_{ge} is the bandgap voltage in the emitter; the exponent $m_e \approx 4$ because at the high doping level of the emitter $s_p \approx s_n \approx a_b \approx 0$. It is furthermore assumed that the emitters are more or less transparent for minority carriers (holes) and that the recombination velocity at the emitter contact is infinite. Eqs. (3.130) and (3.131) together determine the current gain parameter $\beta_f(T)$. The overall result is a moderate increase of β_f with temperature.
For the non-ideal component of the forward base current it follows from Eq. (3.55) that $I_{bf} \propto n_i/\tau$,

$$I_{bf} = \text{const. } T^2 \exp(-qV_{gap}/2k_B T). \tag{3.132}$$

Here V_{gap} pertains to the region around the e-b junction. For the parameter V_{LF} we have

$$V_{LF} = 2\Delta E_t = 2(E_t - E_i) - 2k_B T \ln g_t, \tag{3.133}$$

so V_{LF} varies linearly with T.
Fig. 3.26 shows a typical behaviour of the current gain $h_{FE}(T)$.
The cut-off frequency f_T as a function of I_c also varies with temperature. The rising part of $f_T(I_c)$ is dominated by the term $k_B T/qI_c(C_{T_e} + C_{T_c} + C_{T_{ex}})$ in Eq. (3.116). So f_T decreases here with increasing temperature. The maximum f_T is largely determined by the other factors in Eq. (3.116); some of them increase, others decrease with temperature. As an example we will consider the forward transit time τ_f. From Eq. (3.42) we learn that $\tau_f \propto W_b^2/D_n$. The temperature dependence of D_n is given by Eq. (3.128), but W_b also depends on T: $Q_{b0} \propto g_i W_b$ or $W_b \propto Q_{b0}/g_i \propto (\bar{\mu}_p g_i \rho_\square)^{-1}$. Therefore $\tau_f \propto (D_n \bar{\mu}_p^2 \rho_\square^2 g_i^2)^{-1}$ or $\tau_f = \text{const. } T^{-(1-s_n-2s_p+2a_b)} g_i^{-2}$.

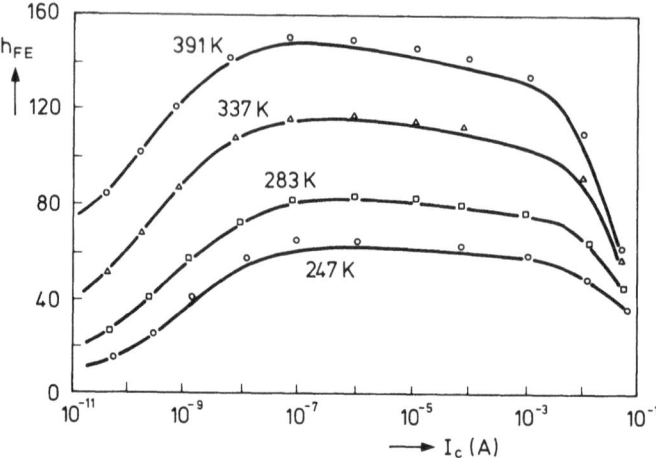

Fig. 3.26 Example of the temperature dependence of the d.c. current gain $h_{FE}(I_c)$

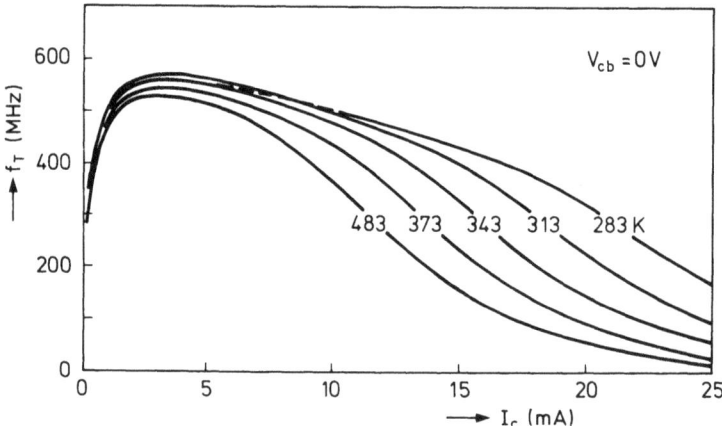

Fig. 3.27 Example of the cut-off frequency f_T as a function of I_c at various temperatures

In the range 250–450 K, g_i can be approximated by $T^{0.3}$ to $T^{0.5}$, with the net result that the forward base transit time τ_f hardly depends on temperature. The fall-off part of $f_T(I_c)$ is due to high injection effects in the base or to quasi-saturation. In both cases f_T has mostly a small negative temperature coefficient. A typical behaviour of $f_T(I_c)$ is shown in Fig. 3.27.

References

[3.1] H. K. Gummel: Measurement of the Number of Impurities in the Base Layer of a Transistor. Proc. I.R.E. 42, 1761 (1954).

References

[3.2] H. C. de Graaff, J. W. Slotboom, A. Schmitz: The Emitter Efficiency of Bipolar Transistors. Solid-State Electr. *20*, 515 (1977).

[3.3] H. Schaber et al.: Process and Device Related Scaling Considerations for Polysilicon Emitter Bipolar Transistors. IEDM Techn. Digest 170 (1987).

[3.4] R. Beaufroy, J. J. Sparkes: The Junction Transistor as a Charge-Controlled Device. Automat. Tel. Eng. J. *13*, 310 (1957).

[3.5] J. te Winkel: Past and Present of the Charge-Control Concept in the Characterization of the Bipolar Transistor. Adv. Electr. and Electr. Phys. *39*, 253 (1975).

[3.6] J. J. Ebers, J. L. Moll: Large Signal Behaviour of Junction Transistors. Proc. I.R.E. *42*, 1761 (1954).

[3.7] J. L. Moll, J. M. Ross: The Dependence of Transistor Parameters on the Distribution of Base Layer Resistivity. Proc. I.R.E. *44*, 72 (1956).

[3.8] H. K. Gummel: A Charge-Control Relation for Bipolar Transistors. Bell Syst. Techn. J. *49*, 115 (1970).

[3.9] H. K. Gummel, H. C. Poon: An Integral Charge-Control Model for Bipolar Transistors. Bell Syst. Techn. J. *49*, 827 (1970).

[3.10] I. Getreu: Modeling the Bipolar Transistor. Tektronix Inc., Beaverton OR (1979).

[3.11] H. C. de Graaff, W. J. Kloosterman: New Formulation of the Current and Charge Relations in Bipolar Transistor Modeling for CACD Purposes. IEEE Trans. Electr. Dev. *ED-32*, 2415 (1985).

[3.12] J. L. Moll: Physics of Semiconductors. McGraw-Hill, New York (1964).

[3.13] H. C. de Graaff: Review of Models for Bipolar Transistors. In: Process and Device Modeling for Integrated Circuit Design (F. van de Wiele, W. L. Engl, P. G. Jespers, eds.). Noordhoff, Leiden (1977).

[3.14] H. C. Poon, H. K. Gummel: Modeling of the Emitter Capacitance. Proc. IEEE *57*, 2181 (1969).

[3.15] C. T. Kirk: A Theory of Transistor Cut-Off Frequency (f_T) Fall-Off at High Current Densities. I.R.E. Trans. Electr. Dev. *ED-9*, 164 (1962).

[3.16] J. M. Early: Effects of Space-Charge Layer Widening in Junction Transistors. Proc. I.R.E. *40*, 1401 (1952).

[3.17] J. W. Slotboom: Iterative Scheme for 1- and 2-Dimensional D.C. Transistor Simulation. Electr. Ltrs. *5*, 677 (1969).

[3.18] J. R. A. Beale, J. A. G. Slatter: The Equivalent Circuit of a Transistor with a Lightly Doped Collector Operating in Saturation. Solid-State Electr. *11*, 241 (1968).

[3.19] J. A. Pals, H. C. de Graaff: On the Behaviour of the Base-Collector Junction of a Transistor at High Collector Current Densities. Philips Res. Rep. *24*, 53 (1969).

[3.20] L. A. Hahn: The Effect of Collector Resistance Upon the High Current Capability of *n-p-v-n* Transistors. IEEE Trans. Electr. Dev. *ED-16*, 654 (1969).

[3.21] D. L. Bowler, F. A. Lindholm: High Current Regimes in Transistor Collector Regions. IEEE Trans. Electr. Dev. *ED-20*, 257 (1973).

[3.22] H. C. de Graaff: High Current Density Effects in the Collector of Bipolar Transistors. In: Process and Device Modeling for Integrated Circuit Design (F. van de Wiele, W. L. Engl, P. G. Jespers, eds). Noordhoff, Leiden (1977).

[3.23] L. J. Turgeon, J. R. Mathews: A Bipolar Transistor Model of Quasi-Saturation for Use in CAD. IEDM Techn. Digest 394 (1980).

[3.24] H. C. de Graaff: Compact Bipolar Transistor Modeling. In: Process and Device Modeling (W. L. Engl, ed.). North-Holland, Amsterdam (1986).

[3.25] G. M. Kull et al.: A Unified Circuit Model for Bipolar Transistors Including Quasi-Saturation Effects. I.E.E.E. Trans. Electr. Dev. *ED-32*, 1103 (1985).

[3.26] S. L. Miller: Ionization Rates for Holes and Electrons in Silicon. Phys. Rev. *105*, 1246 (1957).

[3.27] R. W. Dutton: Bipolar Transistor Modeling of Avalanche Generation for Computer Simulation. I.E.E.E. Trans. Electr. Dev. *ED-22*, 334 (1975).

[3.28] D. A. Divekar, R. E. Lovelace: Modeling of Avalanche Current of Bipolar Junction Transistors for Computer Circuit Simulation. I.E.E.E. Trans. CAD Int. Circ. and Syst. *CAD-1*, 114 (1982).

[3.29] H. C. Poon, J. C. Meckwood: Modeling of Avalanche Effect in Integral Charge Control Model. I.E.E.E. Trans. Electr. Dev. *ED-19*, 90 (1972).

[3.30] S. M. Sze: Physics of Semiconductor Devices, 2nd ed. John Wiley & Sons, New York (1981).

[3.31] J. R. Hauser: The Effects of Distributed Base Potential on Emitter Current Density and Effective Base Resistance for Stripe Transistor Geometries. I.E.E.E. Trans. Electr. Dev. *ED-11*, 238 (1964).

[3.32] J. E. Lary, R. L. Anderson: Effective Base Resistance of Bipolar Transistors. I.E.E.E. Trans. Electr. Dev. *ED-32*, 2503 (1985).

[3.33] G. Rey: Effets de la Défocalisation sur le Comportement des Transistors à Jonctions. Solid-State Electr. *12*, 645 (1969).

[3.34] H. Groendijk: Modeling Base Crowding in a Bipolar Transistor. I.E.E.E. Trans. Electr. Dev. *ED-20*, 329 (1973).

[3.35] H. C. de Graaff: Electrical Behaviour of Lightly Doped Collectors in Bipolar Transistors. Thesis, University of Technology, Eindhoven (1975).

[3.36] H. C. de Graaff: Approximate Calculations on the Spreading Resistance in Multi-Emitter Structures. Philips Res. Rep. *24*, 34 (1969).

[3.37] J. Lindmayer, C. Y. Wrigley: Fundamentals of Semiconductor Devices. Van Nostrand, Princeton (1965).

[3.38] J. te Winkel: Extended Charge-Control Model for Bipolar Transistors. I.E.E.E. Trans. Electr. Dev. *ED-20*, 389 (1973).

[3.39] J. G. Fossum, S. Veeraraghavan: Partitioned-Charge-Based Modeling of Bipolar Transistors for Non-Quasi-Static Circuit Simulation. I.E.E.E. Electr. Dev. Ltrs. *EDL-7*, 652 (1986).

[3.40] H. Klose, A. W. Wieder: The Transient Integral Charge Control Relation—A Novel Formulation of the Currents in a Bipolar Transistor. I.E.E.E. Trans. Electr. Dev. *ED-34*, 1090 (1987).

[3.41] P. B. Weil, L. P. McNamee: Simulation of Excess Phase in Bipolar Transistors. I.E.E.E. Trans. *CAS-25*, 114 (1978).

[3.42] J. J. H. van den Biesen: A Simple Regional Analysis of Transit Times in Bipolar Transistors. Solid-State Electr. *29*, 529 (1986).

[3.43] R. G. Meyer, R. S. Muller: Charge-Control Analysis of the Collector-Base Space-Charge-Region Contribution to Bipolar Transistor Time Constant τ_T. I.E.E.E. Trans. Electr. Dev. *ED-34*, 450 (1987).

[3.44] J. A. Pals: On the Noise of a Transistor with d.c. Current Crowding. Philips Res. Rep. *26*, 91 (1971).

[3.45] J. L. Plumb, E. R. Chenette: Flicker Noise in Transistors. I.E.E.E. Trans. Electr. Dev. *ED-10*, 304 (1963).

[3.46] J. M. C. Stork et al.: High Performance Operation of Silicon Bipolar Transistors at Liquid Nitrogen Temperatures. IEDM Techn. Digest 405 (1987).

[3.47] J. W. Slotboom, H. C. de Graaff: Bandgap Narrowing in Silicon Bipolar Transistors. I.E.E.E. Trans. Electr. Dev. *ED-24*, 1123 (1977).

Compact Models for Vertical Bipolar Transistors 4

In chapter 3 we discussed the most relevant phenomena of bipolar device physics. In this chapter we will give a precise description of the model equations which together define a given model. This description will be limited to vertical *npn* transistors for integrated circuits, including the substrate effects of the parasitic *pnp* transistor. Vertical *pnp* transistors also exist, but they require no new fundamental additions.

In principle one can make various selections of the elements discussed in chapter 3 to define a compact model; we have attempted to make a rough division into three parts, in order of increasing complexity. The simplest models are those of the Ebers-Moll type. Better, but more complex models are the Gummel-Poon model and its modifications. This chapter will be concluded with a recently developed model called Mextram which emphasizes accurate modelling of the device physics, especially in the quasi-saturation.

4.1 Ebers-Moll-Type Models

The original version [4.1] of the Ebers-Moll (E-M) model was an injection model, but the description given here will be for the transport formulation: See the discussion in section 3.1. The network topology of the model is given in Fig. 4.1.

4.1.1 Basic Ebers-Moll Model

We first define the ideal forward and reverse currents I_f and I_r which are already given by Eq. (3.21),

$$\left. \begin{array}{l} I_f = I_s \{\exp(V_{b_1 e_1}/U_T) - 1\}, \\ I_r = I_s \{\exp(-V_{c_1 b_1}/U_T) - 1\}. \end{array} \right\} \tag{4.1}$$

Note, that these currents are functions of the internal nodal voltages $V_{b_1 e_1}$ and $V_{c_1 b_1}$. Such internal nodes emerge when series resistances are added to

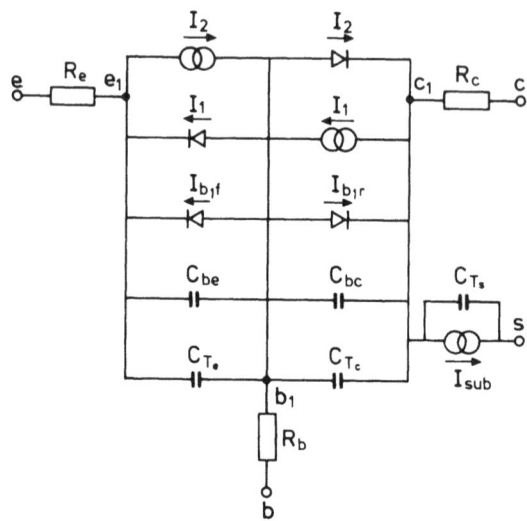

Fig. 4.1 Equivalent circuit diagram of a standard Ebers-Moll model, with additional constant series resistances and bias-dependent depletion capacitances

enhance the performance; they make the numerical evaluation more difficult because they increase the rank of the matrix of the set of equations, or they require the introduction of an iteration scheme.

The saturation current I_s is, physically speaking, determined by the emitter geometry and the doping profile in the base (see section 3.3). Here we will consider it as a parameter. $U_T = k_B T/q$ is the thermal voltage. For a reverse-biased collector-substrate junction the substrate current is defined in a similar way,

$$I_{sub} = I_{ss} \exp(-V_{c_1 b_1}/U_T) - 1, \tag{4.2}$$

with I_{ss} as parameter.

A transport model contains separately added base currents (see section 3.4),

$$I_{b_1 f} = \frac{I_f}{\beta_f}$$

and (4.3)

$$I_{b_1 r} = \frac{I_r}{\beta'_r} + I_{sub} + \frac{I_{sub}}{\beta_s},$$

where β_f, β'_r and β_s can be considered for the moment as parameters.

Fig. 4.2 shows a cross-section of a vertical npn transistor. Base, collector and substrate form a parasitic pnp transistor. The substrate-collector junction usually remains reversely biased, but in the inverse mode of operation of the active transistor the base-collector junction becomes forward-biased and we may have an appreciable substrate current, consisting of holes emitted by

4.1 Ebers-Moll-Type Models

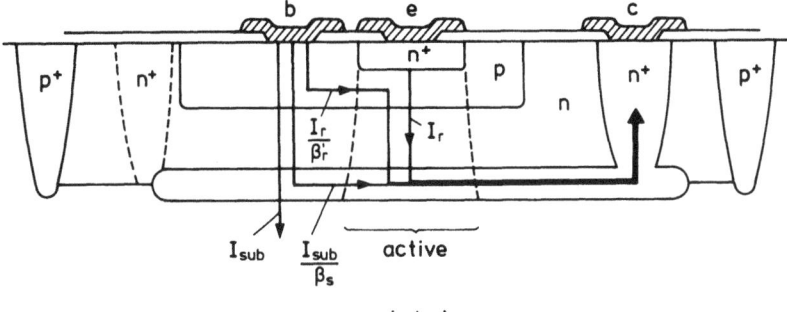

Fig. 4.2 Cross-section of a vertical *npn* transistor with indication of the various current flows in the reverse mode of operation

the base and collected by the substrate (I_{sub}). The component I_{sub}/β_s consists of holes that recombine in the buried layer outside the active part, plus electrons recombining in the base and at the base contact. I_r is the reverse current of the active part, flowing from emitter to collector, and I_r/β_r' arises from recombination in the active part (mainly holes in the buried layer). The above explains the total reverse base current I_{b_1r} in Eq. (4.3),

$$I_{b_1r} = \left\{ \frac{I_s}{\beta_r'} + \left(1 + \frac{1}{\beta_s}\right) I_{ss} \right\} \{\exp(-V_{c_1b_1}/U_T) - 1\}.$$

However, the problem is that we cannot measure the components I_r/β_r' and I_{sub}/β_s separately, but only their sum. Describing this sum by I_r/β_r gives us a measurable parameter β_r,

$$\frac{1}{\beta_r} = \frac{1}{\beta_r'} + \frac{1}{\beta_s} \frac{I_{ss}}{I_s}.$$

Recasting Eq. (4.3) then gives for the ideal base current components

$$I_{b_1f} = \frac{I_f}{\beta_f}$$

and

$$I_{b_1r} = \frac{I_r}{\beta_r} + I_{sub}.$$

(4.4)

If the deep n^+ contact diffusion regions, together with the buried layer, completely surround the *n*-epi layer of the collector, plausible values are $\beta_s \sim 5$, $\beta_r' \sim 10$ and $I_{ss}/I_s \sim 0.1$. If the n^+ contact diffusion is missing on one side (see Fig. 4.2) and the epilayer "sees" the *p*-type substrate directly, I_{ss} and β_s are substantially higher, but β_r will not change much.

If we put $I_1 = I_f$ and $I_2 = I_r$, see Fig. 4.1, Eqs. (4.1), (4.2) and (4.4) define a simple d.c. model with 4 parameters: $I_s - I_{ss} - \beta_f - \beta_r$. The forward current gain h_{FE} is a constant and equals β_f, the reverse current gain h_{RE} is also constant and equals

$$\beta_r \bigg/ \left(1 + \frac{I_{ss}}{I_s}\beta_r\right).$$

With an open emitter the gain of the parasitic *pnp* transistor is $\beta_r I_{ss}/I_s$. The model does not account for high injection in the base, quasi-saturation and Early effect. It can be enlarged for transient and a.c. behaviour by adding the stored charges and using the quasi-static and charge-control principles of sections 3.2 and 3.3,

$$\left.\begin{array}{ll} Q_{be} = \tau_f I_f, & C_{be} = \dfrac{dQ_{be}}{dV_{b_1 e_1}}, \\[2mm] & \\ Q_{bc} = \tau_r I_r, & C_{bc} = -\dfrac{dQ_{bc}}{dV_{c_1 b_1}}. \end{array}\right\} \quad (4.5)$$

This requires two more parameters: the transit times τ_f and τ_r. The cut-off frequency $f_T = 1/2\pi\tau_f = $ constant. This simple model now requires in total 6 parameters. Several other extensions are possible, but at the cost of more parameters and increased computer time.

4.1.2 Extensions of the Basic Ebers-Moll Model

1. We can first add the Early effect by putting

$$\left.\begin{array}{l} I_1 = \dfrac{I_f}{1+q_1}, \\[2mm] I_2 = \dfrac{I_r}{1+q_1}, \end{array}\right\} \quad (4.6)$$

with

$$q_1 = \frac{V_{b_1 e_1}}{V_{ear}} - \frac{V_{c_1 b_1}}{V_{eaf}}.$$

The Early voltages V_{ear} and V_{eaf} (cf. Eq. (3.64)) are parameters. They mainly affect the $I_c - V_{ce}$ characteristics, see Fig. 3.12. If $V_{cb}/V_{eaf} = 1 + V_{be}/V_{ear}$ we have punch-through: The collector depletion charge has "consumed" the total neutral base charge. This gives a numerical instability, which must be tamed by suitable limiting.

2. Addition of fixed series resistances R_e, R_b and R_c. These are considered as parameters. See also Fig. 4.1.

3. Addition of the depletion capacitances C_{T_e}, C_{T_c} and C_{T_s}, as discussed in section 3.5. The singularity at $V_j = V_d$ can be avoided by e.g. a tangential

4.1 Ebers-Moll-Type Models

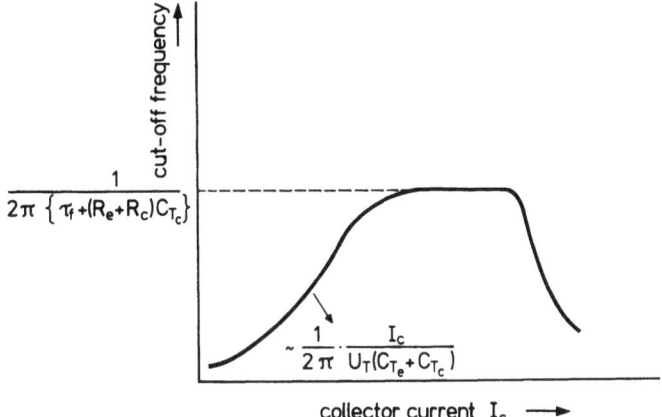

Fig. 4.3 Typical shape of the $f_T(I_c)$ curve as generated by the E-M model. The rising part is dominated by the depletion capacitances C_{T_e} and C_{T_c}, the maximum value by τ_f and $(R_e + R_c)C_{T_c}$. The fall-off is due to saturation (Q_{bc})

continuation. We need here three times 4 parameters (C_0, V_d, p and a). With the extensions sub 1., 2. and 3. the a.c. small-signal circuit becomes as given in Fig. 3.20. The resistance r_{ea} represents the Early effect.
The total transit time $\tau_{tot} = 1/2\pi f_T$ already contains several contributions, also present in Eq. (3.116),

$$\tau_{tot} = \tau_f + (R_e + R_c)C_{T_c} + \frac{U_T}{I_c}(C_{T_e} + C_{T_c}). \tag{4.7}$$

The typical shape of $f_T(I_c)$, as sketched in Fig. 4.3, is already present. The maximum f_T value is $1/2\pi\tau_f$. The fall-off at high currents is due to saturation: dQ_{bc}/dI_c must be added to τ_{tot}.

4. Further additions aim at making the current gain current-dependent [4.2] by introduction of non-ideal base current components as discussed in section 3.4, Eq. (3.54),

$$I_{b_2} = I_{bf}\{\exp(V_{b_1e_1}/m_f U_T) - 1\} + I_{br}\{\exp(-V_{c_1b_1}/m_r U_T) - 1\}. \tag{4.8}$$

We have 4 parameters: I_{bf}–I_{br}–m_f–m_r. The non-ideality coefficients (m_f, m_r) hardly have a physical background, they are fit parameters.
In section 3.3 we have seen that at high injection levels in the base ($n \gg N_a$) the exponential slope of the $I_c - V_{be}$ curve changes into $V_{be}/2U_T$. We can also obtain this by means of curve-fitting [4.2]; we replace I_s in Eq. (4.1) by

$$I'_s = \frac{I_s}{1 + \Theta \exp(V_{b_1e_1}/2U_T)}. \tag{4.9}$$

Together Eqs. (4.8) and (4.9) make h_{FE} current-dependent and $h_{FE}(I_c)$ has the familiar shape, as given in Fig. 3.9. It has, however, a weak physical back-

ground and more of a curve-fitting nature. Better solutions are offered in the Gummel-Poon and Mextram models, see sections 4.2 and 4.3.
5. The fall-off of the $f_T(I_c)$ curve can be improved by making the forward transit time τ_f a function of I_c. We will give two possibilities here. The first is [4.2]

$$\tau_f = \tau_{f0}\left\{1 + A\left(\frac{I_1}{I_\tau} - 1\right)^2\right\} \quad \text{for } I_1 > I_\tau,$$
$$= \tau_{f0} \quad \text{for } I_1 \leq I_\tau.$$
(4.10)

I_1 is given by Eq. (4.6), τ_{f0}, A and I_τ are parameters.
The second is [4.3]

$$\tau_f = \tau_{f0}(1 + q_1)^2 \left[1 + \ln\left\{1 + \exp\left(\frac{I_1 - I_k}{I_\tau}\right)\right\}\right]. \quad (4.11)$$

The parameters are τ_{f0}, I_k and I_τ.
Eqs. (4.10) and (4.11) reflect the fact that the transit time usually increases at high injection levels. Moreover, Eq. (4.11) incorporates the Early effect by means of the $(1 + q_1)^2$ factor and it avoids the if-statement in Eq. (4.10). To summarise, when we use all the mentioned extensions, we need the following parameters:

the d.c. parameters	I_s–I_{ss}–β_f–β_r,
the transit times	τ_{f0}–τ_r,
the Early voltages	V_{eaf}–V_{ear},
the series resistances	R_e–R_b–R_c,
the depletion charge parameters	C_{0e}–p_e–V_{de}–a_e,
	C_{0c}–p_c–V_{dc}–a_c,
	C_{0s}–p_s–V_{ds}–a_s,
non-ideal base current and high injection	I_{bf}–I_{br}–m_f–m_r–Θ
for the f_T fall-off	I_k–I_τ.

This makes altogether 30 parameters.

4.1.3 Temperature Dependence of the Parameters

To give a full description of the temperature dependence of all parameters is very cumbersome and we will limit ourselves here to the most important. The following is based on the theory of section 3.13.
For I_s we use Eq. (3.130), introducing a reference temperature T_r,

$$I_s(T) = I_s(T_r)\left(\frac{T_c}{T_r}\right)^{m_b} \exp\left\{-\frac{qV_{gb}}{k_B}\left(\frac{1}{T} - \frac{1}{T_r}\right)\right\}. \quad (4.12)$$

The temperature parameters are m_b and V_{gb}.
For the base currents we have (see Eq. (3.131))

4.1 Ebers-Moll-Type Models

$$\frac{I_s}{\beta_f}(T) = \frac{I_s}{\beta_f}(T_r)\left(\frac{T}{T_r}\right)^{m_e} \exp\left\{-\frac{qV_{ge}}{k_B}\left(\frac{1}{T} - \frac{1}{T_r}\right)\right\}$$

and (4.13)

$$\frac{I_s}{\beta_r}(T) = \frac{I_s}{\beta_r}(T_r)\left(\frac{T}{T_r}\right)^{m_c} \exp\left\{-\frac{qV_{gc}}{k_B}\left(\frac{1}{T} - \frac{1}{T_r}\right)\right\}.$$

For the substrate current I_{ss} we can write

$$I_{ss}(T) = I_{ss}(T_r)\left(\frac{T}{T_r}\right)^{m_c} \exp\left\{-\frac{qV_{gc}}{k_B}\left(\frac{1}{T} - \frac{1}{T_r}\right)\right\}. \qquad (4.14)$$

Parameters are m_e–m_c–V_{ge}–V_{gc}.
For the non-ideal base currents we use Eq. (3.132) with two more bandgap parameters V_{gje} and V_{gjc}, referring to the eb- and cb-junctions, respectively:

$$I_{bf}(T) = I_{bf}(T_r)\left(\frac{T}{T_r}\right)^2 \exp\left\{-\frac{qV_{gje}}{k_B}\left(\frac{1}{T} - \frac{1}{T_r}\right)\right\},$$

(4.15)

$$I_{br}(T) = I_{br}(T_r)\left(\frac{T}{T_r}\right)^2 \exp\left\{-\frac{qV_{gjc}}{k_B}\left(\frac{1}{T} - \frac{1}{T_r}\right)\right\}.$$

The different bandgap voltages in Eqs. (4.12) to (4.15) arise from the different doping levels in the various regions (see section 2.4).
The zero-bias depletion capacitances depend, according to Eq. (3.124), on the diffusion voltage $V_d(T)$. For each of the three depletion capacitances we then must know the relevant bandgap voltages: V_{gb} for V_{de}, V_{gc} for V_{dc} and a new one, V_{gs}, for V_{ds}. The series resistances may vary according to a power law as in Eq. (3.129), or to a second-order polynomial:

$$R(T) = R(T_r)\{1 + tc_1(T - T_r) + tc_2(T - T_r)^2\}. \qquad (4.15)$$

The latter is more accurate, but requires two parameters (tc_1 and tc_2) instead of one.
For the sake of simplicity we will disregard the weak temperature dependence of the transit times and the Early voltages, but even then we already need at least 13 parameters for the characterisation of the temperature behaviour. Of these the bandgap voltages are of special importance.

4.1.4 Typical Results

Some typical results at 300 K of an E-M model, extended with series resistances and depletion capacitances, are given in Fig. 4.4.
In Fig. 4.4a we see log I_c and log I_b versus V_{be}, the so-called Gummel plot. The increase in V_{be} at high currents is due to the series resistances R_e and R_b. Non-ideal base currents were not modelled, so h_{FE} only shows a slight decrease with increasing current, due to the increase of q_1 with V_{be} (Early effect at the emitter side!); see Fig. 4.4b. The greatest misfit in Fig. 4.4c arises

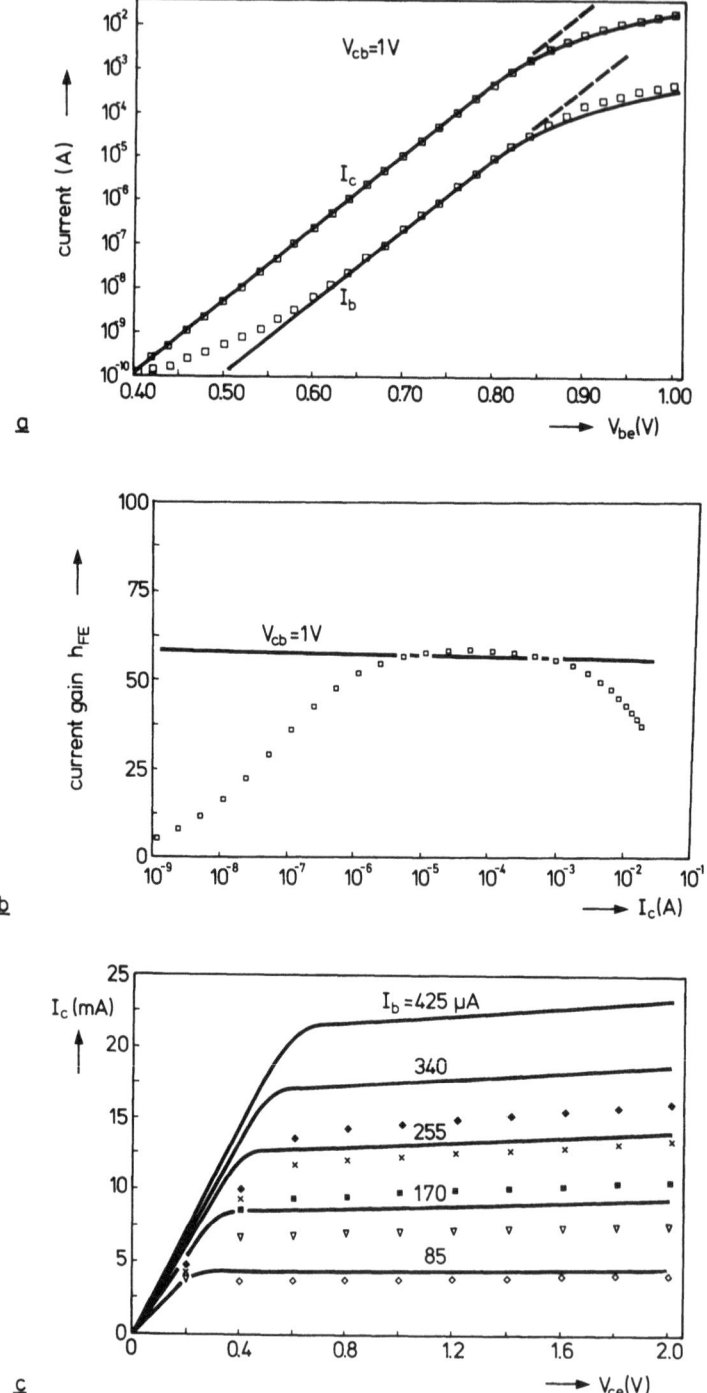

Fig. 4.4 Model calculations with an E-M model (fully drawn lines) compared with measurements: a) I_c and I_b versus V_{be} (Gummel plot). b) Current gain h_{FE} ($=I_c/I_b$) versus I_c. The standard E-M model gives a rather constant h_{FE}. c) (I_c, V_{ce}) characteristics; the misfit at higher currents is due to the nearly constant h_{FE}. d) f_T as a function of I_c and V_{cb}

4.2 Gummel-Poon-Type Models

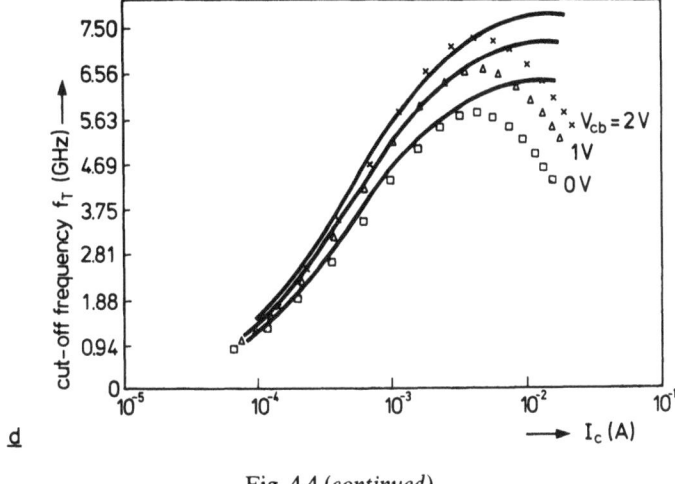

Fig. 4.4 (continued)

from the h_{FE} being nearly constant. Finally, the $f_T(I_c, V_{cb})$ curves in Fig. 4.4 d reflect that in this E-M model the τ_f is taken as a constant; the maximum value of f_T depends on $\tau_f + C_{T_c}(R_e + R_c)$; see Eq. (4.7). This maximum is influenced by V_{cb} via C_{T_c}. Hard saturation and f_T fall-off do not occur in the measured range of I_c and V_{cb} values: $(V_{cb} - I_c R_c + I_b R_b)$ remains positive.

4.2 Gummel-Poon-Type Models

The main attraction of the E-M model is its simplicity (see preceding section), but it has many shortcomings. We can repair the latter by several extensions of a curve-fitting nature, but a more elegant alternative, partly based on physical principles, is offered by the Gummel-Poon (G-P) model [4.4]. The starting points for this model are the Moll-Ross relation of Eq. (3.25) and the Integral Charge Control formulas (3.29), (3.30) and (3.31). The G-P model incorporates in a natural way the effects of high base injection and Early effect. Here too, several extensions can be added, e.g. the base widening in quasi-saturation.

4.2.1 Basic Gummel-Poon Model

An important quantity in the derivation of the G-P model is the total base charge Q_b, as given by Eq. (3.26). From this equation we get

$$Q_b/Q_{b0} = \tfrac{1}{2}(1 + q_1) + \tfrac{1}{2}\sqrt{(1 + q_1)^2 + 4q_2}. \qquad (4.17)$$

Here $(1 + q_1)$ is the Early factor (see section 3.6), which can be written as

$$1 + q_1 = 1 + \frac{V_{b_1 e_1}}{V_{ear}} - \frac{V_{c_1 b_1}}{V_{eaf}}, \quad (4.18)$$

with the Early voltages V_{ear} and V_{eaf} as parameters.
The normalized charge q_2 represents the charge storage in the neutral base region

$$q_2 = \frac{I_f}{I_{kf}} + \frac{I_r}{I_{kr}}, \quad (4.19)$$

where

$$\left. \begin{array}{l} I_f = I_s\{\exp(V_{b_1 e_1}/U_T) - 1\}, \\ I_r = I_s\{\exp(-V_{c_1 b_1}/U_T) - 1\}. \end{array} \right\} \quad (4.19)$$

I_{kf} and I_{kr} are parameters. The equivalent circuit is given in Fig. 4.5.
The main currents are I_1 and I_2,

$$I_1 = \frac{I_f}{Q_b/Q_{b0}}, \quad I_2 = \frac{I_r}{Q_b/Q_{b0}}. \quad (4.20)$$

The stored charges are obtained from the charge-control relations (3.27),

$$Q_{be} = \tau_f I_1, \quad Q_{bc} = \tau_r I_2. \quad (4.21)$$

The forward and reverse transit times are given by $\tau_f = Q_{b0}/I_{kf}$ and $\tau_r =$

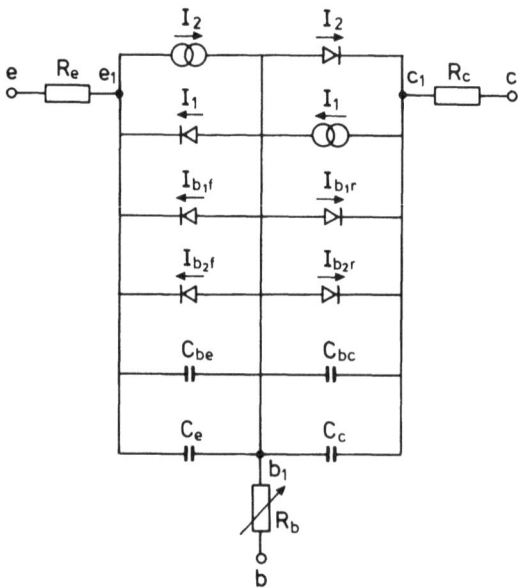

Fig. 4.5 Equivalent circuit diagram of the basic Gummel-Poon model. The base resistance is current-dependent

4.2 Gummel-Poon-Type Models

Q_{b0}/I_{kr}. Connected with these charges are the diffusion capacitances

$$C_{be} = \frac{dQ_{be}}{dV_{b_1 e_1}} \quad \text{and} \quad C_{bc} = -\frac{dQ_{bc}}{dV_{c_1 b_1}}.$$

See also Eq. (3.32).
Together, Eqs. (4.17), (4.18), (4.19), (4.20) and (4.21) give the essential features of the active part of the transistor for the G-P model. The following parameters are required: $I_s - I_{kf} - I_{kr} - V_{ear} - V_{eaf}$.
The equivalent circuit of Fig. 4.5 is completed, as in the E-M model, by the following additions:

— the ideal base current components

$$I_{b_1 f} = I_f / \beta_f \quad \text{and} \quad I_{b_1 r} = I_r / \beta_r,$$

— the non-ideal base current components

$$I_{b_2} = I_{bf} \{\exp(V_{b_1 e_1}/m_f U_T) - 1\} + I_{br} \{\exp(-V_{c_1 b_1}/m_r U_T) - 1\},$$

— depletion capacitances C_{T_e} and C_{T_c}, as discussed in section 3.5 and with the tangential continuation,
— series resistances R_e, R_b and R_c. R_e and R_c are fixed, R_b is split up into a constant part, R_{bc}, of the inactive base region, and a varying part, R_{bv}, of the active region; see Fig. 3.17. R_{bv} is modulated by the base charge Q_b, thus by the depletion charges and the stored charges (conductivity modulation by injected carriers).

So we have for the base resistance

$$R_b = R_{bc} + \frac{R_{bv}}{Q_b / Q_{b0}}. \tag{4.22}$$

Q_b/Q_{b0} follows from Eq. (4.17).
The above-mentioned additions require, of course, extra parameters:

— the current gain parameters β_f and β_r,
— the non-ideal base current parameters I_{bf}, I_{br}, m_f and m_r,
— the depletion capacitance parameters, two times four (C_0, V_d, p and a),
— the resistance parameters R_e, R_c, R_{bc}, R_{bv}.

This brings the total number of parameters so far to 23.
In the standard G-P model, as described above, the quasi-saturation effect (see section 3.7) is not modelled. As far as base widening or push-out is concerned, this is sometimes remedied as follows. Instead of Eq. (4.21) we write for Q_{be}

$$Q_{be} = B\tau_f I_1 = \left(1 + \frac{x_i}{W_b}\right)^2 \tau_f I_1.$$

B is called the base push-out factor, x_i is the width of the injection region as given by Eq. (3.68). From the theory in section 3.7 it follows that in the absence of hot-carrier flow in the collector epilayer ($I_c < I_{hc}$) we have

$$\frac{x_i}{W_c} \approx 1 - \frac{V_{cb} + V_{dc}}{I_c R_{epi}} = \frac{I_c R_{epi} - V_{cb} - V_{dc}}{I_c R_{epi}} = \frac{V_{ci}}{I_c R_{epi}}.$$

Here V_{ci} is an internal junction voltage. This equation has a singularity at $I_c = 0$ and does not always guarantee a non-negative x_i, so we reshape

$$\left(\frac{x_i}{W_c}\right)^2 = \frac{\{\sqrt{V_{ci}^2 + V_1^2} + V_{ci}\}^2}{4\{(I_c R_{epi})^2 + V_1^2\}}.$$

Now the value of x_i/W_c always lies between 0 and 1.
With this the base push-out factor becomes

$$B = \left[1 + \frac{W_c}{4W_b}\frac{\{\sqrt{V_{ci}^2 + V_1^2} + V_{ci}\}^{n_p}}{(I_c R_{epi})^2 + V_1^2}\right]^2. \tag{4.23}$$

The push-out exponent n_p is an extra fit parameter that accounts for such effects as emitter-base crowding and charge storage in inactive regions. As parameters we need the epilayer resistance R_{epi}, the diffusion voltage V_{dc} (can be the same as for the collector depletion capacitance), the "smoothing" voltage V_1, the quantity W_c/W_b and n_p.
By the introduction of the base push-out factor B we have in fact added the stored epilayer charge Q_{epi} of Eq. (3.71) to the forward base charge Q_{be}. A similar, but more elaborate approach is also followed in [4.6].

4.2.2 Extensions

In the circuit simulator SPICE [4.7] the forward transit time parameter τ_f is modified as a function of $V_{c_1 b_1}$ and I_f,

$$\tau_f = \tau_{f0}\left\{1 + x_{\tau_f}\left(\frac{I}{I_f + I_{\tau f}}\right)^2 \exp\left(-\frac{V_{c_1 b_1}}{V_{\tau_f}}\right)\right\}. \tag{4.24}$$

The parameters are here $\tau_{f0}-x_{\tau_f}-I_{\tau_f}-V_{\tau_f}$. Here too, τ_f increases with the forward current and when $V_{c_1 b_1}$ becomes forward-biased; it is more curve-fitting, however, then physical theory.
Another effect of quasi-saturation is the modulation of the collector epilayer resistance. This is modelled in [4.8] as

$$R_{cv} = \frac{R_{epi}}{1 + \exp\left\{-2\left(1 - 2\frac{X_i}{W_c}\right)\right\}} \tag{4.25}$$

with

$$\frac{x_i}{W_c} = \frac{U_T}{I_c R_{epi}}\left[2\left(\frac{n_i}{N_{epi}}\right)^2\left\{\exp\left(-\frac{V_{c_1 b_1}}{U_T}\right) - 1\right\} - 1\right].$$

4.2 Gummel-Poon-Type Models

Only two parameters are needed: R_{epi} and N_{epi}. The collector resistance in Fig. 4.5 now consists of a variable part R_{cv} and a constant part R_{cc}, formed by the buried layer.

An alternative fit for R_{cv} is obtained by making it a function of $V_{c_1 e_1}$ only,

$$\left. \begin{array}{l} R_{cv} = R_{epi} \dfrac{M_{epi} + \sqrt{M_{epi}^2 + \delta_M}}{2}, \\ M_{epi} = 1 - \exp\{(V_{Rc} - V_{c_1 e_1})/U_T\}. \end{array} \right\} \qquad (4.26)$$

R_{epi}, V_{Rc} and δ_M are the parameters here, although the smoothing factor δ_M may just have a default value.

4.2.3 Full Quasi-Saturation Model

The modifications of the G-P model mentioned so far all model only one aspect of the quasi-saturation. A more complete extension to the standard G-P model is offered in [4.9], see Fig. 4.6. The equivalent circuit consists of two transistors: an *npn* (Tr_1) for the active part and a *pnp* (Tr_2) for the parasitic substrate behaviour. Both transistors are modelled by the standard G-P model (without any addition). We have furthermore R_e, R_c and R_{bc} as fixed resistors, R_{bv} and R_{bvp} are variable according to Eq. (4.22).

Quasi-saturation is modelled by the current source I_{epi} and two capacitances where the charges Q_w and Q_o reside. These charges represent the charge storage in the collector epilayer: $Q_{epi} = Q_0 + Q_w$. They are given by [4.9]

$$\left. \begin{array}{l} Q_0 \approx Q_{c0}\{K_1(V_{c_2 b_2}) - 1 - \tfrac{1}{2}\gamma\}, \\ Q_w \approx Q_{c0}\{K_1(V_{c_1 b_2}) - 1 - \tfrac{1}{2}\gamma\} \end{array} \right\} \qquad (4.27)$$

with

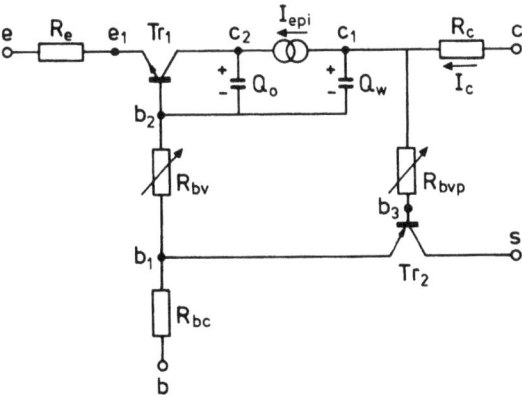

Fig. 4.6 Extended Gummel-Poon model with separate modelling of the quasi-saturation and of the parasitic substrate *pnp* transistor

$$Q_{c0} = \tfrac{1}{4} q A_{em} N_{epi}$$

and

$$\gamma = \left(\frac{2n_i}{N_{epi}}\right)^2.$$

The function K_1 is defined as

$$K_1(V) = \sqrt{1 + \exp(-V/U_T)}. \tag{4.28}$$

The current I_{epi} is the electron current in the collector epilayer,

$$I_{epi} = I_n \approx I_c = \frac{K_1(V_{c_2 b_2}) - K_1(V_{c_1 b_2}) - \ln\left\{\dfrac{1 + K_1(V_{c_2 b_2})}{1 + K_1(V_{c_1 b_2})}\right\} + \dfrac{1}{U_T}(V_{c_2 b_2} - V_{c_1 b_2})}{R_{epi}\left\{1 + \dfrac{|V_{c_2 b_2} - V_{c_1 b_2}|}{V_0}\right\}\dfrac{1}{U_T}} \tag{4.29}$$

Under normal operation conditions $V_{c_1 b_2} > V_{c_2 b_2} > 0$, in quasi-saturation $V_{c_1 b_2} > 0$, $V_{c_2 b_2} < 0$ and in hard saturation $V_{c_1 b_2} < 0$ and $V_{c_2 b_2} < 0$. The voltage V_0 in Eq. (4.29) stands for $V_0 = v_s W_c / \mu_{epi}$, with v_s as the saturated drift velocity.

The internal voltages $V_{c_1 b_2}$ and $V_{c_2 b_2}$ are connected with the pn products at $x = 0$ and $x = W_c$, see Fig. 3.14 and e.g. Eq. (3.80),

$$n(0)p(0) = \{p(0) + N_{epi}\}p(0) = n_i^2 \exp(-V_{c_2 b_2}/U_T),$$
$$n(W_c)p(W_c) = \{p(W_c) + N_{epi}\}p(W_c) = n_i^2 \exp(-V_{c_1 b_2}/U_T).$$

For $I_c < I_{hc}$ the model gives a good physical description of the quasi-saturation, but it requires a large number of parameters: for each of the transistors Tr_1 and Tr_2 20 parameters, plus the fixed resistors R_e, R_c and R_{bc} (3 parameters) and 4 parameters for the quasi-saturation ($R_{epi} - Q_{c0} - \gamma - V_0$), that is 47 parameters in total. Another drawback is the different topology of the equivalent circuit with three extra internal nodes (b_2, b_3 and c_2 in Fig. 4.6) as compared to the standard G-P and E-M models (see Fig. 4.1 and 4.5). This gives rise to extra elements in the network matrix and complicates the numerical solution. It prevents moreover the easy reduction to a simpler model by setting some parameter values to zero, which is indeed possible for models with the same topology.

For the temperature dependence of the parameters of the standard model we can use the same equations as for the E-M model: Eqs. (4.12) to (4.16). A simplification used in the SPICE simulator is that all bandgap voltages are equal: $V_{ge} = V_{gb} = V_{gc} = V_{gje} = V_{gjc} = V_g$.

4.2.4 Typical Results

In Fig. 4.7 we have given a few results of the standard G-P model, extended with modulation of the collector epilayer resistance according to Eq. (4.26).

4.2 Gummel-Poon-Type Models

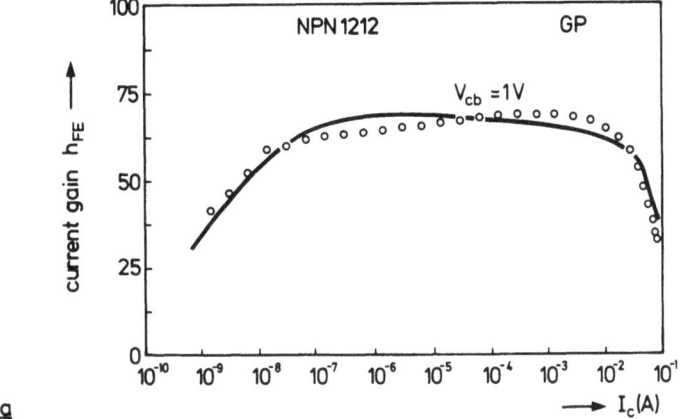

a

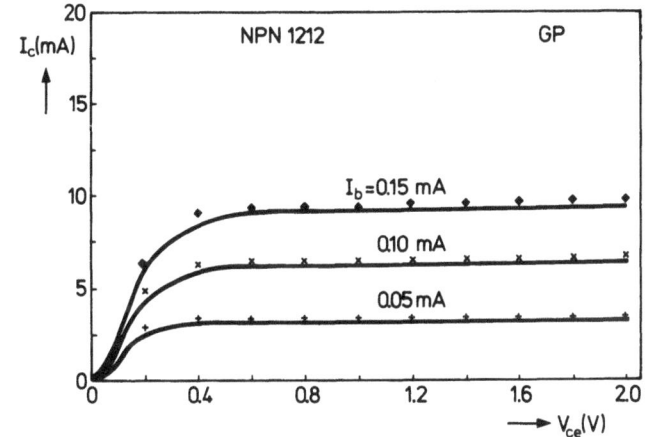

b

Fig. 4.7 Model calculations of the G-P model with modulated epilayer resistance, compared with measurements: a) current gain $h_{FE}(I_c)$, b) (I_c, V_{ce}) characteristics, c) $f_T(I_c, V_{cb})$, d) the polar diagram of the a.c. small-signal power gain parameter S_{21} in grounded emitter configuration. The real part of S_{21} is along the horizontal axis, the imaginary part along the vertical one. For 3 different d.c. currents

In Fig. 4.7a the current gain is shown; it is really current-dependent. Fig. 4.7b gives the (I_c, V_{ce}) characteristics, Fig. 4.7c the $f_T(I_c)$ curves at different values of V_{cb}. Finally Fig. 4.7d gives the small-signal scattering parameter S_{21}, which is a measure for the power gain, at three different values of I_c, one for the rising part of the f_T curve, one near the top and one in the fall-off region. The improvement over the E-M model is illustrated in Fig. 4.8, where the same quantities are given for the same transistor, but now modelled with an E-M model with additional Early effect, depletion capacitances and fixed series resistances.

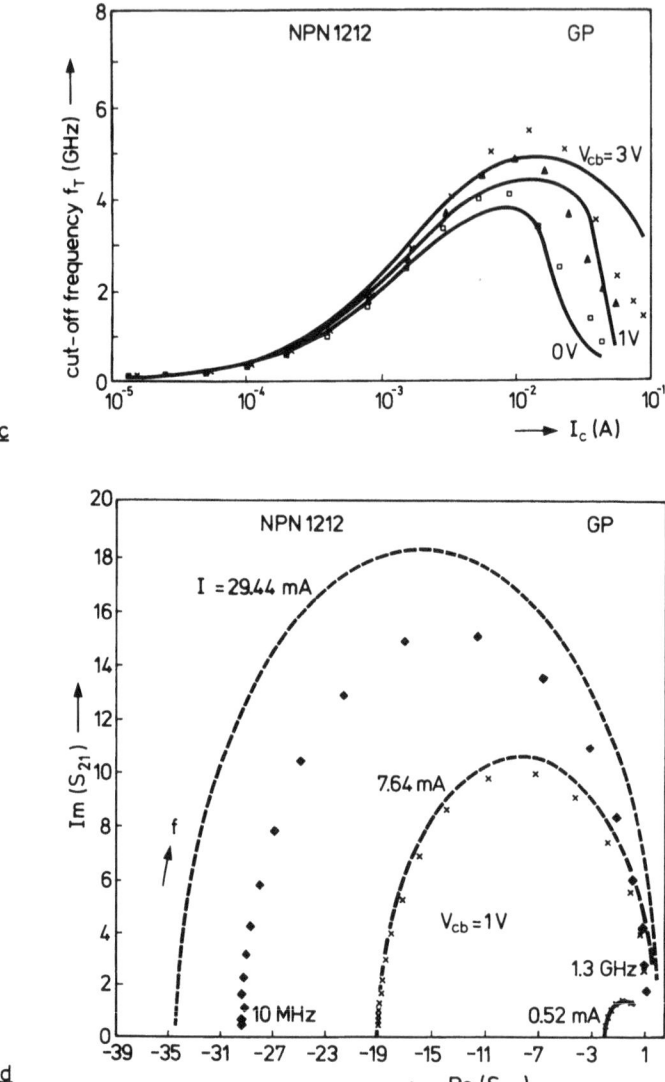

Fig. 4.7 (*continued*)

4.3 The MEXTRAM Model

In this section we will give a description of a rather extensive model called MEXTRAM [4.10]. Its topology resembles the extended G-P model of Fig. 4.6, with the extra internal nodes (c_2 and b_2). It incorporates many physical effects, the theory of which has already been described in chapter 3:

— currents and charges are related to minority carrier concentrations, which, in turn, depend on applied junction voltages;

4.3 The MEXTRAM Model

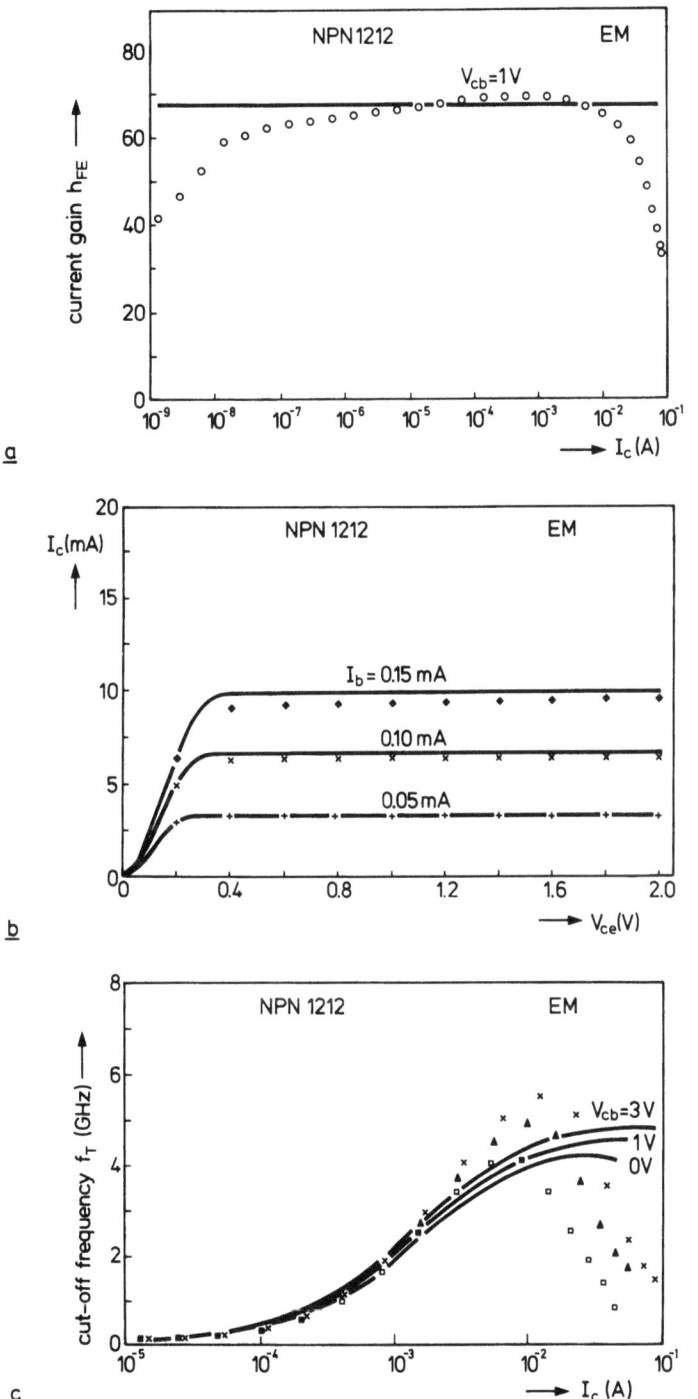

Fig. 4.8 The same measurements of the same transistor as in Fig. 4.7. The model calculations are done here with the E-M model, to show the superiority of the G-P model

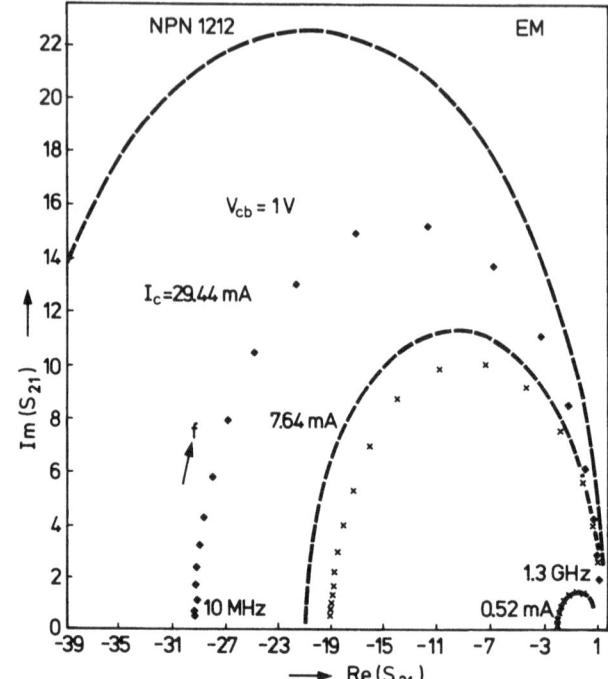

Fig. 4.8 (continued)

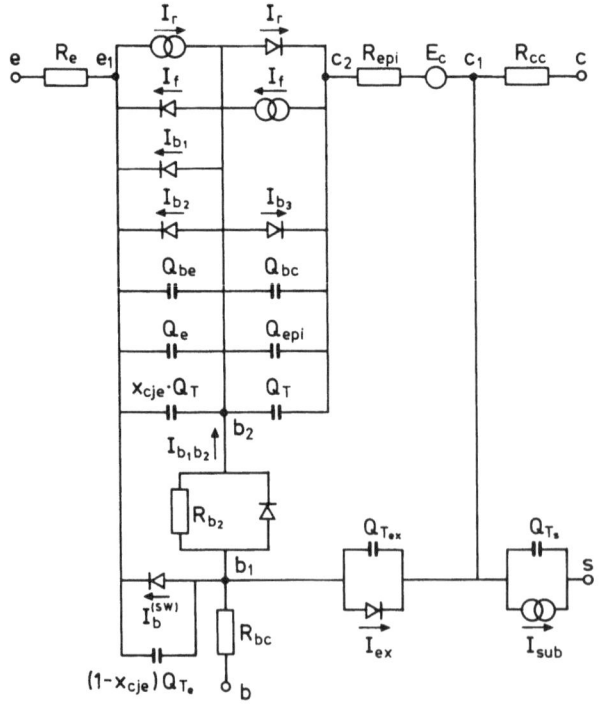

Fig. 4.9 Equivalent circuit diagram of the MEXTRAM model

4.3 The MEXTRAM Model

— the influence of the built-in field in the base is taken into account;
— transit times are bias-dependent;
— non-ideal base currents are modelled, so current gain is current-dependent;
— the influence of emitter-base current crowding and conductivity modulation on the base resistance is taken care off;
— quasi-saturation is extensively modelled, including base widening, Kirk effect and hot-carrier behaviour;
— the inactive part and the parasitic substrate effect are described separately;
— excess phase shift and non-quasi-static behaviour can be taken into account for extension of the model.

In spite of all this, the model requires a smaller number of parameters than the extended G-P model of Fig. 4.6, while its performance is even better in many aspects. The MEXTRAM equivalent circuit is sketched in Fig. 4.9.

4.3.1 Main Currents and Stored Charges

The built-in electric field in the base is $F = -\eta U_T/W_b$, see Eq. (3.34). The parameter η is the field factor. In chapter 3 the following were already defined,

$$f_b = \exp(\eta),$$

$$a_1 = \frac{2}{\eta} \frac{f_b - 1}{f_b},$$

$$a_2 = \frac{(\eta - 1)f_b^2 + f_b}{(f_b - 1)^2},$$

$$a_3 = \frac{f_b(f_b - \eta - 1)}{(f_b - 1)^2}.$$

The collector (electron) current is given by

$$I_c = I_n = \frac{I_f - I_r}{1 + q_1} \qquad (4.30)$$

with

$$q_1 = \frac{Q_{T_e} + Q_{T_c}}{Q_{b0}}.$$

Q_{T_e} and Q_{T_c} are the depletion charges and will be defined later on. The forward and reverse currents depend on minority carrier concentrations (see Eqs. (3.49) and (3.50)),

$$I_f = g_{I_f} \frac{I_k}{a_1} n_0 - I_s$$

with

$$g_{I_f} = \frac{2 + a_1 + 4n_0}{(2 + a_1)a_1 + 4n_0}$$

and

$$I_r = g_{I_r} \frac{I_k}{a_1} n_b - I_s$$

with

$$g_{I_r} = \frac{f_b^{-1} + n_b}{a_1 + n_b}.$$

(4.31)

The normalized concentrations of the minority carriers are given by the junction voltages,

$$n_0 = \frac{n(0)}{\hat{N}_a} = \frac{1}{2} \frac{f_1}{1 + \sqrt{1 + f_1}}$$

with

$$f_1 = 4a_1^2 \frac{I_s}{I_k} \exp(V_{b_2 e_1}/U_T),$$

and

$$n_b = \frac{n(W_b)}{\hat{N}_a} = \frac{1}{2} f_b^{-1} \frac{f_2}{1 + \sqrt{1 + f_2}}$$

with

$$f_2 = 4a_1^2 \frac{f_b^2 I_s}{I_k} \exp(V_{dc}/U_T) \, p_0(p_0 + 1),$$

where

$$p_0 = \frac{p(0)}{N_{epi}}$$

(4.32)

follows from

$$2p_0 + \ln p_0 = -V_{c_2 b_2}/U_T.$$

The stored charges in the base are

4.3 The MEXTRAM Model

$$Q_{be} = Q_{b0}(1 + q_1)g_{Q_f}\frac{1}{a_1}n_0$$

with

$$g_{Q_f} = \frac{2 + a_1 + n_0}{2 + a_1 + a_1 a_2 n_0} a_2,$$

and (4.33)

$$Q_{bc} = Q_{b0}(1 + q_1)g_{Q_r}\frac{1}{a_1}n_b$$

with

$$g_{Q_r} = \frac{a_3 + f_b n_b}{1 + a_1 f_b n_b}.$$

See also Eqs. (3.51) and (3.52).
In addition to these stored charges in the base, we also define a charge, stored in the neutral emitter,

$$Q_e = Q_{e0}\{\exp(V_{b_2 e_1}/m_\tau U_T) - 1\}. \tag{4.34}$$

Q_{e0} and m_τ are the parameters needed here; instead of Q_{e0} we can also use $\tau_{ne} \approx Q_{e0}/I_s$ as parameter.
In itself Q_e is usually not so important, but its contribution to the total transit time is essential in most cases.
In quasi-saturation we may have a charge storage in the collector epilayer

$$Q_{epi} = Q_{b0}\exp\left(\frac{V_{dc}}{U_T}\right)\frac{p_0(p_0 + 1) - p_w(p_w + 1)}{I_c/I_s}. \tag{4.35}$$

In Eq. (4.35) p_w is the normalized hole concentration at the end of the epilayer,

$$p_w = \frac{p(W_c)}{N_{epi}} = \frac{1}{2}\frac{f_3}{1 + \sqrt{1 + f_3}},$$

$$f_3 = 4\exp\{-(V_{c_1 b_2} + V_{dc})/U_T\}. \tag{4.36}$$

See also Eq. (3.80).

4.3.2 Quasi-Saturation and Hot-Carrier Effect in the Epilayer

The voltage source E_c in the collector circuit of the diagram in Fig. 4.9 models the quasi-saturation. From that diagram it follows that $V_{c_2 b_2} = -I_c R_{epi} - E_c + V_{c_1 b_2}$. R_{epi} is the unmodulated epilayer resistance: The voltage drop $I_c R_{epi}$ is usually not the real voltage drop across the epilayer, but E_c is meant to correct this. In chapter 3 it was already derived (Eq. (3.75)) that in the forward mode, in quasi-saturation, we have

$$-V_{c_2b_2} = V_{ci} = I_c R_{\text{epi}} - I_c R_{\text{epi}} \frac{2V}{BW_c + \sqrt{(BW_c)^2 + 4VAW_c^2}} - \frac{1}{2}V_{dc}.$$

In the reverse mode we have, according to Eq. (3.81)

$$-V_{c_2b_2} = I_c R_{\text{epi}} + U_T(2p_w + \ln p_w).$$

We must here define E_c in such a way that it can cover both Eqs. (3.75) and (3.81) in one single, analytical function. So, for the reverse mode E_c should read as

$$E_c = V_{c_1b_2} + U_T(2p_w + \ln p_w)$$
$$= -V_{dc} + U_T\left(2p_w + \ln \frac{2}{1 + \sqrt{1 + f_3}}\right),$$

where use has been made of f_3 in Eq. (4.36).
In the forward mode E_c should read as

$$E_c = V_{c_1b_2} - \frac{1}{2}V_{dc} - I_c R_{\text{epi}} \frac{2V}{BW_c + \sqrt{(BW_c)^2 + 4VAW_c^2}}.$$

In the ohmic case ($I_c < I_{hc}$, $A = 0$) it is then

$$E_c = V_{c_1b_2} - \tfrac{1}{2}V_{dc} - V_{c_1b_2} - \tfrac{1}{2}V_{dc} = -V_{dc}$$

and this smoothly connects with the values in the reverse mode. Combining, we now get

$$E_c = E_{\text{ohmic}} = -V_{dc} + U_T\left(2p_w + \ln \frac{2}{1 + \sqrt{1 + f_3}}\right). \tag{4.37}$$

With the help of Eq. (4.36) E_{ohmic} can be calculated as function of $V_{c_1b_2}$, see Fig. 4.10.
In the saturated velocity case ($I_c > I_{hc}$) we have $AW_c^2 > 0$, so the voltage drop in the epilayer is no longer ohmic and the source E_c must be enlarged with a contribution E_{hc},

$$E_c = E_{\text{ohmic}} + E_{hc}, \tag{4.38}$$

where E_{hc} has the form

$$V\left\{1 - I_c R_{\text{epi}} \frac{2}{BW_c + \sqrt{(BW_c)^2 + 4VAW_c^2}}\right\}.$$

We will now also consider the fact that the drift velocity as a function of the electric field does not abruptly change from ohmic to saturated behaviour, but shows a gradual transition. In lightly doped silicon this transition interval runs from $F_{hc}/\sqrt{6} \approx 2.5$ kV/cm to $\sqrt{6}\, F_{hc} \approx 15$ kV/cm [4.11], see Fig. 4.11.
The consequence is that $BW_c(I_c)$ in Eq. (3.76) also changes, see Fig. 4.12.
Finally, for E_{hc} we get a rather complicated formula:

4.3 The MEXTRAM Model

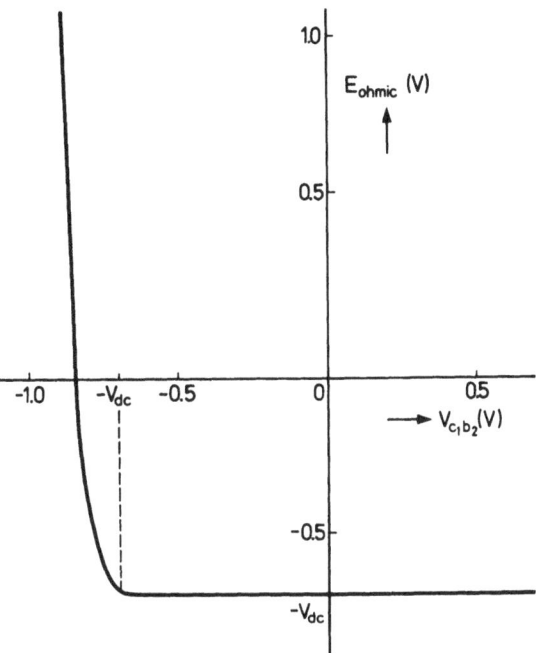

Fig. 4.10 The ohmic part (E_{ohmic}) of the voltage source E_c that models the epilayer modulation in the ohmic region of the quasi-saturation, hard saturation and reverse mode, as a function of the internal collector-base voltage $V_{c_1b_2}$ (see circuit diagram of Fig. 4.9)

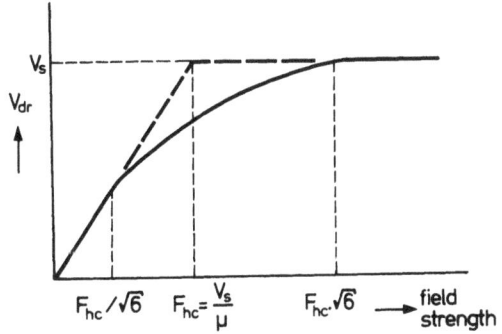

Fig. 4.11 Drift velocity (v_{dr}) as a function of field strength. Dashed line: simplified division into an ohmic and a saturated velocity regime. Fully drawn line: more realistic division into an ohmic, tepid and hot carrier (saturated) regime

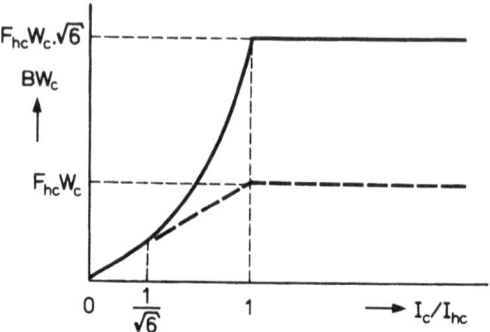

Fig. 4.12 The quantity BW_c as a function of I_c in the simplified version (dashed line) and in the more realistic version (fully drawn line)

$$E_{hc}/(V_{c_1b_2} + \tfrac{1}{2}V_{dc}) = 0, \qquad I_c < \frac{I_{hc}}{\sqrt{6}},$$

$$= 1 - \frac{I_{hc}}{I_c\sqrt{6}}, \qquad \frac{I_{hc}}{\sqrt{6}} < I_c < I_{hc},$$

$$= 1 - \frac{1}{\sqrt{6}}\frac{I_c}{I_{hc}}\frac{2}{1+\sqrt{1+4V\left(\dfrac{I_c}{I_{hc}}-1\right)}}, \qquad I_c > I_{hc}.$$

The voltage V stands for

$$V = \frac{qN_{epi}W_c^2}{12\varepsilon(I_{hc}R_{epi})^2}(V_{c_1b_2} + \tfrac{1}{2}V_{dc});$$

the term $U_T \ln(I_c/I_{hc})$ in Eq. (3.76) has been neglected here.
The quantity $E_{hc}/(V_{c_1b_2} + \tfrac{1}{2}V_{dc})$ is plotted as a function of I_c/I_{hc} for various values of V in Fig. 4.13, the dashed lines. It is clear that in this form $E_{hc}(V_{c_1b_2}, I_c)$ is not an analytical function. So we have to remodel E_{hc} to circumvent these pitfalls. The final result is sketched in Fig. 4.13 by the fully drawn lines; the formula for E_{hc} becomes

$$E_{hc} = \frac{12\varepsilon}{qN_{epi}}\left(\frac{I_{hc}R_{epi}}{W_c}\right)^2 V \cdot \frac{i^4 F(V)}{0.18F^2 + i^4}\cdot\frac{1 - iG + \sqrt{(1-iG)^2 + 0.01}}{2},$$

$$V = \frac{qN_{epi}}{12\varepsilon}\left(\frac{W_c}{I_{hc}R_{epi}}\right)^2 \frac{V_{c_1b_2} + \tfrac{1}{2}V_{dc} + \sqrt{(V_{c_1b_2} + \tfrac{1}{2}V_{dc})^2 + 0.01}}{2},$$

$$F(V) = \frac{1}{1+3.14V} + \frac{0.92V}{1+V},$$

$$G(V) = 0.408\frac{1+0.0237V}{1+1.928V}$$

and

$$i = I_c/I_{hc}.$$

(4.39)

4.3 The MEXTRAM Model

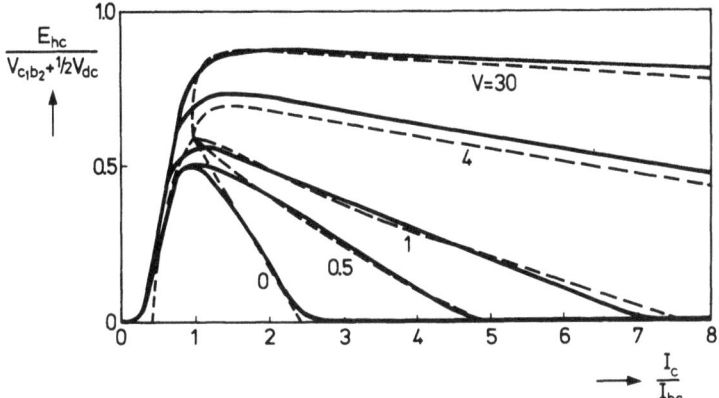

Fig. 4.13 The hot-carrier part E_{hc} of the voltage source E_c, as a function of I_c and $V_{c_1b_2}$. Dashed lines: according to the physical device theory. Fully drawn lines: the mathematical fit with analytical functions (Eq. (4.39))

Modelling of the main current and charge storage, including quasi-saturation and reverse mode of operation so far requires the following parameters: $I_s - I_k - Q_{b0} - \eta - I_{hc} - R_{epi} - \tau_{ne} - m_\tau - V_{dc}$.

4.3.3 Depletion Charges

The depletion charges are formulated according to Eq. (3.58) with avoidance of the singularity at $V_j = V_d$.
The emitter depletion charge becomes

$$Q_{T_e} = \frac{C_{0e} V_{de}}{1 - p_e} \left[(1 + K)^{-\frac{1}{2}p_e} - \frac{1 - V_{b_2e_1}/V_{de}}{\{(1 - V_{b_2e_1}/V_{de})^2 + K\}^{\frac{1}{2}p_e}} \right]. \quad (4.40)$$

If required, this charge is split into a bulk (or bottom) contribution $x_{cje} Q_{T_e}$ and a sidewall contribution $(1 - x_{cje}) Q_{T_e}$. This may be important for very narrow emitter stripes. The parameters are $C_{0e} - V_{de} - p_e - x_{cje}$. K is not really a parameter; it usually has a default value of 0.01.
The collector depletion charge of the active transistor part is

$$Q_{T_c} = x_{cje} \frac{C_{0c} V_{dc}}{1 - p_c} \left[(1 + K)^{-\frac{1}{2}p_c} - \frac{1 - V_{jcd}/V_{dc}}{\{(1 - V_{jcd}/V_{dc})^2 + K\}^{\frac{1}{2}p_c}} \right]. \quad (4.41)$$

C_{0c} is the total capacitance, $x_{cje} C_{0c}$ the active part of it.
This charge has its own internal junction voltage $V_{jcd} = -(V_{c_1b_2} - AW_c^2 - BW_c)$, where AW_c^2 and BW_c are modelled as analytical functions of I_c to represent the internal voltage drop

$$BW_c = I_c R_{epi} \frac{1 + 3(I_c/I_{hc})^2}{1 + (I_c/I_{hc})^3}$$

and

$$AW_c^2 = \frac{qN_{epi}W_c^2}{2\varepsilon} \frac{(I_c/I_{hc} - 1) + \sqrt{(I_c/I_{hc} - 1)^2 + 0.01 I_c/I_{hc}}}{2}.$$

Eq. (4.41) only gives the depletion charge of the collector-base junction in the active transistor part. The remainder belongs to the inactive, extrinsic depletion charge

$$Q_{T_{ex}} = (1 - x_{cjc}) \frac{C_{0c} V_{dc}}{1 - p_c} \left[(1+K)^{-\frac{1}{2}p_c} - \frac{1 + V_{c_1 b_1}/V_{dc}}{\{(1 + V_{c_1 b_1}/V_{dc})^2 + K\}^{\frac{1}{2}p_c}} \right]. \tag{4.42}$$

The parameters are x_{cjc}–C_{0c}–p_c–N_{epi}.
Finally we have the substrate depletion charge

$$Q_{T_s} = \frac{C_{0s} V_{ds}}{1 - p_s} \left[(1+K)^{-\frac{1}{2}p_s} - \frac{1 + V_{c_1 s}/V_{ds}}{\{(1 + V_{c_1 s}/V_{ds})^2 + K\}\}^{\frac{1}{2}p_s}} \right] \tag{4.43}$$

with the parameters C_{0s}–V_{ds}–p_s.

4.3.4 Base Currents

The base currents in this model follow the theory of section 3.4. I_{b_1} is the ideal forward base current of Eq. (3.53),

$$I_{b_1} = (1 - x_{ijb}) \frac{I_s}{B_f} \{\exp(V_{b_2 e_1}/U_T) - 1\}, \tag{4.44}$$

where x_{ijb} is that part, that is attributed to the sidewall component and which depends on $V_{b_1 e_1}$,

$$I_{b_1}^{sw} = x_{ijb} \frac{I_s}{B_f} \{\exp(V_{b_1 e_1}/U_T) - 1\}. \tag{4.45}$$

The non-ideal components follow Eq. (3.55),

$$\left. \begin{array}{l} I_{b_2} = I_{bf} \dfrac{\exp(V_{b_2 e_1}/U_T) - 1}{\exp(V_{b_2 e_1}/2U_T) + \exp(V_{lf}/2U_T)} \\[1em] I_{b_3} = I_{br} \dfrac{\exp(-V_{c_1 b_2}/U_T) - 1}{\exp(-V_{c_1 b_2}/2U_T) + \exp(V_{lr}/2U_T)} \end{array} \right\} \tag{4.46}$$

As parameters B_f–I_{bf}–I_{br}–V_{lf}–V_{lr}–x_{ijb} are needed.

4.3.5 Series Resistances

The resistances R_e, R_{cc} and R_{bc} are constants, the modulation of R_{bv} is due to emitter-base current crowding (Eq. (3.98)) and conductivity modulation,

$$I_{b_1b_2} = \frac{2U_T}{R_{b_2}}\{\exp(V_{b_1b_2}/U_T) - 1\} + \frac{V_{b_1b_2}}{R_{b_2}} \qquad (4.47)$$

with

$$R_{b_2} = \frac{3R_{bv}}{1 + \dfrac{Q_{T_e} + Q_{T_c} + Q_{be} + Q_{bc}}{Q_{b0}}}.$$

The parameters are R_e–R_{cc}–R_{bc}–R_{bv}.

4.3.6 Modelling the Inactive Part and Substrate

The substrate current model includes high injection,

$$I_{\text{sub}} = \frac{2I_{ss}\{\exp(-V_{c_1b_2}/U_T) - 1\}}{1 + \sqrt{1 + 4\dfrac{I_s}{I_{ks}}\{\exp(-V_{c_1b_2}/U_T) - 1\}}}. \qquad (4.48)$$

The current I_{ex} is related to injection currents from the inactive, extrinsic part of the c-b junction. It is modelled similarly to I_f and I_r,

$$I_{ex} = \frac{1}{\beta_{rx}}\left(\frac{f_b^{-1} + n_{bx}\dfrac{I_k}{a_1}}{a_1 + n_{bx}\,a_1}n_{bx} - I_s\right) + I_{\text{sub}}. \qquad (4.49)$$

The minority carrier concentration n_{bx} follows from

$$n_{bx} = \frac{1}{2}f_b^{-1}\frac{f_4}{1 + \sqrt{1 + f_4}}$$

with (4.50)

$$f_4 = 4a_1^2 f_b^2 \frac{I_s}{I_k}\exp(-V_{c_1b_1}/U_T).$$

Related to n_{bx} is also a stored charge Q_{ex},

$$Q_{ex} = \frac{1 - X_{cjc}}{X_{cjc}}Q_{b0}(g_{1x}p_{wx} + g_{2x}n_{bx}),$$

where

$$\left.\begin{aligned}
g_{1x} &= \frac{I_s R_{epi}}{U_T}\exp(V_{dc}/U_T), \\
g_{2x} &= \frac{b_2 f_b^{-1} + n_{bx}}{\dfrac{1}{a_1} f_b^{-1} + n_{bx}}, \\
b_2 &= \frac{1 - (\eta + 1)f_b^{-1}}{(1 + f_b^{-1})^2}, \\
p_{wx} &= \frac{1}{2}\frac{f_5}{1 + \sqrt{1 + f_5}}, \\
f_5 &= 4\exp\{-(V_{c_1 b_1} + V_{dc})\}.
\end{aligned}\right\} \qquad (4.51)$$

Because Q_{b0} is only the fixed base charge in the active part, we must write for the similar quantity in the inactive part $[(1 - x_{cjc})/x_{cjc}]Q_{b0}$. The depletion charges Q_{T_s} and $Q_{T_{ex}}$ have already been mentioned in section 4.3.3. The additional parameters are here $\beta_{rx} - I_{ss} - I_{ks}$.

The required number of parameters of the complete model can be summarized as follows:

for main currents and charges: $I_s - I_k - Q_{b0} - \eta - I_{hc} - R_{epi} - V_{dc} - \tau_{ne} - m_\tau$,
for depletion charges: $C_{0e} - V_{de} - p_e - x_{cje} - C_{0c} - p_c - N_{epi} - x_{cjc} - C_{0s} - V_{ds} - p_s$,
base currents: $\beta_f - I_{bf} - I_{br} - V_{lf} - V_{lr} - x_{ijb}$,
series resistances: $R_e - R_{cc} - R_{bc} - R_{bv}$,
substrate and inactive part: $\beta_{rx} - I_{ss} - I_{ks}$,

in total 33 parameters. This is slightly more than required for the extended E-M model (30 parameters), but much less than required for the extended G-P model.

The MEXTRAM model may also be extended by adding:

1. excess phase shifts to I_f and Q_{be} as outlined in section 3.10,
2. high-frequency current crowding,
3. weak avalanche, see section 3.8,
4. noise sources, see section 3.12,
5. a further splitting-up of the extrinsic, inactive part.

A detailed description of such additions is, however, beyond the scope of this book and will therefore be omitted.

4.3.7 Typical Results

For an illustration of the results that can be obtained with the MEXTRAM model, we must look at Fig. 4.14. In this figure the same quantities are plotted as in Figs. 4.7 (for the G-P model) and 4-8 (for the E-M model), for the same transistor. We see that at low currents the h_{FE} fit of the MEXTRAM

4.3 The MEXTRAM Model

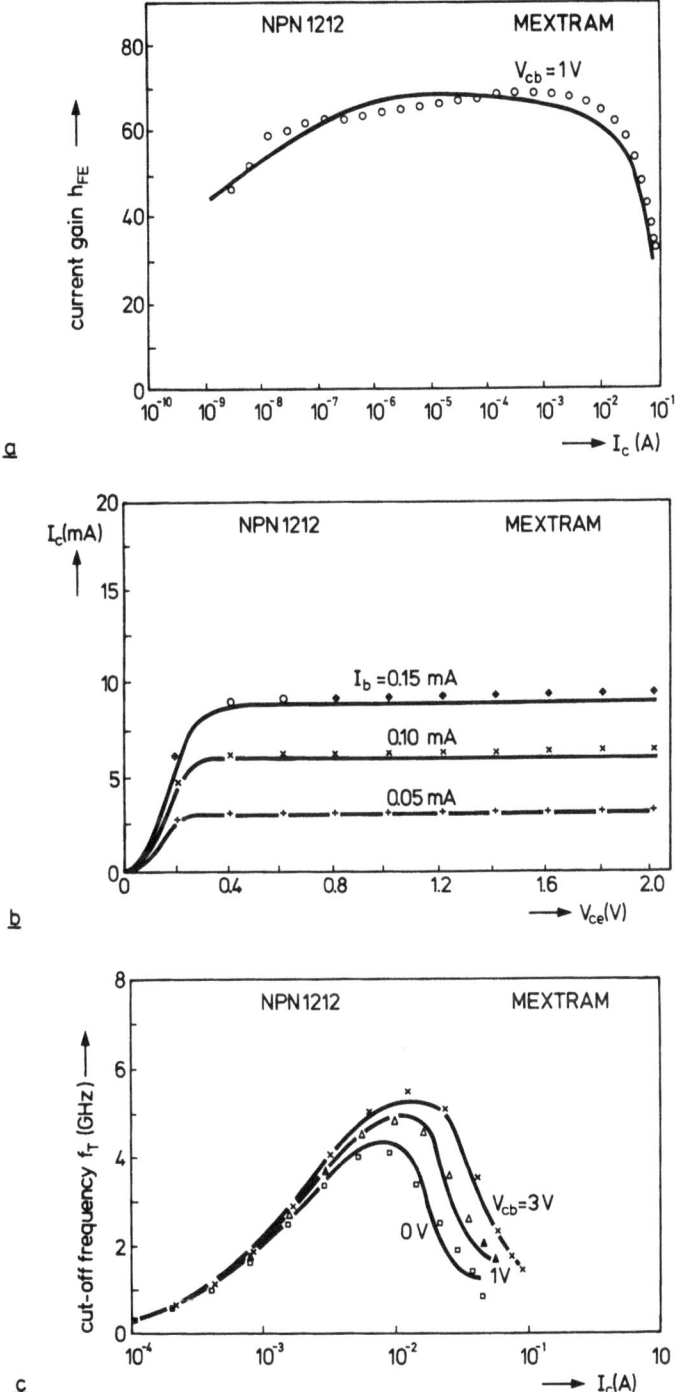

Fig. 4.14 The same measurements of the same transistor as in Figs. 4.7 and 4.8. The model calculations are done with the MEXTRAM model (fully drawn and dashed lines)

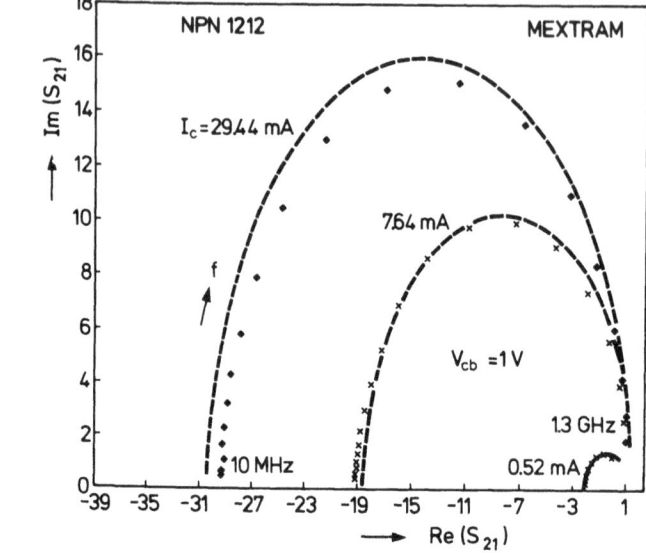

Fig. 4.14 (continued)

model is about as accurate as that of the G-P model, and much better than that of the E-M model. The other characteristics, $I_c(V_{ce})$, $f_T(I_c)$ and the power gain S_{21} are best modelled with MEXTRAM.
The temperature dependence of the main parameters is more or less the same as for the E-M model (see Eqs. (4.12) to (4.15)).
The resistances follow the power law of Eq. (3.129), the hot carrier parameter I_{hc} is taken as temperature-independent.

4.4 Short Review

This chapter will now be concluded with a summing up of the main features of the various models.

4.4.1 Basic Ebers-Moll Model

— Only low injection for base and substrate,
— ideal base current components,
— storage charges with constant transit time.

4.4.2 Extensions to the Ebers-Moll Model

— Early effect with constant Early voltages,
— fixed series resistances,

4.4 Short Review

— depletion capacitances with tangential continuation,
— non-ideal base currents with non-ideality coefficients,
— high injection in the base by means of a fit formula,
— transit time is made bias-dependent by means of a fit formula.

The E-M models are simple, but fast; extensions are realized by means of fit formulas. The basic model requires 6 parameters, the extensions altogether 24 more.

4.4.3 Basic Gummel-Poon Model

— High injection in the base included,
— ideal and non-ideal base current components,
— Early effect with Early voltages,
— charge storage with constant transit time,
— variable base resistance by conductivity modulation in the active part.

4.4.4 Extensions to the Gummel-Poon Model

— Base push-out,
— modulation of the epilayer resistance R_{epi},
— quasi-saturation,
— parasitic *pnp* modelling.

The G-P models rely more on device physics and less on curve fitting than the E-M models. They are more complex, but still fast: CPU time is roughly 1.2 to 1.5 times higher than with E-M models. Quasi-saturation is modelled but remains a weak point. The basic G-P model requires 23 parameters; that number increases to 47 when all extensions are included.

4.4.5 Mextram Models

— High injection in base and substrate,
— influence of built-in field in the base,
— bias-dependent transit times,
— ideal and non-ideal base current components,
— variable base resistance due to conductivity modulation and current crowding,
— current-dependent collector epilayer resistance,
— quasi-saturation and hot-carrier effects,
— parasitic *pnp* and inactive region modelling.

Mextram models stick closely to the device physics, give a good description of the quasi-saturation and require a moderate number of parameters (33).

Table 4.1

	Basic E-M	Extended E-M	Basic G-P	Extended G-P	Mextram
h_{FE}	constant	current-dependent	current-dependent	current-dependent	current-dependent
f_T	constant	current-dependent	rising part current-dependent	fall-off included	bias-dependent
r_b	absent	fixed	conductivity modulation	conductivity modulation	cond. mod. + crowding
r_c	absent	fixed	fixed	bias-dependent	current-dependent
Early effect	absent	Early voltages	Early voltages	Early voltages	depletion charges
quasi-satur.	absent	absent	absent	modelled separately	modelled separately
inact. region	absent	absent	absent	parasitic *pnp*	parasitic *pnp* + inact. region
high inject.	absent	fit formula	physical model	physical model	physical model
number param.	6	30	23	47	33

Because of their complexity a larger CPU time is needed: 2 to 3 times more than with E-M models.
In Table 4.1 the results are summarized. The emitter series resistance is fixed in all models (not present in the basic E-M). The parameters of the temperature dependence are not included in the required number of parameters in this table.

References

[4.1] J. J. Ebers, J. L. Moll: Large Signal Behaviour of Junction Transistors. Proc. I.R.E. 42, 1761 (1954).
[4.2] I. Getreu: Modeling the Bipolar Transistor. Tektronix Inc., Beaverton OR (1979).
[4.3] P. B. Weil: Companion Networks for Advanced Transistor Models. Ph.D. Thesis, UCLA (1976).
[4.4] H. K. Gummel, H. C. Poon: An Integral Charge-Control Model for Bipolar Transistors. Bell Syst. Techn. J. 49, 827 (1970).
[4.5] H. C. de Graaff: Compact Bipolar Transistor Modeling. In: Process and Device Modeling (W. L. Engl, ed.). North-Holland, Amsterdam (1986).

References

[4.6] H. Stübing, H. M. Rein: A Compact Physical Large-Signal Model for High-Speed Bipolar Transistors at High Current Densities. Part I: One-Dimensional Model; H. M. Rein, M. Schröter: Part II: Two-Dimensional Model and Experimental Results. IEEE Trans. Electr. Dev. *ED-34*, 1741 (1987).

[4.7] A. Vladimirescu, A. R. Newton, D. O. Pederson, A. Sangiovanni-Vincentelli, K. Zhang: Spice Version 2 G.5 User's Guide. Berkeley, CA (1981).

[4.8] L. J. Turgeon, J. R. Mathews: A Bipolar Transistor Model of Quasi-Saturation for Use in CAD. IEDM Techn. Digest 394 (1980).

[4.9] G. M. Kull et al.: A Unified Circuit Model for Bipolar Transistors Including Quasi-Saturation Effects. I.E.E.E. Trans. Electr. Dev. *ED-32*, 1103 (1985).

[4.10] H. C. de Graaff, W. J. Kloosterman, T. N. Jansen: Compact Bipolar Transistor Model for CACD, with Accurate Description of Collector Behaviour. Ext. Abstr. 18th Conf. Solid St. Dev. and Materials. Tokyo (1986), p. 287.

[4.11] E. J. Ryder: Mobility of Holes and Electrons in High Electric Fields. Phys. Rev. *90*, 766 (1953).

5 Lateral *pnp* Transistor Models

The use of lateral *pnp* transistors is widespread in linear integrated circuits. They are applied as active load devices, current sources, level shifters etc. [5.1]. Furthermore they are an integral part of integrated injection logic (I^2L) circuits [5.2].

The typical structure of a lateral *pnp* transistor [5.3] is sketched in Fig. 5.1. The lateral *pnp* transistors in a circuit are fabricated together with the vertical *npn* transistors.

The structural difference of the lateral *pnp* with respect to the vertical *npn* has consequences for the electrical behaviour:

— The lateral distances are usually longer than the vertical ones, so the transit times in lateral devices are also longer. This reduces the obtainable f_T compared to vertical transistors.
— The *n* epilayer in a lateral *pnp* is part of the base region and not part of the collector as in a vertical *npn* transistor. So the lightly doped epilayer easily gives rise to high injection effects in the base, but quasi-saturation does not occur and $R_{cv} \equiv 0$. However, extra charge storage does occur, but in the base region under the emitter. This too is detrimental to high f_T values.
— Emitter and collector dope concentrations in lateral *pnp* transistors resemble those of the base in a vertical *npn*, so they are in the range 10^{18}–10^{19}/cm^3 instead of the usual emitter dope around 10^{20}/cm^3. This makes the emitter less efficient because of increased electron injection from the base; it also increases the emitter series resistance and enhances current crowding under the emitter.
— The forward-biased emitter injects holes into the base under the emitter. Most of these holes are stored there, but part of them will be collected by the substrate, giving rise to a substrate current that reduces the gain h_{FE}.
— Recombination in the epilayer bulk is usually negligible, but recombination at the oxide interface may be important.

In general we can say that, in comparison with vertical *npn* transistors, lateral *pnp* transistors have a low current gain (10–50), low maximum f_T

5.1 Model Definitions

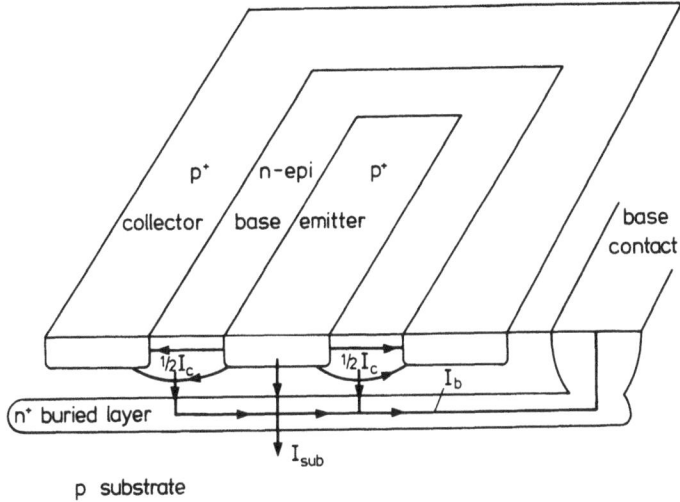

Fig. 5.1 Typical structure of a lateral *pnp* transistor. The arrows indicate the current flow in the forward mode of operation

(10–100 MHz) and a low current-handling capability. The latter is due to the emitter area being only effective at the sidewall and corner regions (see Fig. 5.1). This makes the compatibility with *npn* transistors a problem in most cases, although new, different structures have recently been proposed that possess much higher h_{FE} and/or f_T values [5.3, 5.4].

5.1 Model Definitions

Although the device physics of the lateral *pnp* transistor is rather well established [5.5 to 5.11], literature about compact models for circuit simulation is not very abundant [5.12]. In the following sections we will describe a few examples of the E-M type and the G-P type; for comparison the reader is also referred to chapter 4.

5.1.1 Lateral pnp Models of the Ebers-Moll Type

In Fig. 5.2 the circuit diagram of an extended E-M model is given. The extensions refer to the addition of fixed series resistances, depletion capacitances and Early effect (see also section 4.1.2.). The topology of Fig. 5.2 reveals three internal nodes (e_1, b_1 and c_1).
The main currents in the forward and reverse directions are I_1 and I_2, respectively, the main (hole) current is I_p,

$$\left. \begin{array}{l} I_p = I_1 - I_2, \\ I_1 = \dfrac{I_f}{1+q_1}, \quad I_2 = \dfrac{I_r}{1+q_1} \end{array} \right\} \quad (5.1)$$

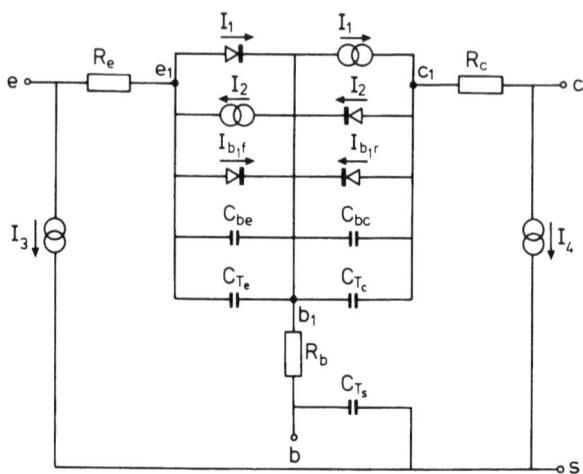

Fig. 5.2 Equivalent circuit diagram of an extended E-M model. The current sources I_3 and I_4 make up the substrate current and are characteristic of the lateral pnp

with

and
$$\left. \begin{array}{l} I_f = I_s\{\exp(V_{e_1 b_1}/U_T) - 1\} \\ I_r = I_s\{\exp(V_{c_1 b_1}/U_T) - 1\}. \end{array} \right\} \quad (5.2)$$

The Early effect is modelled by the term $(1 + q_1)$ with

$$q_1 = \frac{V_{e_1 b_1}}{V_{ear}} + \frac{V_{c_1 b_1}}{V_{eaf}}. \quad (5.3)$$

The saturation current I_s and the Early voltages are parameters. The model contains only ideal base current components

$$I_{b_1 f} = \frac{I_f}{\beta_f}, \quad I_{b_1 r} = \frac{I_r}{\beta_r}. \quad (5.4)$$

The forward and reverse charge storages are, respectively,

$$Q_{be} = \tau_f I_f, \quad Q_{bc} = \tau_r I_r, \quad (5.5)$$

from which are derived the capacitances

$$C_{be} = dQ_{be}/dV_{e_1 b_1} \quad \text{and} \quad C_{bc} = dQ_{bc}/dV_{c_1 b_1}.$$

In the foregoing β_f and β_r are the current gain parameters, τ_f and τ_r the transit time parameters. The depletion capacitances are modelled in the usual way, with a tangential continuation as outlined in section 3.5.
So far the only difference with the E-M models of section 4.1 is the polarity of the applied voltages: Here $V_{eb} > 0$ and $V_{cb} < 0$ in the normal forward

5.1 Model Definitions

mode of operation. Specific additions for the lateral *pnp* behaviour are the current sources I_3 and I_4. They are defined as

$$I_3 = X_{es}I_f, \quad I_4 = X_{cs}I_r \tag{5.6}$$

and they represent the direct injection into the substrate from the emitter and collector: $I_{sub} = I_3 + I_4$.

For the collector and base current we need five parameters ($I_s-\beta_f-\beta_r-V_{eaf}-V_{ear}$), two for the charge storage ($\tau_f-\tau_r$), and two for the substrate currents ($X_{es}-X_{cs}$). We need furthermore three parameters for the series resistances ($R_e-R_b-R_c$) and three times four for the depletion capacitances (C_0-V_d-p-a). This makes altogether 24 parameters.

The high-injection effect is not modelled, so the current gain h_{FE} is almost bias-independent, except for the voltage dependence of the Early factor $(1 + q_1)$. The total transit time is given by Eq. (4.7) and the typical $f_T(I_c)$ curve has already been sketched in Fig. 4.4 d.

A simple modification of the network topology can introduce an h_{FE} fall-off at high currents: See Fig. 5.3.

The ideal base current components now have a different voltage dependence

$$\left. \begin{aligned} I_{b_1f} &= \frac{I_s}{\beta_f} \{\exp(V_{eb_1}/U_T) - 1\}, \\ I_{b_1r} &= \frac{I_s}{\beta_r} \{\exp(V_{cb}/U_T) - 1\}. \end{aligned} \right\} \tag{5.7}$$

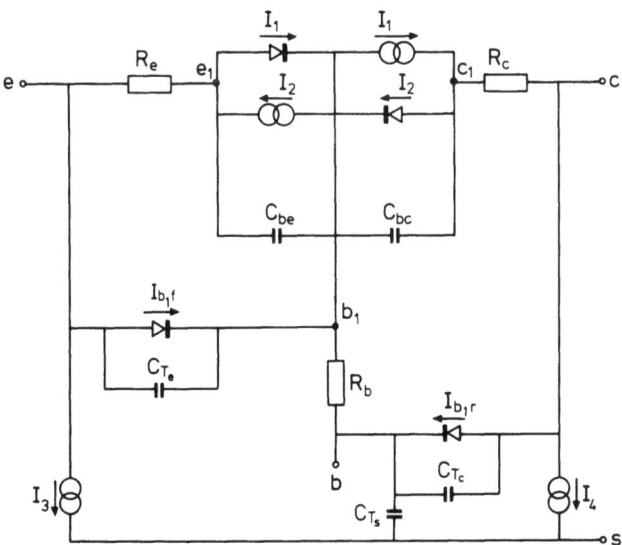

Fig. 5.3 Modification of the circuit diagram of Fig. 5.2. It introduces h_{FE} fall-off at high currents

At high currents we get $V_{eb_1} > V_{e_1b_1}$ and $V_{cb} > V_{c_1b_1}$, which causes a decrease in h_{FE}.

5.1.2 Lateral pnp Models of the Gummel-Poon Type

The foregoing E-M models are rather primitive and not very satisfactory in practice. Many improvements of the basic G-P model (section 4.2.1) are very often required for lateral *pnp* transistors. They are:

— high injection effects in the collector and substrate currents,
— bias-dependent base resistance,
— non-ideal base currents,
— a bias-dependent forward transit time.

Fig. 5.4 gives the circuit diagram; the network topology is similar to the one in Fig. 5.3.

As in section 4.2.1, we first model the base charge ratio Q_b/Q_{b0}:

$$\frac{Q_b}{Q_{b0}} = \frac{1}{2}(1 + q_1) + \frac{1}{2}\sqrt{(1 + q_1)^2 + 4q_2}. \tag{5.8}$$

$(1 + q_1)$ is the Early factor, already given by Eq. (5.3). The quantity q_2 represents the charge storage and is modelled as

$$q_2 = \left(\frac{I_f}{I_{kf}}\right)^{N_f} + \frac{I_r}{I_{kr}}. \tag{5.9}$$

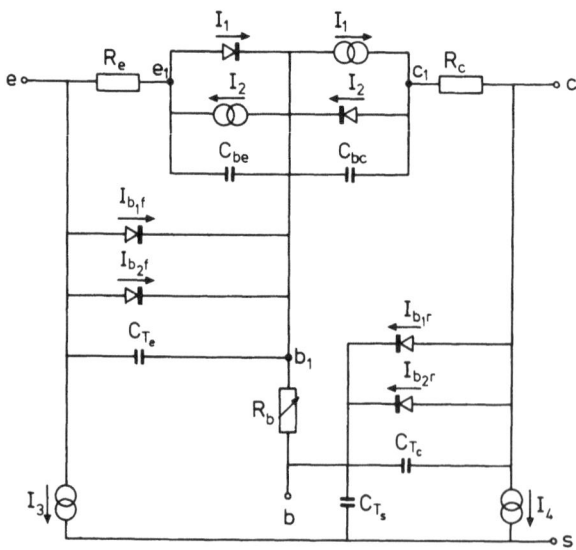

Fig. 5.4 Equivalent circuit diagram of a basic G-P model for lateral *pnp* transistors

5.1 Model Definitions

The difference with Eq. (4.19) is the presence of the fit parameter N_f, which gives an extra degree of freedom in modelling the high-injection effect.
The ideal forward (I_f) and reverse currents (I_r) have already been given by Eq. (5.2). The main current is now given by

$$I_p = I_1 - I_2$$

with

$$I_1 = \frac{I_f}{Q_b/Q_{b0}}, \quad I_2 = \frac{I_r}{Q_b/Q_{b0}}.$$

(5.10)

The ideal base currents are as given by Eq. (5.4), the non-ideal base currents are defined as

$$\begin{aligned} I_{b_2f} &= I_{bf}\{\exp(V_{e_1b_1}/m_f U_T) - 1\}, \\ I_{b_2r} &= I_{br}\{\exp(V_{c_1b_1}/m_r U_T) - 1\}. \end{aligned}$$

(5.11)

I_{bf}, I_{br}, m_f and m_r are model parameters.

Although the base current components in Fig. 5.4 are connected in a similar way as in Fig. 5.3, they depend on internal voltages. This saves computer time (the exponentials are also calculated for I_f and I_r) and Eqs. (5.10) and (5.11) already ensure a bias-dependent h_{FE}.
The substrate currents I_3 and I_4 are provided with a high injection effect, characterized with one parameter for both currents,

$$\begin{aligned} I_3 &= X_{es} \frac{2I_f}{1 + \sqrt{1 + 4I_f/I_{ks}}}, \\ I_4 &= X_{cs} \frac{2I_r}{1 + \sqrt{1 + 4I_r/I_{ks}}}. \end{aligned}$$

(5.12)

I_{ks} is the high injection parameter.
The stored charges are modelled as

$$\begin{aligned} Q_{be} &= \tau_f(I_1 + I_s)(1 + q_1)^2 + \tau_{ne} I_s \left(\frac{I_f + I_s}{I_s}\right)^{1/m_\tau}, \\ Q_{bc} &= \tau_r(I_2 + I_s)(1 + q_1)^2. \end{aligned}$$

(5.13)

The charge storage from the collector injection (Q_{bc}) is modelled quite simply, with τ_r as model parameter. The forward charge storage Q_{be} requires three parameters (τ_f, τ_{ne} and m_τ). The first term on the right-hand side of the Q_{be} formula includes the Early effect (cf. section 3.6), the second term represents the Q_e charge from the Mextram model (Eq. (4.34)).
The depletion capacitances C_{T_e}, C_{T_c} and C_{T_s} are modelled as described in section 5.1.1; of the series resistances, the base resistance shows bias dependence by means of conductivity modulation of a part of it,

$$R_b = R_{bc} + \frac{R_{bv}}{Q_b/Q_{bo}}. \tag{5.14}$$

The total number of required parameters is 35, partitioned as follows:

the main currents	$I_s-I_{kf}-I_{kr}-V_{eaf}-V_{ear}-N_f$,
the base currents	$I_{bf}-I_{br}-m_f-m_r-\beta_f-\beta_r$,
the substrate currents	$X_{es}-X_{cs}-I_{ks}$,
the stored charges	$\tau_f-\tau_r-\tau_{ne}-m_\tau$,
the series resistances	$R_e-R_c-R_{bc}-R_{bv}$,
the depletion capacitances	$3 \times (C_0-V_d-p-a)$.

The temperature dependence does not need specific modifications. It is again based on the theory of section 3.13 and it uses the same formulas as in section 4.13. Because the various doping levels in the lateral *pnp* transistor are lower than those in the vertical *npn*, the respective bandgap voltages are higher in the lateral *pnp*.

5.2 Results

To give an idea of the performance of the G-P model for lateral transistors we have compared the model calculations with measurements. The tested device has a square emitter of 4×4 μm, completely surrounded by the collector; there is one collector and one base contact (see Fig. 5.5).

Fig. 5.6 gives the Gummel plots for the forward and reverse operation. At the higher current levels we can see the influence of the series resistances (especially the base resistance) plus the high injection. The relevant parameter values are

$$I_{kf} = 7.65 \times 10^{-5} \text{ A}, \qquad I_{kr} = 6.95 \times 10^{-5} \text{ A},$$
$$I_{ks} = 4.78 \times 10^{-2} \text{ A}, \qquad R_{bc} = 100 \text{ }\Omega, \qquad R_{bv} = 1933 \text{ }\Omega,$$

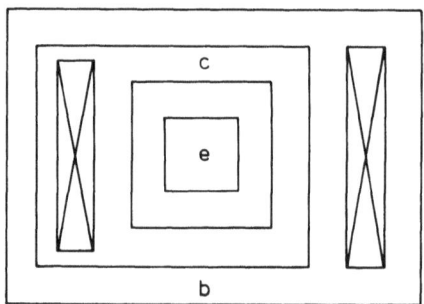

Fig. 5.5 Top view of a transistor layout. The emitter size is 4×4 μm

5.2 Results

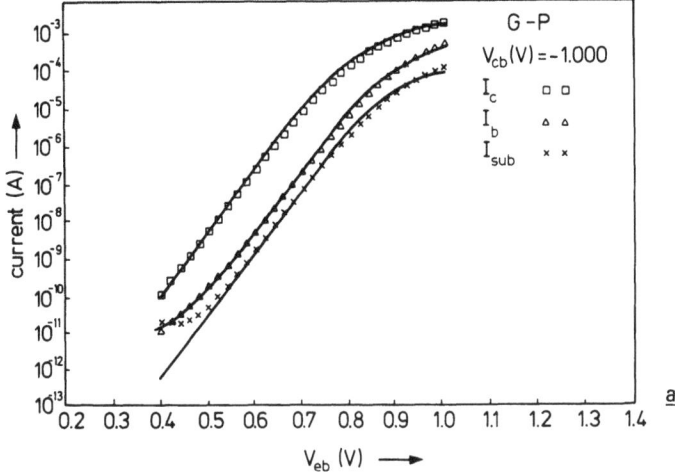

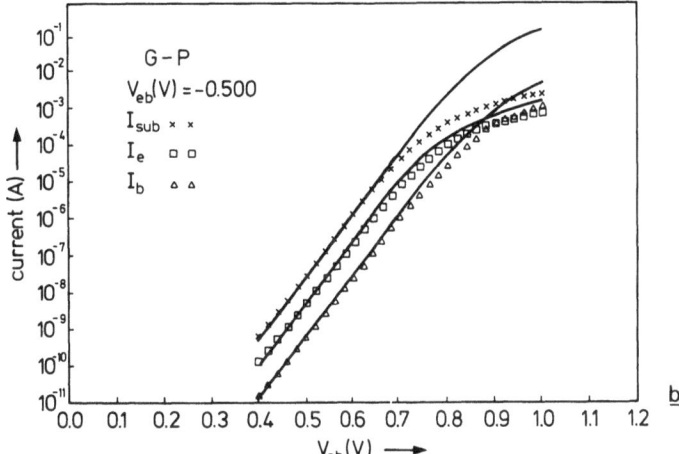

Fig. 5.6 Gummel plots of the transistor in Fig. 5.5. a) forward mode, b) reverse mode. The fully drawn lines are model calculations, the measurements are represented by □ △ ×

$R_e = 10\ \Omega$ and $R_c = 56.6\ \Omega$. The model predictions in the reverse mode are not very accurate, although the current gains (Fig. 5.7) are almost perfectly modelled.

Finally, in Fig. 5.8, we have plotted the measured and calculated f_T curves. The overall impression of these figures is that the lateral G-P model does not perform very accurately; the E-M results are not given here, but are far worse. The reasons for this mediocre performance will be indicated in the next section.

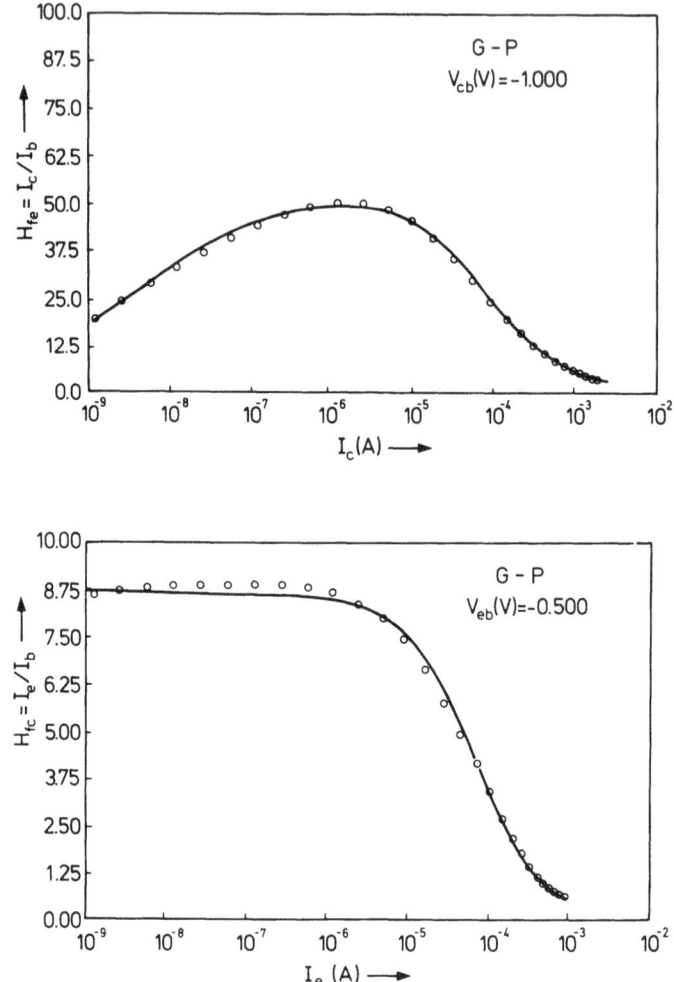

Fig. 5.7 The forward (h_{FE}) and reverse (h_{FC}) current gains of the transistor in Fig. 5.5, as a function of I_c and I_e, respectively. Fully drawn lines are model calculations, the circles represent the measurements

5.3 Shortcomings of Existing Models

The main difficulties in modelling the lateral transistor lie in the essentially two-dimensional nature of the current flow, which means that the series resistances are structurally not well-defined; see Fig. 5.9.
In this figure we can roughly distinguish two different transistors (① and ②) in parallel.
Each has different series resistances in emitter and collector and also different base widths. The latter implies different base charges Q_{b0}, different transit times and different Early effects. At low currents transistor ① may dominate

5.3 Shortcomings of Existing Models

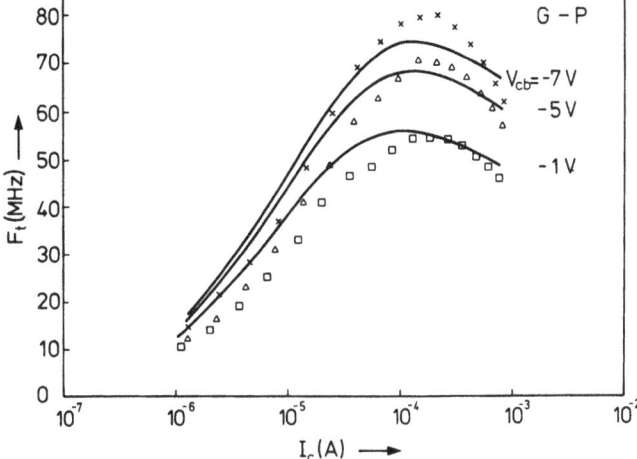

Fig. 5.8 The measured (□△×) and calculated (fully drawn lines) cut-off frequency f_T as a function of I_c for various values of V_{cb}. Same transistor as in Figs. 5.6 and 5.7

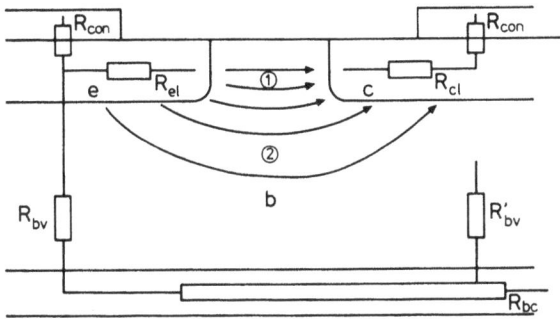

Fig. 5.9 Two-dimensional current flow in a lateral device, constituting two transistors in parallel (① and ②). R_{con} represents the contact resistances, R_{el} and R_{cl} the lateral resistances in the emitter and collector regions, respectively. The lateral base resistance R_{bc} of the buried layer is constant, R_{bv} and $R_{bv'}$ may vary because of conductivity modulation

slightly, but at higher currents transistor ② gradually takes over. Because of its larger base width, number ② has e.g. a higher Early voltage. So the V_{eaf} of the compound device is by no means a constant, as a parameter should be, but it varies strongly with bias [5.13], see Fig. 5.10. Another deficiency is that the substrate currents I_3 and I_4 in Eq. (5.12) have the same parameter I_{ks} for high injection. In reality they may have different parameter values because the emitter and collector areas are not the same.

Because of this complex behaviour of a lateral *pnp* transistor, the results are still not satisfactory, even with a G-P model, which already requires 35 parameters (two more than the Mextram model!). Modelling the lateral *pnp* by two separate G-P transistors in parallel would roughly double the

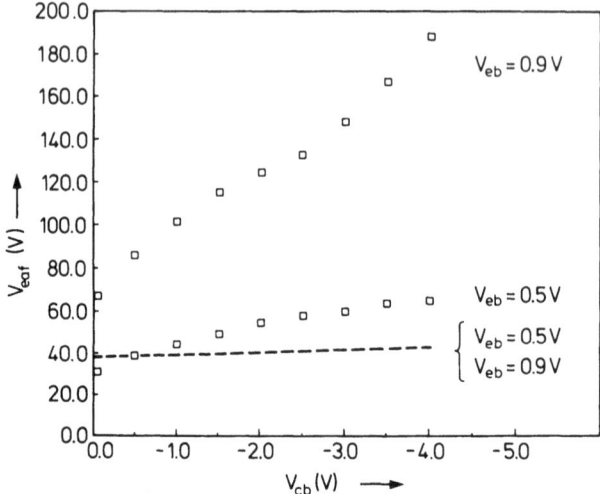

Fig. 5.10 The forward Early voltage V_{eaf}, defined as $V_{eaf} = I_c |dV_{cb}/dI_c|$, measured under various bias conditions. The dotted line represents the predictions of the model in section 5.1.2. The emitter shape was octagonal

number of parameters and would give problems with their determination. In conclusion we can say that, although the models of the lateral *pnp* transistor are more complex than those of the vertical *pnp*, they perform worse.

References

[5.1] H. C. Lin: D. C. Analysis of Multiple Collector and Emitter Transistors in Integrated Structures. IEEE J. Solid-St. Circ. *SC-4*, 20 (1969).
[5.2] K. Hart, A. Slob: Integrated Injection Logic: A New Approach to LSI. IEEE J. Solid-St. Circ. *SC-7*, 346 (1972).
[5.3] J. Sugawara, T. Kamei: A High Performance, High Voltage Self-Aligned Double-Diffused Lateral (SADDL) *pnp* Transistor. IEEE Trans. Electr. Dev. *ED-33*, 23 (1986).
[5.4] K. Nakazato, T. Nakamura, M. Kato: A 3 GHz Lateral *pnp* Transistor. IEDM Techn. Digest (paper 16.3) 416 (1986).
[5.5] J. Lindmayer, W. Schneider: Theory of Lateral Transistors. Solid-St. Electr. *10*, 225 (1967).
[5.6] S. Chou: An Investigation of Lateral Transistors—d.c. Characteristics. Solid-St. Electr. *14*, 811 (1971).
[5.7] S. Chou: Small-Signal Characteristics of Lateral Transistors. Solid-St. Electr. *15*, 27 (1972).
[5.8] D. Seltz, J. Kidron: A Two-Dimensional Model for the Lateral *p-n-p* Transistor. IEEE Trans. Electr. Dev. *ED-21*, 587 (1974).
[5.9] H. H. Berger, U. Dreckmann: The Lateral *p-n-p* Transistor—A Practical Investigation of the d.c. Characteristics. IEEE Trans. Electr. Dev. *ED-26*, 1038 (1979).
[5.10] K-S. Seo, Ch-K. Kim: On the Geometrical Factor of Lateral *p-n-p* Transistors. IEEE Trans. Electr. Dev. *ED-27*, 295 (1980).

References

[5.11] A. A. Eltoukhy, D. J. Roulston: A Complete Analytic Model for the Base and Collector Current in Lateral *p-n-p* Transistors. Solid-St. Electr. *27*, 69 (1984).

[5.12] I. Kidron: Integrated Circuit Model for Lateral *pnp* Transistors Including Isolation Junction Interactions. Int. J. Electr. *31*, 421 (1971).

[5.13] M. C. Schneider, J. A. Bodinaud: A New Formulation of the Early Effect in Lateral *pnp* Transistors. Physica *129B*, 327 (1985).

6 MOSFET Physics Relevant to Device Modelling

6.1 Formation of the Inversion Layer

In order to facilitate the discussion of specific transistor models, which is the subject of the next three chapters, in this chapter we introduce some general device concepts relevant to MOSFET modelling. We deal successively with the following subjects: formation of an inversion layer, the ideal drain current, threshold voltage, carrier mobility in an inversion layer, the saturation mode, dynamic operation and inherent parasitics. Unless stated otherwise, the device structure considered here is that of an n-channel transistor. In most cases, a p-channel device only needs a change of sign to produce its model formulas.

6.1.1 Qualitative Discussion

To understand the basic operation of the MOSFET [6.1, 6.2, 6.3] we assume first that the insulator between gate and substrate is free of defects and additionally that the gate material and the neutral substrate have the same work function. The energy band diagram for a p-type semiconductor without a gate bias applied with respect to the substrate ($V_G = 0$), is given in Fig. 6.1a. Owing to the zero-current condition, the Fermi level E_F in both (semi)conducting materials coincides.

When a positive gate bias is applied, the energy band diagram of Fig. 6.1a changes to that given in Fig. 6.1b. Note that according to the convention the electron potential is plotted in the negative direction. Since the positive charge at the gate forces the holes to move away from the interface, a negative space charge of ionized acceptors is formed at the semiconductor surface. According to Poisson's equation the space charge is only maintained if a potential drop exists across the space charge layer. Consequently, the potential difference V_G is divided across the two nonconducting regions of the MOS structure: the gate insulator and the semiconductor depletion region. Since the insulator still forces the current to remain at zero value, the Fermi level remains flat.

6.1 Formation of the Inversion Layer

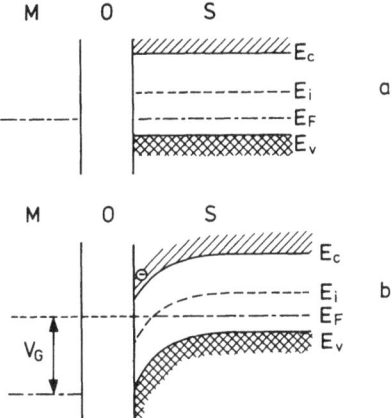

Fig. 6.1 a: Energy band diagram at zero gate bias for an ideal MOSFET with p-type substrate; b: Energy band diagram at positive gate bias for the same situation as Fig. 6.1 a

When the gate bias is further increased the energy bands in the semiconductor become more strongly curved until the difference between the conduction band and the Fermi level at the interface becomes smaller than the corresponding difference between Fermi level and the valence band of the neutral bulk. According to Boltzmann statistics, under the latter condition a considerable amount of electrons can be expected at the interface. These electrons form an inversion layer. When V_G is increased still further, the inversion layer soon shields the underlying depletion layer from the increase of gate charge. Consequently the expansion of the depletion layer comes to a hold and according to Gauss's law the increasing field in the insulator causes the mobile inversion charge to increase almost proportionally with the gate bias. Quantum-mechanical calculations indicate that the inversion layer has a thickness between 30 and 100 Å [6.4, 6.5]. Since the latter value is much smaller than the thickness of the depletion layer (between 0.2 μm and 0.5 μm in practice) at normal values of the gate bias ($V_G < 10$ Volt) the potential drop across the inversion layer can be neglected.
In reality the gate insulator is not perfect, but owing to contamination it usually contains a fixed positive charge. In addition these contaminants and the abrupt change of the crystal structure at the interface give rise to local electron states. Since these states can exchange charges with the semiconductor, their occupation depends on the position in the energy band with respect to the Fermi level. Fortunately, semiconductor technology has advanced to a state, where for most applications fixed charges can be considered of second order and interface state charges even of higher order. Finally, in most practical gate materials the work function ϕ_m differs from the corresponding ϕ_s value in the semiconductor. With a p type substrate we usually have the situation drawn in Fig. 6.2a. The energy diagram corresponds

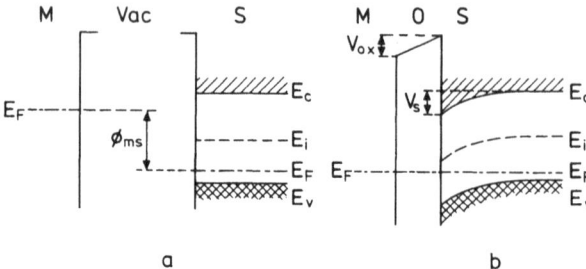

Fig. 6.2 *a*: Energy band diagram of a gate and a semiconductor, which are not in contact and have a different work function; *b*: Energy band diagram, when the gate and the semiconductor form a MOS configuration

to the case where gate and semiconductor are a considerable distance apart.

For the realistic MOST structure, where the Fermi levels have to coincide, the difference in work function causes a contact potential difference between gate and neutral bulk. Again the potential drop is divided across insulator and semiconductor space charge. The latter is formed as the original potential difference between gate and substrate causes net charge to appear on both sides of the insulator. Formally we may write

$$(V_{0x} + V_s)_{V_{G=0}} = -\phi_{ms}, \tag{6.1}$$

where V_{0x} and V_s are the potential drop across the insulator and the semiconductor depletion layer, respectively, and

$$\phi_{ms} = \phi_m - \phi_s. \tag{6.2}$$

In a situation like that shown in Fig. 6.2 a ϕ_{ms} has a negative value.

Taking into account the above insulator charge and the contact potential difference, the real MOS structure of Fig. 6.2 b can be forced into the state of Fig. 6.1 a (flat band condition), by applying a negative gate bias. Assuming that the built-in charge can be represented by an effective charge Q_{0x} located close to the interface [6.6], an equivalent negative gate charge is required to compensate for the effect of Q_{0x}. In addition, a gate voltage ϕ_{ms} is required to eliminate the contact potential difference given by (6.1). The total voltage, which causes the space charges to virtually disappear on both sides, is called the flat band voltage V_{FB} [6.7].

Consequently we have

$$V_{FB} = \phi_{ms} - Q_{0x}/C_{0x}. \tag{6.3}$$

According to the definition Q_{0x} is given by

$$Q_{0x} = \frac{1}{t_{0x}} \int_0^{t_{0x}} y \rho_{0x}(y) \, dy, \tag{6.4}$$

where $\rho_{0x}(y)$ is the charge distribution in the insulator.

6.1.2 Quantitative Analysis

For any applied gate voltage V_G different from V_{FB} a space charge $\rho(y)$ is present in the semiconductor. Here y denotes the normal direction. Formally, for a p-type substrate the above space charge is given by

$$\rho(y) = q[p(y) - n(y) - N_A], \tag{6.5}$$

where N_A is the net doping density.
Assuming the semiconductor to be nondegenerate the electron and hole density is given by Maxwell-Boltzmann statistics (see section 2.2)

$$p(y) \approx N_A \exp[-\psi(y)/U_T], \tag{6.6}$$

$$n(y) \approx \frac{n_i^2}{N_A} \exp[\psi(y)/U_T], \tag{6.7}$$

where n_i is the intrinsic carrier density, ψ is the potential with respect to the neutral bulk at y and $U_T = kT/q$. Furthermore the doping density N_A is usually expressed in terms of the Fermi potential ϕ_F (compare (2.16))

$$N_A = n_i \exp(\phi_F/U_T). \tag{6.8}$$

Using the above definitions the Poisson equation is written as follows,

$$\frac{d^2\psi}{dy^2} = -\frac{qN_A}{\varepsilon_s}[\exp(-\psi/U_T) - 1$$
$$- \exp(-2\phi_F/U_T)\{\exp(\psi/U_T) - 1\}]. \tag{6.9}$$

In addition, deep into the neutral bulk the following boundary conditions apply,

$$\left.\begin{array}{l} \psi = 0, \\ \dfrac{d\psi}{dy} = 0. \end{array}\right\} \tag{6.10}$$

Integrating Poisson's equation once, we obtain for the total semiconductor charge per unit area (Q_s)

$$Q_s = -\varepsilon_s \left(\frac{d\psi}{dy}\right)_{y_s}, \tag{6.11}$$

where the index s refers to the surface (interface). Evaluating Eq. (6.11) we then have [6.7, 6.8]

$$Q_s = -(2\varepsilon_s q N_A)^{1/2} \cdot [\psi_s + U_T\{\exp(-\psi_s/U_T) - 1\}$$
$$+ \exp(-2\phi_F/U_T)\{U_T \exp(\psi_s/U_T) - U_T\}]^{1/2}. \tag{6.12}$$

Furthermore when Gauss's theorem is applied at the interface the above charge can be related to the applied gate bias

$$Q_s = -C_{ox}(V_G - V_{FB} - \psi_s). \tag{6.13}$$

In this relation ψ_s is the surface potential. Note that ψ_s also equals the total potential drop (V_s) across the semiconductor space-charge layer.

Until now the above analysis has been quite general and not restricted to the formation of an inversion layer. Formally we may distinguish the following modes of operation:

— accumulation mode $\psi_s < 0$,
— depletion mode $0 \leqslant \psi_s \leqslant \phi_F$,
— inversion mode $\psi_s > \phi_F$.

Here we are only interested in the latter case. Under the condition $\psi_s > \phi_F$ Eq. (6.12) can be further approximated,

$$Q_s = -(2q\varepsilon_s N_A)^{1/2} \cdot [\psi_s + U_T \exp\{(\psi_s - 2\phi_F)/U_T\}]^{1/2}. \tag{6.14}$$

Equating (6.13) and (6.14) an implicit relation for $\psi_s(V_G)$ is found.

Fig. 6.3 gives ψ_s as a function of applied gate bias for two different device structures. The first one typically represents MOS IC processes with a structural dimension of 5 µm upwards, the other one is typical of a process with submicron dimensions. Two distinct regions of operation can be observed. The weak inversion region ($\phi_F < \psi_s < 2\phi_F$) is characterized by a nearly constant slope $d\psi_s/dV_G$; in strong inversion ($\psi_s > 2\phi_F$), however, this slope drops to small values. The latter result is due to the fact that the mobile

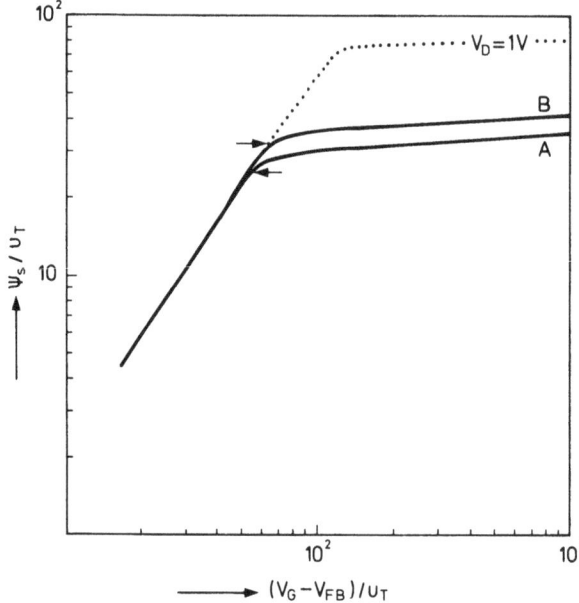

Fig. 6.3 Surface potential as a function of applied gate bias (fully drawn lines). A and B represent a device with $t_{ox} = 100$ nm, $N_A = 5 \times 10^{15}/\text{cm}^3$ and $t_{ox} = 20$ nm, $N_A = 1 \times 10^{17}$ cm^3, respectively. Arrow indicates point where ψ_s equals $2\phi_F$. Dashed line represents B in case of a reverse drain bias

6.1 Formation of the Inversion Layer

carriers of the inversion layer screen the underlying depletion region from an increase in gate charge.

Since the inversion layer is almost two orders of magnitude less thick than the depletion layer, it has been proposed that its thickness be neglected. This concept is known as *charge sheet approximation* [6.9, 6.10, 6.11]. It implies that practically all of the depletion region is assumed to be free of mobile carriers and consequently all of the potential difference is dropped across the depletion region. To some extent this concept is equivalent to the so-called depletion approximation in a *p-n* junction. Solving Poisson's equation under this assumption we obtain for the depletion charge

$$Q_d \approx -(2\varepsilon_s q N_A)^{1/2} \cdot \psi_s^{1/2}. \tag{6.15}$$

Using this result the inversion charge Q_i is easily obtained [6.11] by subtracting Eq. (6.15) from Eq. (6.14),

$$Q_i \approx -(2\varepsilon_s q N_A)^{1/2} \cdot [\{\psi_s + U_T \exp(\psi_s - 2\phi_F)/U_T\}^{1/2} - \psi_s^{1/2}]. \tag{6.16}$$

Nevertheless the result for Q_i remains an implicit function in the variable V_G. Fortunately for two important conditions Q_i can be further approximated. In strong inversion ($\psi_s > 2(\phi_F + U_T)$),

$$\psi_s \approx \psi_{sm} = 2(\phi_F + U_T).$$

Then from Eq. (6.13) it follows that

$$Q_s \approx -C_{ox}[V_G - V_{FB} - \psi_{sm}], \tag{6.17}$$

and therefore

$$Q_i \approx -C_{ox}[V_G - V_{FB} - \psi_{sm} - \gamma \psi_{sm}^{1/2}], \tag{6.18}$$

where

$$\gamma = \frac{(2\varepsilon_s q N_A)^{1/2}}{C_{ox}}. \tag{6.19}$$

Furthermore the quantity

$$V_{T0} = V_{FB} + \psi_{sm} + \gamma \psi_{sm}^{1/2} \tag{6.20}$$

is known as the threshold voltage. Usually it is obtained by extrapolating the $Q_i(V_G)$ curve.

In weak inversion ($\phi_F < \psi_s < 2\phi_F$) the exponential in the right-hand term of Eq. (6.14) becomes much smaller than ψ_s. Therefore the square-root term can be approximated by the first two terms of its Taylor expansion,

$$Q_s \approx -\gamma C_{ox} \left[\psi_s^{1/2} + \left(\frac{U_T}{2\psi_s^{1/2}}\right) \exp\{(\psi_s - 2\phi_F)/U_T\} \right]. \tag{6.21}$$

Consequently

$$Q_i \approx \left(\frac{U_T \gamma C_{ox}}{2\psi_s^{1/2}}\right) \exp\{(\psi_s - 2\phi_F)/U_T\}. \tag{6.22}$$

Since under this condition the charge Q_i is much smaller than the charge Q_d, we obtain by equating Eqs. (6.13) and (6.15)

$$\gamma \psi_s^{1/2} \approx V_G - V_{FB} - \psi_s. \tag{6.23}$$

Solving this result for ψ_s we obtain [6.12]

$$\psi_s = \left[-\frac{\gamma}{2} + \left\{ \frac{\gamma^2}{4} + (V_G - V_{FB}) \right\}^{1/2} \right]^2. \tag{6.24}$$

Substituting Eq. (6.23) in Eq. (6.22) we may express Q_i in terms of the gate bias. However, bearing in mind that the function $dV_G/d\psi_s$ remains almost constant in the region considered, it is useful to develop Eq. (6.22) near the middle of the weak-inversion region ($\psi_s \approx 1.5\phi_F$) [6.13, 6.14]. Denoting the slope as

$$m = \left(\frac{dV_G}{d\psi_s} \right)_{1.5\phi_F},$$

the latter value can be obtained by differentiating Eq. (6.23),

$$m = 1 + \frac{\gamma}{2(1.5\phi_F)^{1/2}}. \tag{6.25}$$

In addition, by making use of the corresponding value of the gate bias

$$V_{Gr} = V_{FB} + 1.5\phi_F + \gamma(1.5\phi_F)^{1/2}, \tag{6.26}$$

we write

$$\psi_s - 1.5\phi_F \simeq \frac{(V_G - V_{Gr})}{m}.$$

Substituting this result into Eq. (6.22) we finally obtain

$$Q_i \approx Q_{i0} \exp\left(\frac{V_G - V_{Gr}}{mU_T} \right), \tag{6.27}$$

where

$$Q_{i0} = \frac{(2\varepsilon_s q N_A)^{1/2}}{2.45\phi_F^{1/2}} \cdot U_T \exp\left(\frac{-\phi_F}{2U_T} \right).$$

Now the inversion charge follows a purely exponential relation in V_G.
In Fig. 6.4 the approximated results (6.18) and (6.27) are compared with the more accurate result (6.16). Except for a small region around V_{T0} (with a width 10 U_T) the approximations are sufficiently accurate. Obviously the transition region, where an explicit relation is lacking, can be narrowed by developing Eq. (6.22) around some higher value of V_G (at the expense of accuracy at deep depletion). For the realistic MOST structure with source and drain, almost all MOST circuit models have been developed on the basis of either Eq. (6.18) or Eq. (6.27), and using a curve-fitting routine for the transition region.

6.2 The Ideal MOS Transistor Current

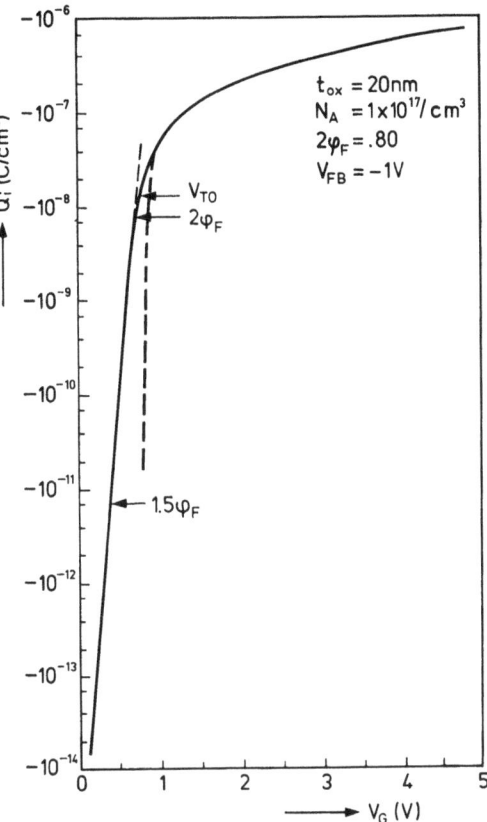

Fig. 6.4 Inversion layer charge as a function of applied gate bias. Only in a narrow range around the threshold voltage V_{TO} do the broken line approximations (6.18) and (6.27) differ from the general result (6.16). Arrows indicate specific points of interest

6.2 The Ideal MOS Transistor Current

Although modelling of the drain current is strictly the subject of another chapter, we consider here the ideal MOST current to serve as a reference to more practical MOST models, which have to take into account numerous degrading effects. This implies that we neglect these effects here, an assumption which is more or less allowed for large devices made on lightly doped substrates. Our basic structure is slightly more complicated than in the previous section, as we have to add source and drain junctions. In addition these junctions can be biased at different potentials with respect to the substrate (compare Fig. 6.5).

If, for instance, the n^+ drain is biased at a reverse voltage V_R, keeping the substrate grounded, the minority carriers in the depletion layer close to the drain obey a law different from Eq. (6.7). In this case Eq. (6.7) has to be changed into

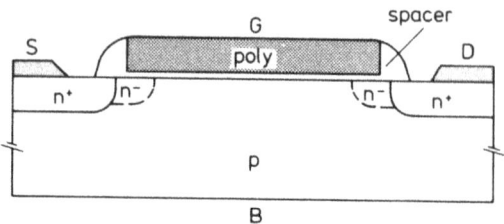

Fig. 6.5 Cross-section of a practical *n*-type MOS transistor. In the conventional type the heavily doped junctions (fully drawn lines) overlap the gate; in the LDD-type the overlap is achieved by means of a lightly doped extension

$$n = N_A \exp\{(\psi - 2\phi_F - V_R)/U_T\}.$$

Consequently, when solving the Poisson equation with a bias V_G applied to the gate, the depletion charge close to the drain differs from Eq. (6.14). Now we have

$$Q_s = -\gamma C_{ox}[\psi_s + U_T \exp\{(\psi_s - 2\phi_F - V_R)/U_T\}]^{1/2}. \tag{6.28}$$

On the other hand from Gauss's theorem we conclude that Eq. (6.13) can be maintained. When combining Eq. (6.13) with the above expression for Q_s, we obtain the following implicit relation for the surface potential at the drain end of the channel,

$$V_G = V_{FB} + \psi_s + \gamma[\psi_s + U_T \exp\{(\psi_s - 2\phi_F - V_R)/U_T\}]^{1/2}. \tag{6.29}$$

In Fig. 6.3 the dotted line gives $\psi_s(V_G)$ for $V_R = 1$ Volt. Qualitatively the same behaviour as for $V_R = 0$ Volt is observed. Like Eq. (6.13) for the inversion charge, Eq. (6.18) too can be maintained.

Naturally, for a situation with $V_S = 0$, $V_D > 0$, and with close to the source end $\psi_s(V_G) > \phi_F$, sufficient mobile carriers are present in the channel to cause a drain current to flow. On the basis of the charge-sheet approximation a current expression has been derived in several references [6.11, 6.15, 6.16]. Here we follow an approach, that combines simplicity with sufficient accuracy [6.17]. Formally the channel current is written (compare Eq. (2.43))

$$I(x) = -\mu_n W Q_i \frac{d\psi_s}{dx} + \mu_n U_T \frac{dQ_i}{dx}, \tag{6.30}$$

where μ_n is the electron mobility, W is the channel width and x is the lateral direction. In this equation the first term represents transport by drift and the second one transport by diffusion. Note that in the latter term Einstein's relation has been used.

Integrating the above expression from the source end of the channel ($x = 0$) to the drain end ($x = L$), and considering that I is independent of x, then by making use of Eq. (6.18) we obtain

$$I_{DS} = \frac{W\mu_n C_{ox}}{L}[f(\psi_{sL}) - f(\psi_{s0})], \tag{6.31}$$

6.2 The Ideal MOS Transistor Current

where

$$f(\psi_s) = [(V_G - V_{FB})\psi_s - \tfrac{1}{2}\psi_s^2 - \tfrac{2}{3}\gamma\psi_s^{3/2}] + U_T(\psi_s + \gamma\psi_s^{1/2}). \quad (6.32)$$

The latter formula has been presented such that the drift and diffusion parts are easily recognized. The values of the surface potential ψ_{s0} and ψ_{sL} can be related to the external voltages by substituting either $V_R = 0$ or $V_R = V_D$ into Eq. (6.29). The resulting equations can be solved for ψ_{s0} and ψ_{sL} by iteration. Usually only a few steps are sufficient [6.16].

For the case considered in Fig. 6.4 the current is plotted as a function of gate bias in Fig. 6.6. In strong inversion the drift term is dominant, in weak inversion the current is mainly due to the presence of diffusion. However, in the transition region both drift and diffusion play a role. This result is not surprising. When evaluating the inversion charge $Q_i(V_R)$ on the basis of the difference between relations (6.28) and (6.15), it is found that Q_i varies exponentially with V_R in the weak inversion region, but Q_i follows a low

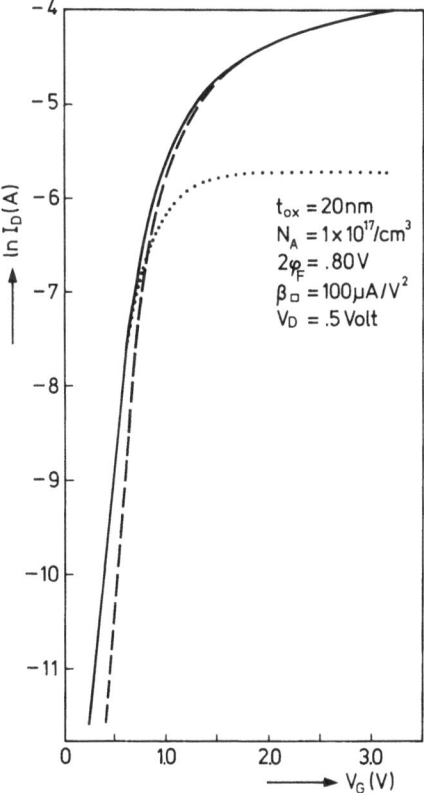

Fig. 6.6 Drain current as a function of gate bias at a fixed value of drain bias (fully drawn line). Dotted line represents the diffusion component; broken line represents the drift component

power law in V_R in strong inversion. The first result is usually associated with a diffusion transport mechanism, but the low power law implies diffusion to be much less probable.

When the charge sheet approximation is skipped, the evaluation of the current relation (6.30) is cumbersome. Generally the result is given in terms of a double integral in ψ and y [6.18], the lengthy evaluation of which is out of proportion to the increase in accuracy.

Since the general result (6.31) is less suitable for circuit modelling, approximate explicit expressions have been derived for the cases of strong inversion and weak inversion. In fact these approaches are only a consequence of the procedures considered in the last part of the previous section.

Assuming that in the case of *strong inversion* V_S is kept zero, the inversion charge Q_i at the source end of the channel is given by Eq. (6.18). However, when applying a drain voltage $V_D > 0$, the current flow causes a potential distribution $V(x)$ to be settled across the entire channel path. This additional potential $V(x)$ adds to the original surface potential ψ_{sm} across the depletion layer. Consequently at a position x in the channel the charge Q_i is now given by

$$Q_i = -C_{ox}[V_G - V_{FB} - \psi_{sm} - V - \gamma(\psi_{sm} + V)^{1/2}],$$

which relation can be rewritten with the help of (6.20) as

$$Q_i = -C_{ox}[V_G - V_{T0} - V - \gamma(\psi_{sm} + V)^{1/2} + \gamma\psi_{sm}^{1/2}]. \quad (6.33)$$

Since the result of Fig. 6.6 indicates that the current is dominated by drift, we obtain by integrating the channel current expression [6.19, 6.20] $I_D = -\mu_n W Q_i (dV/dx)$ from $x = 0$ to $x = L$

$$I_{DS} = \frac{\mu_n C_{ox} W}{L}[g(V_G, V_D) - g(V_G, V_S)], \quad (6.34)$$

where

$$g(V_G, V) = [(V_G - V_{T0} + \gamma\psi_{sm}^{1/2})V - \tfrac{1}{2}V^2 - \tfrac{2}{3}\gamma(\psi_{sm} + V)^{3/2}]. \quad (6.35)$$

In principle this result is only valid if a strong inversion layer exists along the channel. When the drain voltage is gradually increased, the inversion charge Q_i according to Eq. (6.33) diminishes at the drain end. From a mathematical viewpoint the drain voltage V_{DSAT}, at which Q_i approaches zero value, is given by

$$V_{DSAT} = -\psi_{sm} + \left[-\frac{\gamma}{2} + \left\{\frac{\gamma^2}{4} + V_G - V_{T0} + \gamma\psi_{sm}^{1/2} + \psi_{sm}\right\}^{1/2}\right]^2. \quad (6.36)$$

Correspondingly the expression (6.35) as a function of V_D passes through a maximum. In reality Q_i can never become zero owing to the continuity of current flow. In fact, when Q_i approaches zero, the electric field increases to a value at which velocity saturation occurs. This mechanism causes the

6.2 The Ideal MOS Transistor Current

current to saturate. Although the physical drain saturation voltage is slightly lower than the value given by (6.36), for long channel devices the difference can be neglected. Therefore the expression (6.36), although strictly not valid, can be used somewhat beyond its physical limit.

For values V_D exceeding V_{DSAT} the expression (6.34) can be used again, provided that the variable V_D is replaced by $V_{DSAT}(V_G)$. However, when considering the measured $I_D - V_D$ characteristics, in particular for devices with shorter channel length, a small increase of the current in the saturation region is observed. One possible explanation of this undesirable effect is the assumption that upon an increase of V_D the point at which a low value of Q_i leads to saturation is slowly moving towards the source. This implies that the value of L in Eq. (6.34) is reduced. The above mechanism will be discussed in detail in section 6.5. In short-channel devices a reduction of the threshold voltage with drain bias might also cause an increase of drain current. This effect is discussed in the next section. Finally, in short-channel devices marked deviations from (6.34) occur owing to the fact that the mobility is not a constant and that series resistances become important. These subjects will be considered in sections 6.4 and 6.6.

Since diffusion is the dominant transport mechanism in the *weak-inversion region*, it is sufficient to base the calculation on the most right-hand term of Eq. (6.30).

After integration of this term from $x = 0$ to $x = L$, the above current is given by

$$I_{DS} = \frac{\mu_n U_T W}{L}(Q_{iD} - Q_{iS}).$$

Taking into account that the extra effect of V_D in the value of Q_i is given by Eq. (6.22), we easily obtain

$$I_{DS} = -\frac{\mu_n U_T W Q_i(\psi_{s0})}{L}\left\{1 - \exp\left(\frac{-V_D}{U_T}\right)\right\}.$$

Similar to the treatment in the previous section, the latter result can be further evaluated by developing the expression (6.22) near the middle of the weak inversion region.

Then we finally obtain [6.14, 6.21, 6.13]

$$I_{DS} = I_0 \exp\left(\frac{V_G - V_{Gr}}{mU_T}\right)\left\{1 - \exp-\left(\frac{V_D}{U_T}\right)\right\}, \tag{6.37}$$

where

$$I_0 = \frac{\mu_n U_T W Q_{i0}}{L}, \tag{6.38}$$

and Q_{i0}, V_{Gr} and m are given in (6.27), (6.26) and (6.25), respectively.

In contrast to the strong-inversion case, now the current starts to saturate with V_D at a value

$$V_{DSAT} \approx 3U_T,$$

which is independent of V_{GS}.
In practice values of the slope factor m may be observed that differ from Eq. (6.25). This can be caused by the presence of surface states, which can exchange charges with the substrate. Further analysis shows that this mechanism affects the derivative $d\psi_s/dV_G$ [6.13], and therefore the value of m too.

6.3 The Threshold Voltage

In practice the strong-inversion regime is the most important mode of operation. Furthermore the threshold voltage as the onset of strong inversion plays a major role. Since this quantity depends on many parameters related to the construction of the device, this subject will be discussed in detail in this section.

6.3.1 The Body Effect

In the previous discussions it was assumed that the source was biased at the same potential as the underlying substrate. When the MOS transistor is part of an integrated circuit, this assumption has to be dropped. When a reverse bias voltage V_S is applied between the source and the substrate, the same arguments apply to the depletion charge near the source end of the channel region as given in the previous section for the drain end. Therefore Eq. (6.29) is valid, provided V_R is replaced by V_S. When ψ_s is plotted as a function of V_G, a graph qualitatively similar to Fig. 6.3 is obtained. However, it is observed that the limiting value of the strong inversion condition is shifted to a value [6.2]

$$\psi_s \approx 2\phi_F + 2U_T + V_S = \psi_{sm} + V_S.$$

Therefore in the strong inversion region Eq. (6.18) can be generalized to

$$Q_i = -C_{ox}(V_G - V_T),$$

where the threshold voltage is given by

$$V_T = V_{FB} + \psi_{sm} + V_S + \gamma(\psi_{sm} + V_S)^{1/2}. \tag{6.39}$$

Note that the threshold voltage given here refers to the gate-bulk voltage. When the source potential is used as a reference, the term V_S is dropped from (6.39) and this expression may be rewritten in the practical form

$$V_{TS} = V_{T0} + \gamma[(V_S + \psi_{sm})^{1/2} - \psi_{sm}^{1/2}], \tag{6.40}$$

where V_{T0} is already given in (6.20).
The increase of the threshold voltage with the source-substrate bias is known as the substrate or body effect [6.19]. Furthermore the parameter γ, which

6.3 The Threshold Voltage

is proportional to the gate insulator thickness and to the square root of the substrate doping, is called the body-effect coefficient. According to Eq. (6.40) V_{TS} is determined by three process parameters.

Since the body effect is undesirable in integrated-circuit design, attempts have been made to minimize its magnitude [6.22, 6.23]. As a simple reduction of t_{ox} and N_A would lead to a low breakdown voltage and an undesired low value of V_{T0} as well, a solution is sought for n-channel devices in reducing the substrate doping and simultaneously implanting a shallow layer of acceptors. Since the latter layer can be made to become largely depleted at the onset of inversion, a fairly high value of V_{T0} can be realized in combination with a low body-effect coefficient. Owing to the fact that the substrate is no longer doped uniformly, Eqs. (6.39) and (6.40) are not generally valid.

6.3.2 Effect of Implants Additional to the Substrate Doping

When a shallow layer of the same dopant type is implanted into the substrate, a doping profile results as given in Fig. 6.7 a.

In order to maintain the strong-inversion condition the depletion charge per unit area has to satisfy the following relations,

$$Q_d = \int_0^{y_d} q N_A(y) \, dy \tag{6.41}$$

and

$$\psi_{sm} = \frac{q}{\varepsilon_s} \int_0^{y_d} y N_A(y) \, dy. \tag{6.42}$$

Strictly, the above relations are only valid if the depletion boundary y_d is located in the substrate area with constant dope. If this is not the case, the

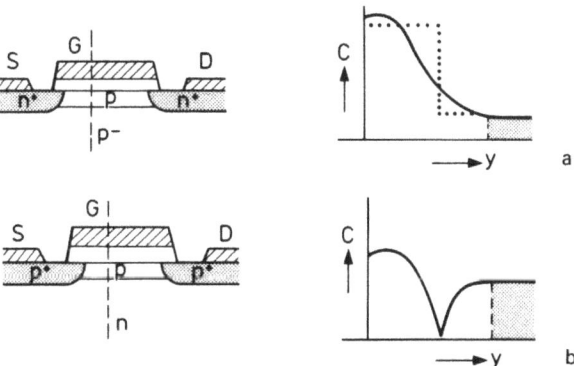

Fig. 6.7 Cross-section and doping profile (across dashed line) of implanted enhancement (a) and buried-channel MOSFET (b)

dope gradient causes a built-in field, which has to be taken into account when integrating Poisson's equation [6.24]. Usually y_d is located in the tail of the doping distribution and therefore the field term only leads to higher order corrections. In addition ψ_{sm} is not simply related to the uniform dope value N_B of the substrate. It has been argued [6.24] that in this case ψ_{sm} is given by

$$\psi_{sm} = U_T \ln\left\{\frac{N_A(y_d)N_B}{n_i^2}\right\} + V_S. \tag{6.43}$$

Equating the relations (6.42) and (6.43), y_d and ψ_{sm} can be solved numerically. Since ψ_{sm} is not a sensitive function of the doping variation or, alternatively, may be treated as a parameter in MOST models, we will not follow the above procedure here, but will assume that ψ_{sm} is known.

Usually the implanted profile is best represented by a Gaussian distribution. Unfortunately in this case the evaluation of Eqs. (6.41) and (6.42) is cumbersome [6.25].

Since, however, V_T is determined by the integral of the doping distribution rather than by its actual shape, it is reasonable to approximate the varying dope by a step profile (dotted line in Fig. 6.7a) with step height N_i and thickness d_i [6.26, 6.27]. Generally, when the surface potential reaches the strong-inversion value, the implanted layer is only partly depleted at low values of V_S. In this case the threshold voltage is still given by Eq. (6.40), where

$$\gamma_i = \frac{(2\varepsilon_s q N_i)^{1/2}}{C_{ox}}. \tag{6.44}$$

This relation remains valid until the implanted layer becomes fully depleted at a source-bulk voltage given by

$$V_{SX} = \frac{q N_i d_i^2}{2\varepsilon_s} - \psi_{sm}. \tag{6.45}$$

When V_S is further increased, the depletion layer now expands in the low-doped substrate region. Solving Eqs. (6.41) and (6.42) in this case we obtain, via

$$y_d = \left(\frac{2\varepsilon_s}{qN_B}\right)^{1/2} \left\{\psi_{sm} + V_S - \frac{q(N_i - N_B)d_i^2}{2\varepsilon_s}\right\}^{1/2},$$

$$V_{TS} = V_{T0} - \gamma_i \psi_{sm}^{1/2} + \frac{q(N_i - N_B)d_i}{C_{ox}} + \gamma(\psi_{sm} + V_S - \Delta\phi)^{1/2}, \tag{6.46}$$

where

$$\Delta\phi = \frac{q(N_i - N_B)d_i^2}{2\varepsilon_s}$$

and

6.3 The Threshold Voltage

$$\gamma = \frac{(2\varepsilon_s q N_B)^{1/2}}{C_{ox}}.$$

Now the threshold voltage increases with V_S according to the smaller body-effect coefficient $\gamma(N_B)$. Neglecting the correction term $\Delta\phi$, Eq. (6.46) only differs from the basic relation (6.39) by

$$\Delta V_T = \frac{q(N_i - N_B)d_i}{C_{ox}},$$

which is often called the threshold shift. Obviously from a viewpoint of application the most ideal situation would be to have

$$V_{SX} \leq 0 \text{ Volt}.$$

Unfortunately for n-channel devices, where B implants have to be used, the above condition rarely occurs owing to high-temperature process steps. Since Eqs. (6.40) and (6.46) and their derivatives have to be continuous at $V_S = V_{SX}$, it can be shown that $\Delta\phi$ and ΔV_T are related to the other parameters:

$$\Delta V_T = \gamma_i (V_{SX} + \psi_{sm})^{1/2} \left\{ 1 - \left(\frac{\gamma}{\gamma_i}\right)^2 \right\}, \tag{6.46 a}$$

$$\Delta\phi = \left\{ 1 - \left(\frac{\gamma}{\gamma_i}\right)^2 \right\} (V_{SX} + \psi_{sm}). \tag{6.46 b}$$

Therefore compared to Eq. (6.40) we need two additional parameters: γ_i and V_{SX}.

In practice the set of equations (6.40) to (6.46) shows excellent agreement with experimental results. This can be seen in Fig. 6.8, where V_{TS} has been plotted as a function of the variable

$$u_S = (V_S + \psi_{sm})^{1/2} - \psi_{sm}^{1/2}. \tag{6.47}$$

For small and large values of V_S, V_{TS} increases with a constant slope factor γ_i and γ, respectively. Only in a small voltage range beyond V_{SX} the slope varies.

Despite the agreement shown, the above equations have some shortcoming. The actual profile might differ too much from the step profile, causing the V_T slope to vary slowly in a large voltage range. A practical solution to the above problem is to split the V_S range into three regions, so that in the outer two regions V_T varies with constant slope. Only in the middle region the slope varies linearly with the variable u_S.

Introducing the boundary values

$$\left. \begin{array}{l} u_{S1} = (V_{SX} - \Delta V_{SX} + \psi_{sm})^{1/2} - \psi_{sm}^{1/2}, \\ u_{S2} = (V_{SX} + \Delta V_{SX} + \psi_{sm})^{1/2} - \psi_{sm}^{1/2}, \end{array} \right\} \tag{6.48}$$

we define V_{TS} as follows:

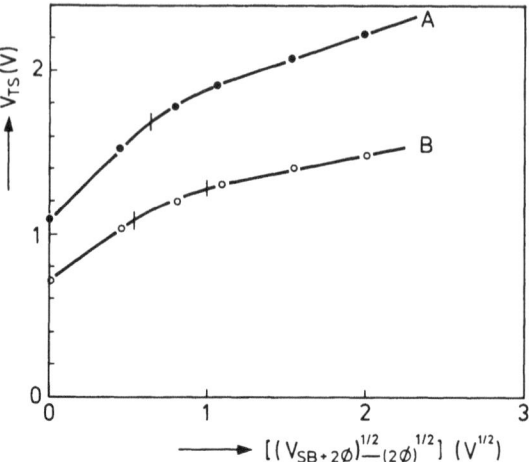

Fig. 6.8 Comparison of measured values of the threshold voltage with calculated data (fully drawn lines) according to different models. Relevant parameter values for model A are: $V_{SX} = 1.5$ V, $\psi_{sm} = 0.75$ V, $\gamma_i = 0.93$ V$^{1/2}$, $\gamma = 0.26$ V$^{1/2}$, and for model B: $V_{SX} = 1.95$ V, $\Delta V_{SX} = 0.70$ V, $\gamma_i = 0.63$ V$^{1/2}$, $\gamma = 0.20$ V$^{1/2}$

When $u_S < u_{S1}$,

$$V_{TS} = V_{T0} + \gamma_i u_S; \qquad (6.49)$$

when $u_{S1} \leqslant u_S \leqslant u_{S2}$,

$$V_{TS} = V_{T0} + \gamma_i u_{S'} - \frac{(\gamma_i - \gamma)(u_S - u_{S1})^2}{2(u_{S2} - u_{S1})}; \qquad (6.50)$$

when $u_S > u_{S2}$,

$$V_{TS} = V_{T0} + \gamma u_S + \frac{(\gamma_i - \gamma)(u_{S1} + u_{S2})}{2}. \qquad (6.51)$$

In Fig. 6.8 the above expressions have also been compared with experimental values (curve B). The associated parameters are given in the subscript. Usually the transition range ΔV_{SX} is small and for most cases can be taken as a constant. Now only two extra parameters are needed.

6.3.3 Effect of Implants of Opposite Type to the Substrate Doping

In order to realize n-channel MOSFETS with a negative threshold-voltage value (depletion-type transistor), usually a shallow n-type layer is implanted in a p-type substrate [6.28, 6.29]. Furthermore in order to increase the threshold voltage of p-channel MOSFETS, which would otherwise have a too negative value in a CMOS process, a shallow p-type layer is implanted

6.3 The Threshold Voltage

in an n-type substrate or an n-type well (see Fig. 6.7 b). Although both types are commonly known as buried-channel devices, their V_T behaviour is generally different and therefore needs careful analysis [6.30].

The main reason for the specific threshold voltage behaviour lies in the potential distribution associated with the built-in junction. For a p-type buried layer device with an n^+-type Si gate this distribution in the off-state condition is sketched in Fig. 6.9 as a dotted line. Typically the potential ψ passes through a minimum ψ_m, located in the p-type layer. The potential distribution cannot decrease to a lower value owing to the small difference in work function (ϕ_{ms}) between gate and substrate. To keep the analysis simple, we assume that the buried layer (N_i) and the substrate (N_B) have uniform doping. In this case within the semiconductor $\psi(y)$ can be described by a combination of parabolas. In order to put the device at the onset of ohmic conduction, a more negative voltage V_{TP} has to be applied to the gate to decrease ψ_m to a value sufficiently negative to allow mobile holes to flow (compare the fully drawn line for ψ in Fig. 6.9 a). Similar to the previous condition for enhancement-type devices, the definition of ψ_m is somewhat arbitrary. However, in practice an identical definition appears to be satisfactory,

$$\psi_m \approx -2[\phi_F(N_B) + U_T] + V_S = \phi_m + V_S. \tag{6.52}$$

Note that V_S is negative for a p-type source region. For reasons similar to those leading to Eq. (6.20) the threshold voltage generally reads

$$V_T = V_{FB} + \psi_s - \frac{q}{C_{ox}}(N_B W - N_i d_i). \tag{6.53}$$

Here the right-hand term is the total depletion charge, which consists of the negative charge of the fully depleted p-layer (thickness d_i) and the positive charge of the partly depleted substrate. ψ_s is the value of the potential at the interface.

The value of W is determined from the condition (6.52) for the potential minimum ψ_m. In fact the $\psi(y)$ distribution sketched in Fig. 6.9 a is subject to the following boundary conditions,

$$\left.\begin{array}{l} N_i(d_i - y_m) = N_B W, \\[6pt] \psi_m = -\dfrac{qN_B}{2\varepsilon_s} W^2 - \dfrac{qN_i}{2\varepsilon_s}(d - y_m)^2, \\[6pt] \psi_m - \psi_s = -\dfrac{qN_i}{2\varepsilon_s} y_m^2, \end{array}\right\} \tag{6.54}$$

where y_m is the location of ψ_m.

Solving ψ_s and W from (6.54) and substituting the result in (6.53) we obtain for the threshold voltage

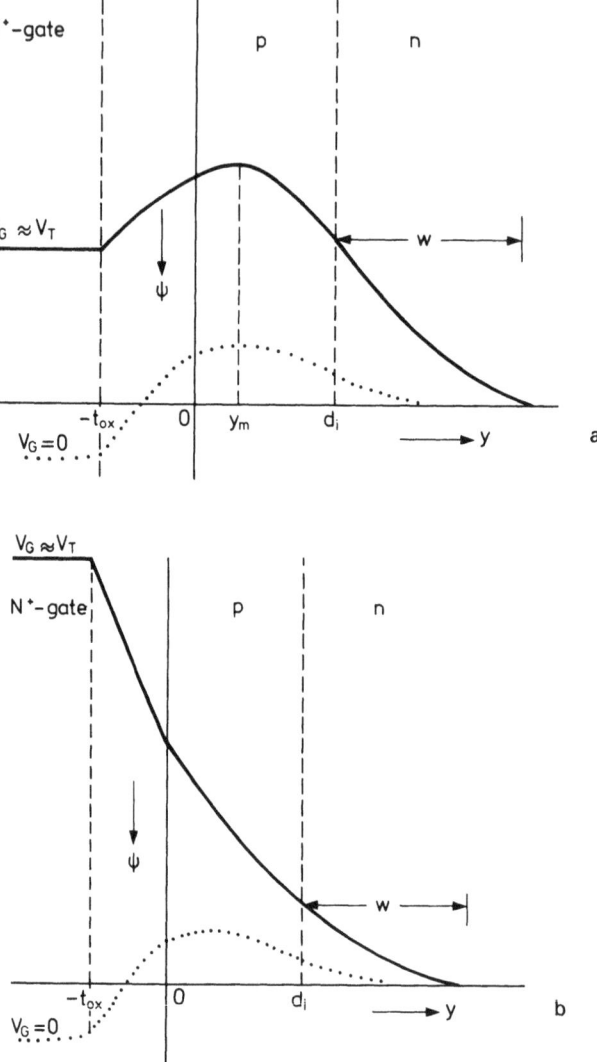

Fig. 6.9 a: Electron potential distribution at threshold condition (fully drawn curve) compared with zero bias condition (broken curve) for depletion-type buried-channel MOSFET; b: Potential distribution at threshold condition compared with zero bias condition for pseudo-enhancement buried-channel MOSFET

$$V_{TS} = V_{FB} - \phi_m + \frac{N_B}{N_B + N_i}(\phi_m - V_S)$$
$$+ qN_i d_i \left(\frac{1}{C_{ox}} + \frac{d_i}{2\varepsilon_s}\right) + \gamma(\phi_m - V_S)^{1/2}, \quad (6.55)$$

where the body-effect coefficient is given by

6.3 The Threshold Voltage

$$\gamma = \left(\frac{1}{C_{ox}} + \frac{d_i}{\varepsilon_s}\right)\left[\frac{2q\varepsilon_s N_i N_B}{N_i + N_B}\right]^{1/2}. \tag{6.56}$$

It has to be remarked that Eq. (6.55) in this form is rarely applied in practice. Usually p-type depletion devices are not made. More important is the n-type depletion device with a negative threshold voltage (at $V_S = 0$). In this case Eq. (6.55) can be applied, provided that, except for V_{FB}, all right-hand terms reverse sign [6.31, 6.32]. Usually in this type of device $N_i \gg N_B$. This causes the middle term of (6.55) to become unimportant. In addition (6.56) can be further simplified. Therefore the practical relation (6.40) also applies to depletion-type devices. Uptil now it has been tacitly assumed that ψ_m is located in the implanted layer ($y_m > 0$). In an n-type depletion transistor this is always the case. However, in most p-type buried layers with an n-type silicon gate this is not true. From (6.54) it is easily obtained that

$$y_m = d_i - \left[\frac{2\varepsilon_s \psi_m N_B}{qN_i(N_i + N_B)}\right]^{1/2}.$$

To suppress punch-through currents at $V_G = 0$ Volt [6.33, 6.34], it is required to keep d_i and N_i to the smallest possible values, and consequently y_m moves towards the interface when $|\psi_m|$ is increased via a decrease of V_{GS}. In fact at $V_{GS} \approx V_{TS}$, in most cases y_m becomes located in the gate insulator and the potential minimum becomes a virtual minimum. Within the semiconductor the potential distribution has the form sketched as a fully drawn line in Fig. 6.9 b. A practical consequence is that in this case hole conduction does not start somewhere in the buried layer (as in the previous case), but right at the interface. Now the buried-layer MOSFET operates in the *quasi-enhancement mode*. Basically Eq. (6.53) can be used again, but for ψ_s we now have

$$\psi_s = -\phi_m + V_S.$$

The depletion width W can be calculated by solving Poisson's equation and making use of Gauss's theorem at the interface. In this way we obtain

$$W = d_i\left[\left\{1 + \frac{N_i}{N_B} + \frac{2\varepsilon_s(\phi_m - V_S)}{d_i^2 qN_i}\right\}^{1/2} - 1\right].$$

Substituting the latter result into Eq. (6.53) we finally obtain [6.30]

$$V_{TS} = V_{FB} - \phi_m + \frac{q(N_i + N_B)d_i}{C_{ox}} - \gamma(\phi_m - V_S + \Delta\phi)^{1/2}, \tag{6.57}$$

where

$$\gamma = \frac{(2\varepsilon_s qN_B)^{1/2}}{C_{ox}}$$

and

$$\Delta\phi = \frac{q(N_i + N_B)d_i^2}{2\varepsilon_s}. \tag{6.58}$$

For implanted structures, the term $(N_i + N_B)d_i$ equals the dose. Consequently V_{TS} is proportional to the implanted dose.

Note that the result (6.58) is quite similar to the relation (6.46) for the n-channel enhancement transistor. However, there is one exception. The correction voltage term $\Delta\phi$ is now added to the diffusion potential value ϕ_F. It can be shown, however, that in practical devices (with V_{TP} values < -0.60 V), where N_i and N_B are of the same order of magnitude,

$$\Delta\phi < 0.10 \text{ Volt}.$$

This implies that for most p-type MOSFETS the earlier result (6.40) is applicable.

6.3.4 Temperature Dependence

Since MOSFET integrated circuits often operate at temperatures far from room temperature it is necessary to consider the temperature dependence of V_T. Looking at the general expressions for V_T (6.46), (6.55) and (6.58), we see that the only factors that may cause a temperature dependence of V_T are ϕ_{ms} and the diffusion potential ϕ_F. Therefore the temperature coefficient of V_T is generally given by [6.35]

$$\frac{dV_T}{dT} = \frac{d\phi_{ms}}{dT} + \left[2 + \frac{\gamma}{\{|V_S| + 2\phi_F - \Delta\phi\}^{1/2}}\right]\frac{d\phi_F}{dT}. \qquad (6.59)$$

Assuming that $n_i^2 = CT^3 \exp(-E_G(0)/kT)$, where $E_G(0)$ is the extrapolated zero-degree bandgap (compare Eq. (2.15)), the temperature coefficient of ϕ_F is given by

$$\frac{d\phi_F}{dT} = \frac{1}{T}\left[\phi_F - \frac{1}{2}\left\{\frac{E_G(0)}{q} + 3U_T\right\}\right]. \qquad (6.60)$$

For a silicon gate with the opposite type of doping to the substrate, the contact potential is determined by the pn-product too. Therefore we have

$$\frac{d\phi_{ms}}{dT} = \frac{1}{T}\left[\phi_{ms} - \left\{\frac{E_G(0)}{q} + 3U_T\right\}\right]. \qquad (6.61)$$

However, when the dope of the gate is of the same type as that of the substrate, ϕ_{ms} is only determined by the difference in majority-carrier density. In this case we have simply

$$\frac{d\phi_{ms}}{dT} = \frac{\phi_{ms}}{T}. \qquad (6.62)$$

Since all terms between brackets vary much less with T than the variable T, a nearly linear variation of V_T with temperature is expected [6.36, 6.37]. In addition, for the various types of MOSFET devices differences in the

6.3 The Threshold Voltage

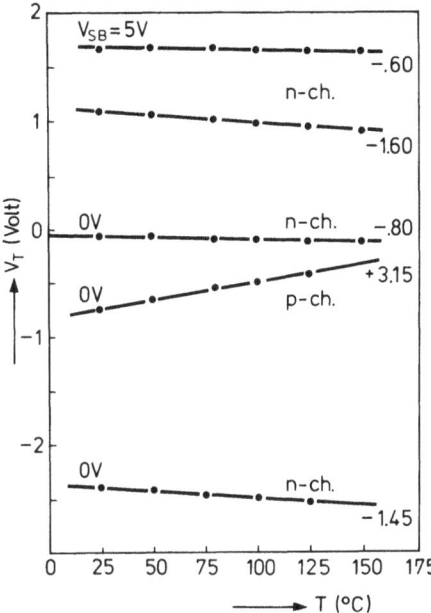

Fig. 6.10 Threshold voltage versus temperature for common MOSFET types. The temperature coefficient is given in mV/degree

temperature coefficient will occur on account of different values of the parameters ϕ_{ms} and γ.

This is demonstrated in Fig. 6.10, where the measured temperature dependence of a number of different MOSFETS has been plotted [6.35].

6.3.5 Short-Channel Effect

Up to now the threshold voltage has been calculated by integrating the 1-D Poisson equation across the depleted region. When the channel becomes shorter, this procedure needs correction. In fact part of the depletion charge is also shared by the source and drain junctions. According to one approach [6.38, 6.39, 6.40] the charge in the grey area of Fig. 6.11 has to be omitted from the charge balance, which determines the value of the threshold voltage. Consequently it becomes easier to turn on the device and the threshold voltage has a smaller value. This is known as the short-channel V_T effect. Formally the net depletion charge per unit width can be written

$$Q'_{dL} = Q_d L - \Delta Q_d, \tag{6.63}$$

where Q_d is the 1-D depletion charge and ΔQ_d is twice the charge of one grey area. Using a trapezoidal form for the effective net depletion region, approxi-

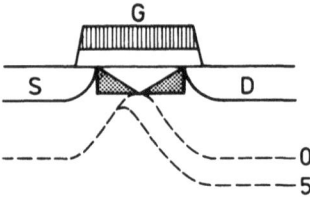

Fig. 6.11 Shape of the depletion region in a short-channel MOSFET at zero and high drain bias. The grey areas indicate the charge shared by the gate and the junctions at zero drain bias

mate expressions for ΔQ_d have been obtained in the case of a uniform substrate doping [6.38, 6.41, 6.42],

$$\Delta Q_d = Q_d d_j \left[\left(1 + \frac{2y_d}{d_j}\right)^{1/2} - 1 \right],$$

where d_j is the junction depth and y_d the thickness of the 1-D depletion layer. In order to keep the effect as small as possible, y_d should be kept small. In that case

$$\Delta Q_d = Q_d y_d.$$

Substituting the latter result into (6.63) and (6.15), the threshold voltage is lowered via a reduction of the body-effect coefficient

$$V_T(L) = V_T(L \to \infty) - \frac{\gamma y_d}{L}(\psi_{sm} + V_S)^{1/2}. \tag{6.64}$$

Despite the pragmatic analysis the latter result is often confirmed experimentally. However, when the short-channel effect is no longer of second order, considerable deviations are found. In this case a rigorous 2-D solution of Poisson's equation is required. By solving this equation in terms of a Fourier series, it has been shown [6.43] that, for extremely short channels, the threshold voltage is lowered by an amount

$$\Delta V_T = \left(\frac{12\phi_F t_{ox}}{y_d}\right) \exp\left(\frac{-\pi L}{4 y_d}\right). \tag{6.65}$$

Although this approach is physically sound, the result (6.65) is still approximate because of slow convergence of the Fourier series and the assumption of uniform substrate doping. In addition the result has been criticized [6.44, 6.45], since for short channels y_d is no longer equal to the long channel value, but increases with decreasing channel length.

Although it can be understood qualitatively from Fig. 6.11 that V_T decreases further upon an increase of the drain bias voltage, the increase of y_d with V_D is the main reason for the fact that attempts to generalize the results (6.64) and (6.65), on account of this V_D effect, have been quite unsatisfactory. For

6.3 The Threshold Voltage

circuit modelling purposes, therefore simple empirical models have mainly been used [6.46, 6.47, 6.48]. In a recent approach the cumbersome procedure to calculate $\Delta Q_D(L, V_D)$ via a 2-D solution of Poisson's equation

$$\frac{\partial^2 \psi}{\partial x^2} + \frac{\partial^2 \psi}{\partial y^2} = \frac{q}{\varepsilon_s} N_s(y) \tag{6.66}$$

has been circumvented by assuming that the potential distribution along the channel direction varies quadratically with the distance [6.49, 6.50]. The latter assumption has been confirmed by means of a numerical simulation. In this case we may write

$$\psi(x, y) = \psi_1(x) + \psi_2(y).$$

Using the boundary conditions

$$\left.\begin{array}{l}\psi_1(0) = V_{bi} + V_S, \\ \psi_1(L) = V_{bi} + V_D,\end{array}\right\} \tag{6.67 a}$$

where $V_{bi} = U_T \ln(N_j N_s/n_i^2)$ and N_j is the junction doping concentration, and defining the minimum value of ψ_1 as

$$\psi_{1m}(y) = \phi_c(y), \tag{6.67 b}$$

it can be proved that

$$\frac{\partial^2 \psi}{\partial x^2} = 2 \frac{V_{DS}^*}{L^2},$$

where

$$V_{DS}^* = (V_D - V_S) + 2(V_{bi} + V_S - \phi_c) \\ + 2[(V_{bi} + V_S + \phi_c)(V_D + V_{bi} - \phi_c)]^{1/2} \tag{6.68}$$

and $\phi_c = \psi_{sm} + V_S$.
Substitution of the latter result into (6.66) yields

$$\frac{\partial^2 \psi}{\partial y^2} = \frac{q}{\varepsilon_s}\left[N(y) - \frac{2\varepsilon_s V_{DS}^*}{qL^2}\right]. \tag{6.69}$$

Essentially the 2-D Poisson equation has been reduced to a 1-D form via a voltage-doping transformation. Physically this means that the influence of the lateral drain-source field on the potential is equivalent to and can be replaced by a reduction of the doping concentration according to (6.69).
Furthermore, assuming as in section 6.3.2. a box-type doping profile, the associated potential distribution $\psi(y)$ and the depletion charge are calculated easily by integrating (6.69). The result obtained for $\psi(y)$ not only agrees reasonably well with simulated results, but also shows that y_d increases upon a decrease of L. Using the calculated value of Q_D, the threshold voltage is obtained in close analogy to Eqs. (6.40) and (6.46).
When $y_d \leqslant d_i$,

$$V_{TS} = V_{FB} + \psi_{sm} + \frac{\left[2\varepsilon_s q \left(N_i - \frac{2\varepsilon_s V_{DS}^*}{qL^2}\right)\right]^{1/2}}{C_{ox}} (\psi_{sm} + V_S)^{1/2}. \quad (6.70)$$

When $y_d > d_i$,

$$V_{TS} = V_{FB} + \psi_{sm} + \frac{qd_i \left(N_i - N_B - \frac{2\varepsilon_s V_{DS}^*}{qL^2}\right)}{C_{ox}}$$

$$+ \gamma \left[\psi_{sm} + V_S - \frac{d_i^2 \left(N_i - N_B - \frac{2\varepsilon_s V_{DS}^*}{qL^2}\right)}{2\varepsilon_s}\right]^{1/2}. \quad (6.71)$$

When the short-channel effect has been made small by making use of scaling rules [6.51, 6.52, 6.53], it is possible to expand the square-root term of Eq. (6.70) in a Taylor series. In this case we obtain for the threshold-voltage short-channel effect at zero back bias ($V_S = 0$)

$$\Delta V_{T0} \approx -\frac{\gamma \varepsilon_s (2\varphi_F)^{1/2}}{qN_i L^2} (2\varphi_F + V_{DS}). \quad (6.72)$$

According to this result, the threshold voltage lowering is inversely proportional to the substrate doping near the interface and the square-power of the channel length, and it is proportional to the drain-source bias voltage. This had already been found earlier by experiment [6.47, 6.54].

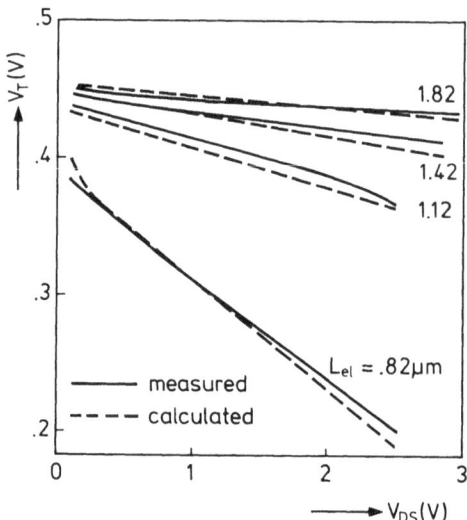

Fig. 6.12 Threshold voltage versus drain bias for MOSFETs with different channel length. Fully drawn lines represent measured data, broken curves have been calculated according to [6.50]

6.3 The Threshold Voltage

Fig. 6.12 gives a comparison between the above calculated results and experimental data [6.50]. A fair agreement is observed. However, it should be remarked that, owing to the assumptions used, the weak dependence of the short-channel effect on junction depth d_j has been left out of account. In normal enhancement MOSFETS this effect is negligible, but unfortunately for buried-channel devices it is rather important.

6.3.6 Narrow-Width Effect

In order to isolate a MOSFET electrically from other devices, the LOCOS concept is widely used in IC technology. Fig. 6.13 gives a cross-section. Generally the channel area is bounded by an insulator, whose thickness rapidly increases within several tenths of a micron. In order to suppress undesired (punch-through) currents between active devices, the substrate concentration under the thick insulator is usually increased by a so-called channel-stop implantation (self-aligned to the insulator). Also drawn in the figure is the shape of the depletion layer corresponding to the onset of strong inversion. Since some of the field lines emanating from the gate charges terminate at ionized acceptors on the sides, the depletion is not limited to just the area below the thin gate insulator. If the channel width (W) is large, this effect can be neglected, but at small values of W the charge in the side parts becomes an essential part of the total depletion charge. Therefore, a higher value of V_{GS} is required to deplete that amount of charge before an inversion layer can be formed. Consequently V_T and the associated body-effect coefficient γ increase with decreasing gate width W (narrow-channel effect).

Assuming that the side parts of the depletion region have a quarter-circle cross-section [6.55, 6.56], it is easily verified that the excess depletion charge is given by

$$\Delta Q_d = Q_d \frac{\pi y_d}{2}, \tag{6.73}$$

where Q_d is the depletion charge per unit area beyond the thin gate insulator,

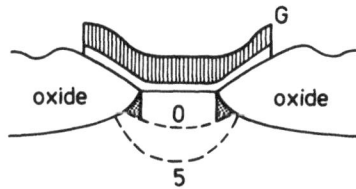

Fig. 6.13 Shape of the depletion region in a narrow-width MOSFET at zero and high back-bias. The grey areas indicate excess charge, which is formed under the LOCOS region at the onset of channel formation

and y_d is the 1-D depletion layer thickness. Since y_d increases with an increase of the back bias V_S, the latter result would imply that the narrow-width effect would become more important with larger back bias. Usually this is not observed experimentally. The above discrepancy can be solved by considering that with increasing V_S the depletion layer of the side parts moves into a more highly doped region under the thicker insulator. Consequently the depletion boundary deviates strongly from a circle. Actually in practice it is better to replace $\frac{1}{2}\pi y_d$ by an adjustable parameter ΔW. In this case we obtain for the narrow-width effect

$$\Delta V_T(W) = \gamma \frac{\Delta W}{W} (\psi_{sm} + V_S)^{1/2}. \tag{6.74}$$

Now the V_T lowering is inversely proportional to the gate width. This result is confirmed experimentally (see section 10.3).

It should be remarked that if there is a strong lateral diffusion of the channel-stop implant into the active device region, the threshold voltage behaviour of a narrow-width device will be described poorly by (6.74). In such a situation it is preferable to split-up the device into three sections operating in parallel (with different V_T and γ values). However, with this approach it is difficult to obtain correct parameters.

6.4 Carrier Mobility in Inversion Layers

6.4.1 Bias Dependence of the Carrier Mobility

In the previous section 6.2 current expressions have been derived under the assumption that the carrier mobility in the conducting channel is constant. In fact, the electric field normal to the surface restricts the channel to a sheet, in which two-dimensional confinement effects and scattering cause the mobility to depend on bias conditions [6.3, 6.57]. In addition, velocity saturation effects associated with a high lateral field cause the mobility to decrease at increased drain bias, in particular for short-channel devices [6.58, 6.59].

Based on theoretical analysis and experimental measurement, the channel mobility is usually expressed in terms of the normal and lateral field (compare section 2.4). Unfortunately, when the expression given in Eq. (2.62) is taken into account, further evaluation of the MOST current expressions leads to complicated implicit relations. Since the effect of field mobility is usually small, it is often sufficient to approximate the mobility in terms of the first and second power of the field. As a first approximation for the normal field effect we then have [6.60]

$$\mu = \frac{\mu_0}{1 + aE_{n,\text{eff}}}, \tag{6.75}$$

where μ_0 and a are constants and the effective normal field is given as the

6.4 Carrier Mobility in Inversion Layers

average of the surface field at the front of the inversion layer and the depletion field at the back of the inversion layer. Thus

$$E_{n,\text{eff}} = \frac{E_{sn} + E_{dpl}}{2}.$$

With zero drain bias the surface field is given by

$$E_{sn} = \frac{\varepsilon_{0x}(V_{GS} - V_{FB} - 2\phi_F)}{\varepsilon_s t_{0x}}$$

and the depletion field is given by

$$E_{dpl} = \frac{\varepsilon_{0x}(V_T - V_{FB} - 2\phi_F)}{\varepsilon_s t_{0x}}.$$

The dependence of channel mobility on gate and substrate bias is therefore given by [6.61]

$$\mu = \frac{\mu_0}{1 + \theta[V_{GS} + V_T - 2(V_{FB} + 2\phi_F)]}, \qquad (6.76)$$

where

$$\theta = \frac{aC_{0x}}{2\varepsilon_s}.$$

Since the current is usually expressed in terms of the conductance at the extrapolated threshold voltage (V_{T0}) at zero back bias ($V_S = 0$), it is useful to introduce a surface mobility μ_{s0} in the above bias conditions. Hence

$$\mu_{s0} = \frac{\mu_0}{1 + \theta[2V_{T0} - 2(V_{FB} + 2\phi_F)]}. \qquad (6.77)$$

When a drain voltage V_{DS} is applied, the resulting channel potential $V(x)$ reduces the surface field considerably. Taking this into account, using the above definition (6.77) and considering that θ is usually small, we can rewrite the general relation (6.76) to

$$\mu_s = \frac{\mu_{s0}}{1 + \theta_A(V_{GS} - V - V_{TS}) + \theta_B\{(V_S + 2\phi_F)^{1/2} - (2\phi_F)^{1/2}\}}, \qquad (6.78)$$

where θ_A and $\theta_B = \gamma\theta_A$ can be considered as process-dependent parameters. The above replacement of normal field by the terminal voltage V_{GS} and V_S is backed up by results of numerical calculations of E_{sn}, given in Fig. 6.14. For n-channel devices E_{sn} has already a high value at threshold and the increase of E_{sn} with V_{GS} and V_{SB} is relatively moderate. Consequently μ_{s0} is much smaller than the bulk value, but the values of θ_A and θ_B are small. In p-channel devices E_{sn} is small at threshold owing to charge-compensation effects in the depletion layer [6.35], but the increase of E_{sn} with V_{GS} and V_{SB} is relatively large. Consequently θ_A and θ_B are relatively large.

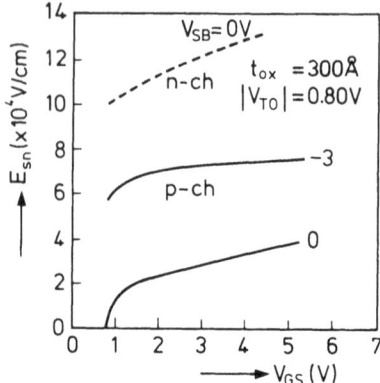

Fig. 6.14 Normal field at the interface as a function of the applied gate voltage for enhancement-type *n*-channel devices and buried-channel-type *p*-channel devices

Until now the above result appears to describe the mobility lowering rather well even for micron-size devices. However, owing to the small gate insulator thickness required for useful devices in the deep submicron range, an additional reduction mechanism has to be taken into account. This is the surface roughness effect, which has been introduced in 2.4. Since this effect is proportional to E_{sn}^2 [6.57] and only affects the conductance at high values of V_{GS}, it suffices to add just a term

$$\theta_D V_{GS}^2$$

to the denominator of Eq. (6.78). Note that the latter term may cause a negative slope in the conductance plot at high values of V_{GS}. This is shown in Fig. 6.15, where the conductance $(\partial I_D/\partial V_{DS})_{V_{DS}\to 0}$ has been plotted for a MOST with a gate insulator of 10 nm.

For short-channel devices the effect of velocity saturation has to be taken into account. In (2.67) a velocity versus field relation has been introduced, based on numerous empirical data [6.62, 6.63]. Unfortunately, for $\beta \neq 1$, the use of this relation in the MOSFET current expressions leads to complicated implicit relations. However, several authors [6.59, 6.64] have shown that the use of (2.67), taking $\beta = 1$, results in satisfactory modelling results for short-channel devices. In addition it should be remarked that the stronger bending of μ_s with lateral field for $\beta = 2$ leads to a stronger bending of the MOSFET drain conductance around saturation voltage than usually observed. Consequently we take for the effect of lateral field on mobility

$$\mu_s = \frac{\mu_s(E_n)}{1 + \theta_c L \dfrac{dV}{dx}}. \tag{6.79}$$

Since there are conflicting data for the saturated velocity [6.65, 6.66],

6.4 Carrier Mobility in Inversion Layers

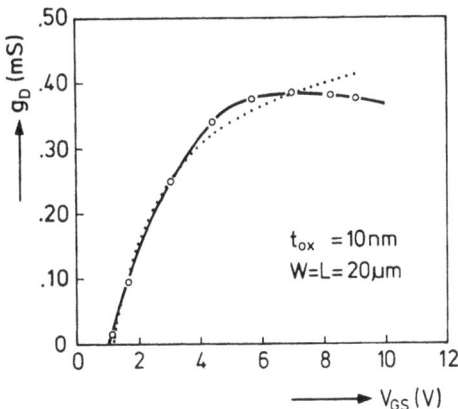

Fig. 6.15 Drain conductance as a function of applied gate bias for a MOSFET with a thin gate insulator. Fully drawn line shows modelled values when taking surface roughness into account; dotted curve shows modelled values taking into account θ_A-parameter only

$$\theta_C = \frac{\mu_s(E_n)}{v_s L} \tag{6.80}$$

has to be considered as a parameter.

As an alternative to the above mobility-lowering model expression the single expression

$$\mu_s = \frac{\mu_{s0}}{1 + \theta_A(V_{GS} - V_{TS}) + \theta_B V_S + \theta_C \dfrac{dV}{dx}} \tag{6.81}$$

has sometimes been used [6.41, 6.67]. This is allowed as long as the θ-parameters are small.

6.4.2 Temperature Dependence

Although the surface mobility differs considerably from the bulk value, empirically the temperature dependence has been found to be well described by the relation (2.65)

$$\mu_s = \mu_s(T_0)\left(\frac{T}{T_0}\right)^s,$$

valid for bulk values. In practice, values of s between 1.5 and 2.0 are found [6.60, 6.67]. Together with the temperature dependence of V_T, the mobility causes several electrical MOSFET parameters to depend on the temperature. The gain constant $\beta_0 = \mu C_{ox}$ and the velocity saturation parameter θ_C are affected directly, and the parameter θ_A may be affected indirectly via a

174   6 MOSFET Physics Relevant to Device Modelling

resistance in series with the source. The latter effect will be explained next.

6.4.3 Modelling of Effects Other than Mobility Via the θ-Parameters

In practical MOSFETS the value of the θ-parameters can often be determined by other effects too. For instance, it will be shown in section 7.2.3 that the presence of a resistor R_s in series with the source leads to a current versus terminal-voltage expression with generalized parameter values [6.68], such as

$$\theta'_A = \theta_A + \frac{W}{L} \beta_0 R_s.$$

In particular in submicron devices with a so-called LDD (lightly doped drain) configuration the increase of θ_A may be significant.

Depending on the isolation technique used for narrow-width devices, a much increased value of θ_B is often measured. This can be explained with the help of Fig. 6.13, which gives a cross section of such a device. Owing to the fact that the depletion layer cannot expand in the higher-doped channel-stop region, at high values of the back-bias voltage V_{SB} the depletion region boundary shows a very curved shape (dashed line no. 2). Because of this the field lines no longer cross perpendicular to the interface in a large part of the channel. Consequently the inversion layer charge decreases from the centre of the channel towards the edges and causes the electrical width effectively to decrease with back-bias. This is shown in Fig. 6.16, where the result of a numerical calculation of the effective channel width has been plotted. The above effect is observed in the characteristics of narrow-width

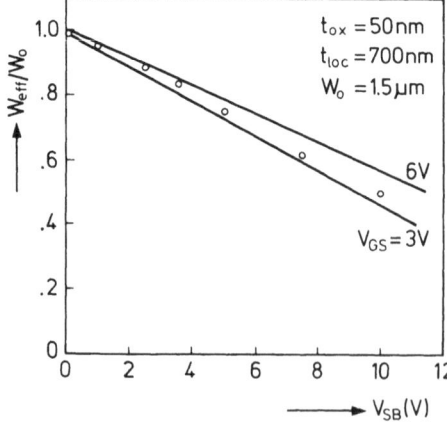

Fig. 6.16 Effective channel width versus applied back-bias. Dots show measured values. Fully drawn lines represent values calculated numerically

6.5 Saturation Mode

devices as a decreases of the conductance slope with V_{GS} at higher back-bias and can be satisfactorily modelled via the parameter θ_B [6.69].

6.5 Saturation Mode

6.5.1 Static Feedback

When the average distance between the conducting drain and the channel becomes small, an increase of the drain bias beyond saturation induces some excess mobile charge in the channel. To some extent this phenomenon is similar to the threshold voltage-lowering effect. Physically, however, it is different since mobile carriers are induced instead of modifying the fixed depletion charge. Naturally the drain end of the channel near the pinch-off region is most affected. The increase of mobile charge has been estimated analytically by assuming that the field lines originating from the drain follow a cylindrical path [6.70]. The result is that

$$\Delta Q_i = \eta C_{0x} V_{DS},$$

where

$$\eta \sim \frac{1}{L^{3/2} N_B^{1/4} C_{0x}}.$$

Of course, the above approach is far too simplified. However, the dependence of ΔQ_i on major process parameters is confirmed by results of numerical simulation of a realistic device structure [6.54]. Fig. 6.17 gives the

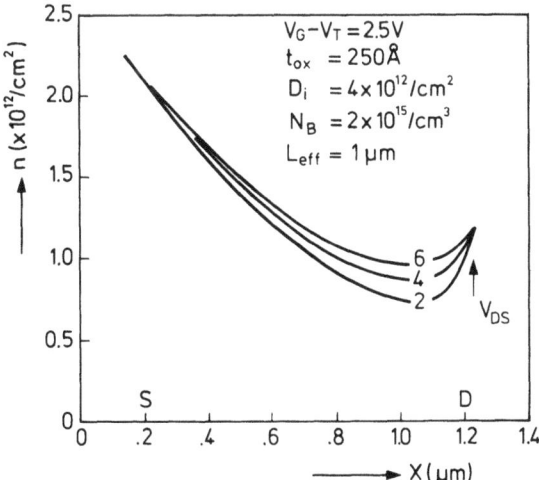

Fig. 6.17 Carrier density distribution along the channel at several values of the applied drain-source voltage, showing the static drain feedback effect

number of mobile carriers in the channel as a function of the position x in the channel for a short-channel MOSFET operating in saturation. From this figure and similar plots obtained at different gate and back-bias voltages, we observe that in first order the carrier charge distribution increases linearly with the applied drain bias. As such, this increase contradicts the results of conventional analysis, which predict a channel charge independent of drain voltage under saturation condition. Naturally the increase ΔQ_i near the pinch-off region enhances the saturation voltage. In order to take the latter increase into account quantitatively, we interpret the above effect as an apparent increase of the effective gate driving voltage [6.47],

$$V_{GT} = V_{GS} - V_{TS} + \eta V_{DS}. \tag{6.82}$$

Generally η depends on the oxide thickness, the drain doping profile and the substrate doping. Consequently this quantity has to be considered as a process parameter. The above gate-driving voltage causes the saturated drain current to vary slightly with drain bias (static-feedback effect). Experimentally it is found that η is inversely proportional to the channel length L [6.54].

6.5.2 Channel-Length Modulation

In addition to the above effect the saturated drain current may increase with V_{DS} by a slight shift of the saturation point towards the source. This effect is known as channel length shortening and it is caused by the fact that the pinched-off space charge region has to expand a little since it has to accommodate more charge upon an increase of the drain voltage. Actually an exact calculation of the above effect is not easy, since it requires a two-dimensional solution of Poisson's equation for the channel saturation region (see Fig. 6.18). In addition to the depletion charge in the above region the mo-

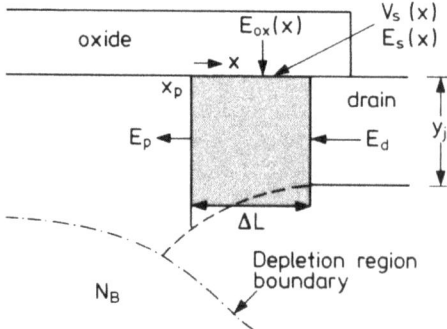

Fig. 6.18 Schematic of the pinch-off depletion region, showing major parameters that determine the channel shortening value ΔL

6.5 Saturation Mode

bile carriers also have to be taken into account. Owing to field inversion in the gate insulator (for a distance $x > x_p$) the latter carriers spread more or less uniformly into the bulk. Consequently the Poisson equation reads

$$\frac{\partial^2 V}{\partial x^2} + \frac{\partial^2 V}{\partial y^2} = \frac{qN_s}{\varepsilon_s} + \frac{I_{DSS}}{qWv_s \Delta y}, \tag{6.83}$$

where Δy is the mean depth of the current spread. A rigorous two-dimensional analysis has been given in implicit form [6.71, 6.72], but the result is limited to long channels and uniformly doped substrates. For short-channel devices more approximate solutions have been proposed [6.64, 6.54, 6.73, 6.74, 6.75]. These approximations are based either on the assumption

$$\frac{\partial^2 V}{\partial y^2} \approx \frac{C_{ox}(V_G - V)}{\varepsilon_s \Delta y} \tag{6.84}$$

for the whole region between x_p and the drain, or on the application of Gauss's law to an area enclosed by the interrupted lines, disregarding some minor effects. Actually both assumptions are complementary and, as numerical results show [6.67], are largely valid for $x > x_p$. Since the latter approach leads to more practical results in the case of high current densities, it is applied here.

Using the assumption (6.84) and expressing I_{DSS} in terms of the surface potential V_s, Eq. (6.83) can be rewritten to a differential equation of the form [6.75]

$$\frac{\partial E_s(x)}{\partial x} = A^2 [V_s(x) - V_{DSS}] + B, \tag{6.85}$$

where the index s refers to the interface and A and B are constants, which depend on junction depth, gate oxide thickness and substrate doping. Defining $E_s(x_p) = E_p$, where E_p is the lateral field value of the pinch-off condition, and considering that $V_s(x_p) = V_{DSS}$ and $V_s(x_p + \Delta L) = V_{DS}$, where $\Delta L = L - x_p$, the length of the space charge region ΔL can be expressed in terms of the lateral surface field $E_{sx} = E_s(x)$ by solving Eq. (6.85). For the field we obtain

$$E_{sx}(V) = [\alpha^2 (V - V_{DSS})^2 + E_p^2]^{1/2}. \tag{6.86}$$

The channel-length shortening term is now given by

$$\Delta L = \alpha^{-1} \ln \left[\frac{\alpha(V_{DS} - V_{DSS}) + E_{sx}(V_{DS})}{E_p} \right]. \tag{6.87}$$

Although α can be expressed in terms of the insulator thickness and junction depth [6.75], we prefer in view of several simplifications made, to consider this factor as a parameter. Another reason is that in practical situations an additional voltage drop occurs over the drain region.

Taking the above value of ΔL and the expressions for the saturated drain current I_{DSS}, like those given in Eqs. (6.34) to (6.36), the varying drain current is given by

$$I_{DS} = \frac{I_{DSS}}{1 - (\Delta L/L)}, \tag{6.88}$$

where L is the original electrical channel length determined by the metallurgical junctions. As will be shown later on, the combination of Eqs. (6.87) and (6.88) results in a calculated value of the drain conductance which, for long-channel devices, shows fair agreement with measured values.

For digital circuit design, where $g_{DS} = \partial I_{DSS}/\partial V_{DS}$ in saturation is not an important quantity, the above result is often considered to be too complicated and has been avoided for reasons of computational speed. In fact, several models have been published [6.76, 6.77, 6.61], where (6.88) is replaced by a form

$$I_{DS} = I_{DSS}\left(1 + \frac{V_{DS} - V_{DSS}}{V_A}\right), \tag{6.89}$$

where V_A is a parameter.

Note that the above saturation model predicts non-zero slope at V_{DSS}, whereas the non-saturation equations predict zero slope at the same point. Such discontinuous slope behaviour is undesirable in the numerical algorithms used in the computer simulation of circuits. In the following chapter it will be shown that this problem can be avoided by adapting the value of the saturation voltage.

As an alternative to the above simplified modelling, the ΔL effect may be totally ignored [6.67] and the saturation behaviour is then modelled using only Eq. (6.82). For short-channel devices this approach is quite satisfactory. This is less valid for long-channel devices, but now η is very low.

6.6 Dynamic Operation

6.6.1 Quasi-Static Operation

When one or more terminal voltages of the MOSFET are changed in time or excess current is supplied (either as a complete transient or as a small a.c. signal), a delay in the source or drain current is observed. Generally this delay is due to two physical mechanisms. A new equilibrium situation requires a change of the internal charge distribution in the transistor, which has to be supplied by the terminal currents. Furthermore due to transit-time effects a terminal current cannot react instantaneously upon a rapid change of a terminal voltage. Fig. 6.19 gives a definition of currents and charges in the presence of varying terminal voltages.

6.6 Dynamic Operation

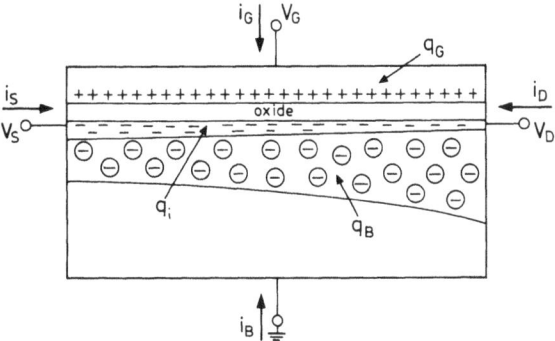

Fig. 6.19 Definition of MOSFET currents and charges in the presence of terminal voltages varying in time

Although the MOSFET does not differ in principle in this respect from the bipolar transistor, it has taken more time to develop a current charge model for circuit analysis like the well-known bipolar charge control concept (see chapter 3). In fact in the past many simulators used a MOSFET model taking into account only a few small signal capacitances (mostly derived for long channels and disregarding the body effect). This approach did not always give rise to large errors since in most applications the MOSFET delays are eclipsed by parasitic circuit capacitances. In some applications, however, large errors did occur owing to the fact that no charge is conserved in these models. In fact these errors can only be corrected by introducing an appropriate charge model [6.78, 6.79, 6.80].

Since phase shifts are difficult to handle, owing to transit-time effects and require a correct calculation of internal charge redistributions, which can only be carried out numerically [6.81, 6.82, 6.83, 6.84, 6.85, 6.86], the concept of quasi-static operation is usually introduced. This implies that the variation of terminal voltages is assumed to be sufficiently slow to allow all terminal transport currents to vary instantaneously with the terminal voltages. Applying this concept to the source, for instance we assume that the transient source current is given as the sum of a time-dependent transport current and a charging current,

$$i_S(t) = I_S(V(t)) + \frac{dQ_S}{dt}. \tag{6.90 a}$$

For the drain a similar relation exists,

$$i_D(t) = I_D(V(t)) + \frac{dQ_D}{dt}. \tag{6.90 b}$$

Since no transport current is flowing to the gate and the substrate, the situation is less complicated for these terminals,

$$i_G(t) = \frac{dQ_G}{dt}, \tag{6.90 c}$$

$$i_B(t) = \frac{dQ_B}{dt}. \tag{6.90 d}$$

In practice the assumption of quasi-static operation does not impose severe limitations on the application of the model. When the channel is not too short, the channel transit time is given by [6.1]

$$\tau_t \approx \frac{(1 + \theta_C V_{DS})L^2}{\mu_s V_{DS}}, \tag{6.91}$$

in which equation V_{DS} equals the saturation voltage V_{DSS} during the corresponding mode of operation. For $L = 0.7$ μm, $\mu_s = 500$ cm^2/Vs and $V_{DSS} = 1.0$ Volt, we have $\theta_C = 0.5$ and $\tau_t = 15$ ps. For this device Fig. 6.20 gives the carrier distribution in the channel $\Delta N(x)$, which was calculated numerically immediately after the source bias had been increased by an amount $\Delta V_S = 0.5$ Volt within $t_1 = 10$ ps.

From the change of $\Delta N(x)$ as a function of time, we see that a new quasi-equilibrium situation is reached after 20 ps. Hence we conclude that transit-time effects are only important within a time frame of a few units of τ_t. In almost all applications, system or circuit constraints cause the externally driven device terminals to vary at least one order of magnitude more slowly.

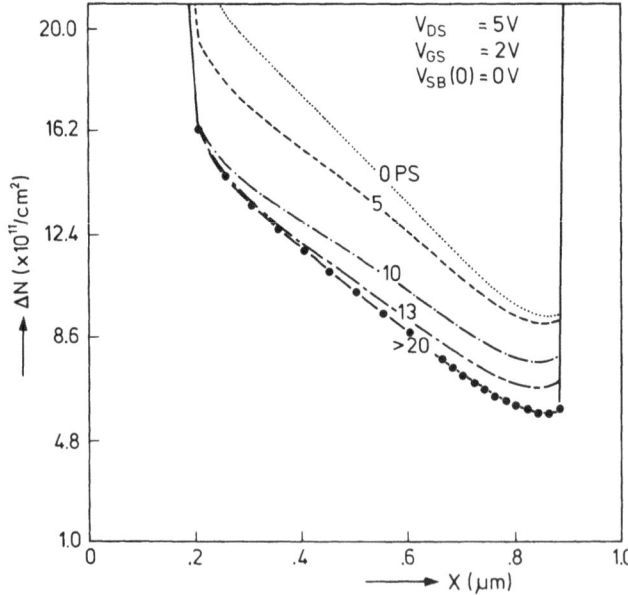

Fig. 6.20 Channel charge distribution at several time intervals after applying a voltage step ΔV_S to the source terminal

6.6 Dynamic Operation

Nevertheless in the next section an approach is discussed which, although conceptually quasi-static, promises to correct to some extent the limitations discussed above.

6.6.2 Charges, Charge Distribution and Capacitances in the Active Region

Although we have already defined in Eq. (6.90 a) the transient current flowing through the source terminal, we tacitly assumed that the charge at the source terminal is known. This is in fact a problem, since only the total channel charge can be calculated from the basic MOSFET equations. For an n-channel device the total inversion charge is given by

$$Q_i = W \int_0^L qn(x)\,dx. \tag{6.92}$$

With the help of the current equations (like Eqs. (6.33) and (6.34)) the above integration can be easily transformed into an integration with respect to the surface potential V. Thus we obtain

$$Q_i = Q_i(V_G, V_D, V_S, V_B). \tag{6.93 a}$$

Similarly we obtain for the charge at the bulk and gate terminal, respectively,

$$Q_B = Q_B(V_G, V_D, V_S, V_B) \tag{6.93 b}$$

and

$$Q_G = -(Q_i + Q_B). \tag{6.93 c}$$

Note that we formally express the charges in four terminal voltages. In practical circuit applications we usually have only three or fewer independent voltages. Since a further evaluation of the above charge expressions requires specific details of modelling the channel current, we will postpone this to the next chapter.

Returning now to the original problem, we can only conclude that a time variation of the inversion layer charge is distributed between the source and drain terminals according to

$$i_D(t) + i_S(t) = \frac{dQ_i}{dt}.$$

However, for circuit calculations we need expressions for $i_D(t)$ and $i_S(t)$ separately.

Several published attempts have been made to divide this channel charge. They vary from an equal division across both terminals [6.79, 6.87] to a weighted distribution on account of physically relevant capacitance values, which can be derived from such a distribution [6.67, 6.88, 6.89, 6.90, 6.91]. Here we discuss an approach [6.90] that is based on a more rigorous derivation and additionally promises to be applicable in a range wider than just quasi-static operation. See also section 3.10.4.

This derivation starts from the well-known current continuity equation (2.50) applied to the inversion layer,

$$\frac{\partial I(x, t)}{\partial x} = -Wq \frac{\partial n(x, t)}{\partial t}. \tag{6.94}$$

Integrating this equation with respect to x from the source ($x = 0$) to a position x on the channel, we obtain

$$i(x, t) - i(0, t) = -Wq \int_0^x \frac{\partial n(x', t)}{\partial t} dx'.$$

Integrating again from source to drain ($x = L$) and solving for $i(0, t) = -i_S(t)$ gives

$$i_S(t) = \frac{1}{L} \int_0^L i(x, t)\, dx + \frac{d}{dt}\left[\frac{W}{L} \int_0^L \int_0^x qn(x', t)\, dx\, dx'\right].$$

Integrating the latter term of the above equation in parts, we finally obtain

$$i_S(t) = \langle i(t)\rangle + \frac{dQ_S}{dt}, \tag{6.95}$$

where

$$\langle i(t)\rangle = \frac{1}{L} \int_0^L i(x, t)\, dx \tag{6.96}$$

and

$$Q_S = -qW \int_0^L n(x, t)\left(1 - \frac{x}{L}\right) dx. \tag{6.97}$$

In a similar way we obtain for the drain current

$$i_D(t) = \langle i(t)\rangle + \frac{dQ_D}{dt}, \tag{6.98}$$

where

$$Q_D = -qW \int_0^L n(x, t)\frac{x}{L}\, dx. \tag{6.99}$$

In this manner both terminal currents are written as the sum of a transient transport current and a charging current. In quasi-static operation the transport current (6.96) is dependent only on the instantaneous values of the terminal voltages and these can be found from the steady-state solution. The source and drain charging currents are simply the time derivatives of the associated charges, which are found by multiplying the instantaneous mobile charge distribution by a linear weighting factor. Exactly the same

6.6 Dynamic Operation

result has been obtained by splitting up the MOSFET channel into an RC network and extracting from this the equivalent time-dependent terminal currents [6.91].

It has been argued [6.92] that the charges Q_D and Q_S defined above do not necessarily represent the actual (discrete) charges, and that the time derivatives therefore represent more than charging currents. In addition, since in first order $\langle i(t) \rangle$ is virtually independent of transit-time, it could well be that the time derivatives of (6.97) and (6.99) include some correction for transit time. No proof has yet been given, but preliminary results for a.c. small-signal y-parameters, which have been obtained from (6.96) and (6.98), agree with exact results obtained by directly solving the continuity equations [6.93].

Since we can express all charges in terms of the terminal voltages (compare (6.93)) we obtain, by applying the chain rule of differentiation, for the charging currents

$$\frac{\partial Q_S}{\partial t} = \frac{\partial Q_S}{\partial V_D}\frac{dV_D}{dt} + \frac{\partial Q_S}{\partial V_G}\frac{dV_G}{dt} + \frac{\partial Q_S}{\partial V_S}\frac{dV_S}{dt} + \frac{\partial Q_S}{\partial V_B}\frac{dV_B}{dt}$$

with similar relations for the other terminal currents. Defining [6.79, 6.3]

$$C_{ij} = \delta_{ij}\frac{\partial Q_i}{\partial V_j}, \qquad (6.100)$$

where

$$\delta_{ij} = \begin{cases} 1, & i = j, \\ -1, & i \neq j, \end{cases}$$

the term C_{ij} can be interpreted as a mutual capacitance. In the next chapter it will be demonstrated that reciprocity does not generally hold. Taking into account that charge is conserved

$$Q_S + Q_D + Q_G + Q_B = 0$$

and that at most only three out of the four terminal voltages can be chosen independently, we conclude that a MOSFET is characterized by nine independent capacitances. Since an evaluation of the capacitances requires specific modelling details, we will leave this to the next chapter.

6.6.3 Charges in the Off-State Region

In weak inversion the only important intrinsic charge left is the depletion region charge. This charge has already been calculated for an n-channel device in section 6.1.2 and is given by

$$Q_B = -\gamma C_{ox} \psi_s^{1/2}, \qquad (6.15)$$

where the surface potential is given by

$$\psi_s = \left[-\frac{\gamma}{2} + \left\{ \frac{\gamma^2}{4} + (V_G - V_{FB}) \right\}^{1/2} \right]^2. \quad (6.24)$$

Since Q_S and Q_D may be neglected, we have

$$Q_G = -Q_B.$$

When $V_G < V_{FB}$ in the off-state region, the depletion region has disappeared and accumulation charge is built up. Consequently in this case

$$Q_B = -C_{ox}(V_G - V_{FB}). \quad (6.101)$$

6.6.4 Parasitic Contributions

In addition to the intrinsic charges several parasitics contribute to the delays met in MOSFET circuits. Naturally these parasitics correspond to and are determined by the integrated circuit process steps.

Fig. 6.21 gives a cross-section of a realistic device showing the origin of the major parasitics.

In order to avoid unduly high lateral electric fields in saturated operation, the drain junction needs some overlap with the gate. Owing to this and stray field lines originating from the gate side-wall and the junction corner, parasitic drain-gate and source-gate capacitances (C_{DGO} and C_{SGO}, respectively) arise, which in particular become manifest in the cut-off mode. The value of these capacitances is discussed in detail in section 6.7.2.

Since the source and drain form a p-n or n-p junction to the underlying substrate or well, the corresponding capacitances also have to be taken into account. For a discussion of the voltage dependence of the unit area capacitance we refer to chapter 3. Generally we have

$$C_T = \frac{C_{T0}}{(1 - V/V_d)^p}, \quad (6.102)$$

where in particular the grading coefficient p depends on the process used. Owing to the fact that the channel-stop area may have a doping value different from the substrate, we have to distinguish between an areal and a peripheral contribution. Thus we write

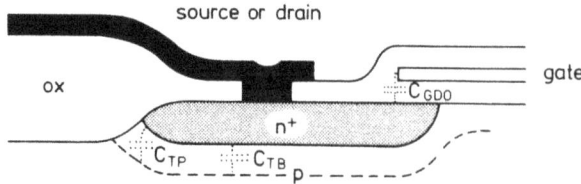

Fig. 6.21 Cross-section of a practical MOSFET, showing the origin of major parasitic capacitances

6.7 Intrinsic Parasitics 185

$$C_T = L_T W_T C_{TB}(V) + (2L_T + W_T)C_{TP}(V), \qquad (6.103)$$

where L_T and W_T are the length and width of the junction, respectively, and C_{TP} is the peripheral capacitance per unit length.

In addition to the above capacitances, it is necessary in circuit speed calculations to take a number of interconnect capacitances into account as well. Since these quantities, in contrast to the previous parasitics, are not at all related to the specific MOSFET structure, a further discussion [6.93, 6.94] is beyond the scope of this book.

6.7 Intrinsic Parasitics

6.7.1 Series Resistance

With the continuing miniaturization of MOSFET devices, the parasitic series resistance has recently become an important issue. The problem arises since, as opposed to the device itself, this resistance is less amenable to scaling. This is mainly due to technical limitations like a contact resistance between different materials and the nonabrupt p-n junction profile of source and drain. Fig. 6.22 illustrates schematically the current pattern in the regions considered. From this figure we conclude that the source/drain series resistance consists of four components: a contact resistance beyond the contact window, a sheet resistance where the current flows along parallel lines, a spreading resistance due to current crowding in the vicinity of the channel end, and an accumulation layer resistance.

The contact resistance R_{co} has been calculated, applying a transmission line model to the interface between metal and semiconductor, and the junction

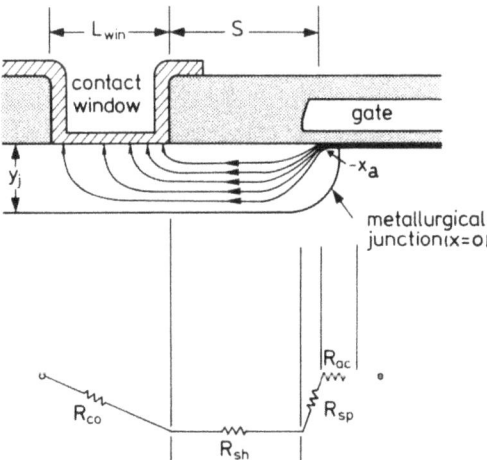

Fig. 6.22 Schematic diagram showing current pattern in the source/drain region and its associated resistance components

6 MOSFET Physics Relevant to Device Modelling

sheet resistance. The result is that [6.96, 6.97]

$$R_{co} = \frac{(\rho_c \rho_\square)^{1/2}}{W} \coth\left[L_W \left(\frac{\rho_\square}{\rho_c}\right)^{1/2}\right], \quad (6.104)$$

where ρ_c is the contact resistance (Ωcm^2), $\rho_\square$ is the sheet resistance (Ω), L_W is the contact window length and W is the MOSFET channel width. In this equation the effects of current crowding at the leading edge of the contact window are included.

The sheet resistance is given simply by

$$R_{sh} = \frac{\rho_\square S}{W}, \quad (6.105)$$

where S is given in Fig. 6.22.

Using the assumption (based on numerical simulation) that the current converges within an angle of 1 radian from the full junction depth (y_j) to the thin accumulation sheet (thickness y_c), the spreading resistance is given by [6.98]

$$R_{sp} = 0.64 * \int_{0.64 y_c}^{0.64 y_j} \frac{\rho \, dx}{xW}, \quad (6.106)$$

where the constant of 0.64 comes from $1/\tan 1$. In the case of uniform junction doping (specific resistance ρ) the above integral yields [6.99]

$$R_{sp} = \frac{0.64 \rho}{W} \ln\left(\frac{y_j}{y_c} - 1\right). \quad (6.107)$$

Using the following values, which are typical of n-type layers, $\rho = 7.5 \times 10^{-4}$ Ωcm, $\rho_c = 25 \times 10^{-8}$ Ωcm^2, $y_j = 0.25$ µm, $y_c = 5$ nm and $L_W = S = 1$ µm, we obtain for $W = 1$ µm

$$R_s = (35 + 30 + 55) = 120\Omega.$$

While the above value is a reasonable estimate of the series resistance of devices with a channel length larger than several microns, experimental values for devices in the (sub)micron range indicate a considerable underestimation by relation (6.107), in particular for lightly doped drain devices. Assuming an exponential junction doping profile in the vicinity of the channel end

$$N = N_0 \exp kx,$$

the integral (6.106) has been recalculated [6.100]. In addition, assuming that the current finally flows through an accumulation sheet with length x_{ac}, the competition between current flow in the accumulation sheet and the underlying junction layer has been solved via a minimization procedure.
Here we mention only that two extra terms have to be added to (6.106)

$$\Delta R_{sp}(k, N_0)$$

6.7 Intrinsic Parasitics

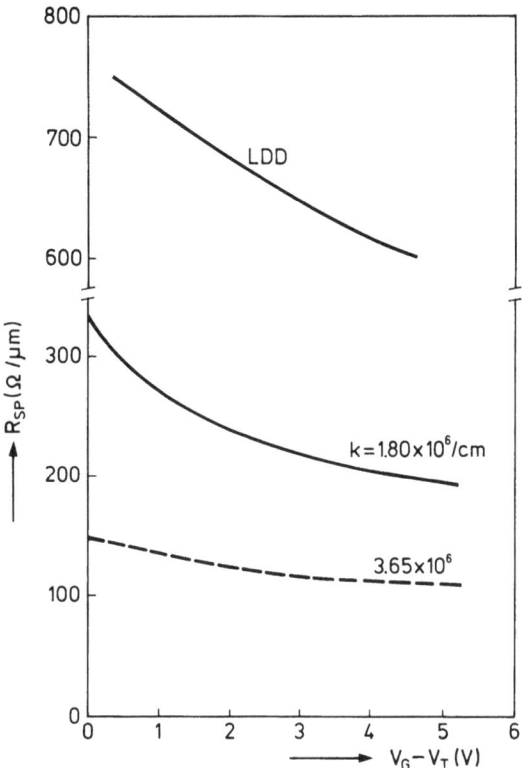

Fig. 6.23 Calculated source/drain spreading resistance as a function of applied gate voltage at two values of the dope gradient coefficient. For comparison the resistance typical of LDD devices is given too

and

$$R_{ac}(k, N_0, N_{ac}),$$

where

$$N_{ac} = \frac{(V_G - V_{FB})C_{ox}}{q}. \qquad (6.108)$$

Fig. 6.23 gives the total value of R_{sp} as a function of gate voltage for two values of the doping gradient k. Other relevant data are: $t_{ox} = 25$ nm, $N_0 = 5 \times 10^{16}/\text{cm}^3$, $\rho = 7.5 \times 10^{-4}$ Ωcm. Owing to the term R_{ac}, the total series resistance decreases slightly upon an increase of the gate bias.

Since the use of an exponential profile for the tail of an LDD configuration leads to an overestimation of R_s, it is better to evaluate the integral (6.106) assuming a linear doping profile [6.68]. A value of R_s typical of LDD devices is also given in Fig. 6.23. Owing to the fact that at larger drain bias values the channel current already spreads into the bulk in the pinch-off region, the spreading resistance at the drain end decreases with V_{DS}. This effect requires special care when modelling the characteristics of submicron LDD MOSFETs.

6.7.2 Gate-Junction Capacitance

When the MOSFET operates in the off-state mode, an active channel carrier charge is no longer present. Nevertheless parasitic gate-drain and gate-source capacitances remain. This is due to the necessary overlap between gate and junctions and the finite gate electrode thickness, which causes a fringing field to exist at the gate perimeter.
The overlap capacitance is simply given by

$$C_{Ov} = \frac{\varepsilon_{0x} \Delta L_G W_G}{t_{0x}}, \qquad (6.108)$$

where ΔL_G is the gate overlap.

The fringing capacitance has been calculated [6.101, 6.102] by transforming a rectangular electrode and ground plane (compare Fig. 6.24) into the complex potential plane by means of Schwartz-Christoffel transformation and using a logarithmic potential transformation to calculate flux. As a result of this the fringing capacitance C_f is found to contain three components: (1) C_1, representing the contribution bounded by flux lines ϕ_1 and ϕ_2; (2) C_2, the contribution bounded by flux lines ϕ_1 and ϕ_3; and (3) C_3, the contribution bounded by flux lines ϕ_3 and ϕ_4. The final result is given by

$$C_f = \gamma_c \varepsilon_{0x} W_G, \qquad (6.109)$$

where

$$\gamma_c = (2 - \ln 4 + \ln a + \ln u/a)/\pi, \qquad (6.110)$$

with

$$a = 2K(K^2 - 1)^{1/2} + 2K^2 - 1 \quad \text{and} \quad K = 1 + t_G/t_{0x},$$

where t_G is the thickness of the gate electrode. The value of u has to be determined from a transcendental equation. Fig. 6.25 gives the factor γ_c as a function of the normalized gate electrode thickness t_G/t_{0x} for two relevant values of the ratio L_G/t_{0x}.

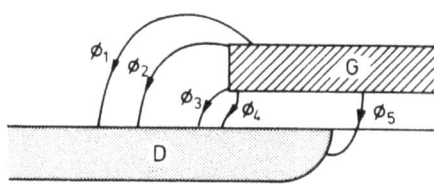

Fig. 6.24 Flux lines representing the three components of the gate-drain fringing capacitance

6.7 Intrinsic Parasitics

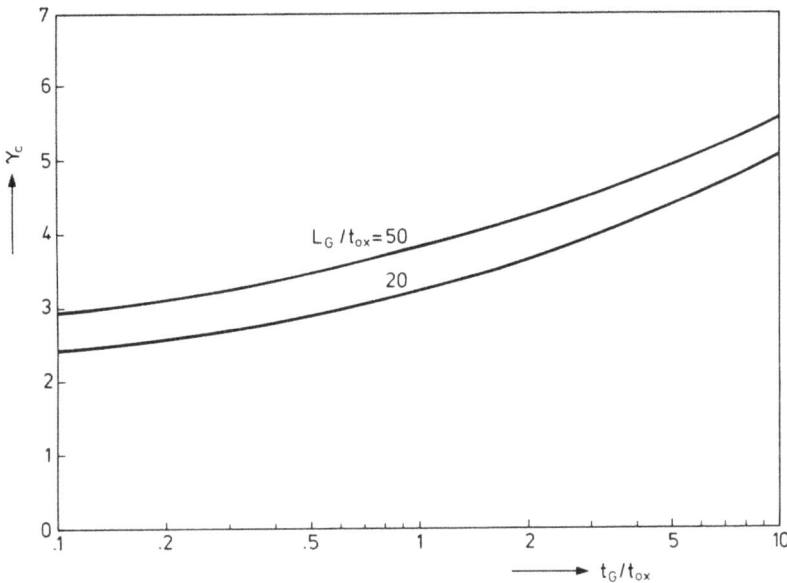

Fig. 6.25 Fringing parameter as a function of normalized gate electrode thickness

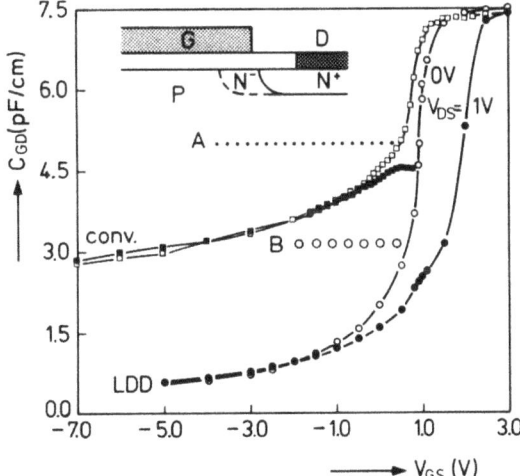

Fig. 6.26 Gate-drain fringing capacitance of a conventional and an LDD MOSFET. Curves are numerical results; dots represent analytical results

Considering that usually $\Delta L_G/t_{0x} \approx 4$, we observe that the correction by C_f can no longer be neglected when $t_G/t_{0x} > 1$. Since in practice $t_G/t_{0x} > 10$ for a 1-µm minimum feature process, C_f is even larger than C_{0v}.
Although the above result is exact, nevertheless a numerical simulation has shown that the situation is more complex, in particular for short-channel devices [6.95]. Fig. 6.26 gives the gate-drain capacitance as a function of gate bias for a conventional and an LDD MOSFET with $L_G = 0.75$ µm. When the device operates in the off-state ($V_G < 0$ Volt) C_{GD0} is not constant, but decreases slightly with increasing negative gate bias. This is caused by a second fringing component not considered before: the capacitance between the drain junction edge and the gate (see Fig. 6.24). This capacitance only becomes negligible when an accumulation layer is formed along the entire channel. This is only the case for large negative values of V_G.
Owing to the above complex situation, it is advisable to extract the off-state model parameters C_{GD0} and C_{GS0} from experimental test configurations.

References

[6.1] S. M. Sze: Physics of Semiconductor Devices. John Wiley & Sons, New York (1981).
[6.2] E. H. Nicollian, J. R. Brews: MOS Physics and Technology. John Wiley & Sons, New York (1982).
[6.3] Y. P. Tsividis: Operation and Modelling of the MOS Transistor. McGraw-Hill, New York (1987).
[6.4] F. Stern, W. E. Howard: Properties of Semiconductor Surface Inversion Layers in Quantum Limit. Physical Reviews *163*, 816–835 (1967).
[6.5] J. A. Pals: Quantization Effects in Semiconductor Inversion and Accumulation Layers. Ph.D. Thesis, Eindhoven University of Technology (1972).
[6.6] A. S. Grove, E. H. Snot, B. E. Deal, C. T. Sah: Simple Physical Model for the Space-Charge Capacitance of MOS Structures. Journal of Applied Physics *35*, 2458–2460 (1964).
[6.7] R. H. Kingston, S. F. Neustadter: Calculation of Space Charge, Field and Free Carrier Concentration. Journal of Applied Physics *26*, 718–720 (1955).
[6.8] C. G. Garrett, W. H. Brittain: Physical Theory of Semiconductor Surfaces. Physical Review *99*, 376–387 (1955).
[6.9] J. A. Geurst: Theory of IGFETs Near and Beyond Pinch-Off. Solid-State Electronics *9*, 129–142 (1966).
[6.10] H. W. Loeb, R. Andrew, W. Love: Application of 2-D Solutions of the Poisson Equation to MOST Devices. Electronics Letters *4*, 352–354 (1968).
[6.11] J. R. Brews: A Charge-Sheet Model of the MOSFET. Solid-State Electronics *21*, 345–355 (1978).
[6.12] J. R. Brews: Subthreshold Behaviour of Uniformly and Non-Uniformly Doped Long-Channel MOST. IEEE Transactions Electron Devices *ED-26*, 1282–1292 (1979).
[6.13] R. J. v. Overstraeten, G. J. Declerck, P. A. Muls: Theory of the MOST in Weak Inversion. IEEE Transactions Electron Devices *ED-22*, 282–288 (1975).
[6.14] R. M. Swanson, J. D. Meindl: MOS Transistors in Low-Voltage Circuits. IEEE Journal of Solid-State Circuits *SC-7*, 146–153 (1972).
[6.15] G. Baccarani, M. Rudan, G. Spadini: Analytical IGFET Model Including Drift and Diffusion. IEE Journal of Solid-State Devices *2*, 62–68 (1978).
[6.16] F. v.d. Wiele: A Long-Channel MOSFET Model. Solid-State Electronics *22*, 991–997 (1979).

References

[6.17] Y. P. Tsividis: reference [6.3], p. 106–123.

[6.18] H. C. Pao, C. T. Sah: Effects of Diffusion Current on Characteristics of MOS Transistors. Solid-State Electronics 10, 927–937 (1966).

[6.19] C. T. Sah, H. C. Pao: Effects of Fixed Bulk Charge on the Characteristics of MOS Transistors. IEEE Trans. Electron Devices ED-13, 393–409 (1966).

[6.20] J. A. van Nielen, and O. W. Memelink: Influence of the Substrate Upon the DC Characteristics of MOS Transistors. Philips Research Reports 22, 55–71 (1967).

[6.21] M. B. Baron: Low-Level Currents in IGFET Transistors. Solid-State Electronics 15, 293–302 (1972).

[6.22] M. R. MacPherson: The Adjustment of MOST Threshold Voltage by Ion Implantation. Applied Physics Letters 18, 502–504 (1971).

[6.23] L. C. Parillo: VLSI Process Integration. In: VLSI Technology. McGraw-Hill, New York (1984).

[6.24] G. Doucet, F. v. d. Wiele, P. Jespers: Theoretical and Experimental Study of MOST Doped by SILOX Technique. Solid-State Electronics 19, 191–199 (1976).

[6.25] A. Das Gupta, S. K. Lahiri: An Analytical Solution of Poisson's Equation for a MOSFET with a Gaussian Doped Channel. Solid-State Electronics 29, 1205–1206 (1986).

[6.26] V. L. Rideout, F. H. Gaensslen, A. LeBlanc: Device Design Considerations for Ion-Implanted MOSFETs. IBM Journal of R&D 19, 50–59 (1975).

[6.27] E. Demoulin, F. v. d. Wiele: Ion-Implanted MOS Transistors. In: Process and Device Modelling for IC Design. Noordhoff, Leyden (1977), pp. 617–676.

[6.28] J. R. Edwards, G. Mar: Depletion-Mode IGFET Made by Deep Implantation. IEEE Transactions Electron Devices ED-20, 283–289 (1973).

[6.29] J. S. T. Huang: Characteristics of a Depletion-Mode IGFET. IEEE Transactions Electron Devices ED-20, 513–515 (1973).

[6.30] F. M. Klaassen, W. Hes: Compensated MOSFET Devices. Solid-State Electronics 28, 359–373 (1985).

[6.31] G. Merckel: Ion-Implanted MOS Transistors. In: Process and Device Modelling for IC Design. Noordhoff, Leyden (1977), pp. 677–688.

[6.32] J. S. T. Huang, G. W. Taylor: Modelling of a Depletion-Mode IGFET. IEEE Transactions Electron Devices ED-22, 995–1001 (1975).

[6.33] S. Chiang, K. M. Cham, R. D. Rung: Optimization of Sub-Micron p-Channel FET. Technical Digest IEDM, Washington D. C. (1983), pp. 534–537.

[6.34] F. M. Klaassen, J. J. Bastiaens, W. Hes, M. Sprokel: Scaling of Compensated MOSFETs. Technical Digest IEDM, San Francisco (1984), pp. 613–616.

[6.35] F. M. Klaassen, W. Hes: On the Temperature Coefficient of the MOSFET Threshold Voltage. Solid-State Electronics 29, 787–789 (1986).

[6.36] G. Giralt, B. Andre, J. Simon, D. Esteve: Influence de la Température sur les Dispositifs Semiconducteurs du Type MOS. Electronics Letters 1, 185–186 (1965).

[6.37] R. Wang, J. Dunkley, T. A. De Massa, J. Jelsma: Threshold Voltage Variations with Temperature in MOS Transistors. IEEE Transactions Electron Devices ED-18, 386–388 (1971).

[6.38] L. D. Yau: A Simple Theory to Predict the Threshold Voltage of Short-Channel IGFETs. Solid-State Electronics 17, 1059–1063 (1974).

[6.39] D. J. Coe, H. E. Brakman, K. H. Nicholas: A Simple Approach for Accurate Modelling the MOST Threshold Voltage. Solid-State Electronics 20, 993–998 (1977).

[6.40] G. W. Taylor: Subthreshold Conduction in MOSFETs. IEEE Transactions on Electron Devices ED-25, 337–350 (1978).

[6.41] G. Merckel: Short Channel MOSFETs. In: Process and Device Modelling for IC Region. Noordhoff, Leyden (1977), 705–724.

[6.42] W. Fichtner, H. W. Pötzl: MOS Modelling by Analytical Approximations, Part I. International Journal of Electronics 46, 33–55 (1979).

[6.43] K. N. Ratnakumar, J. D. Meindl, D. L. Scharfetter: New IGFET Short-Channel Threshold Voltage Model. IEEE Journal of Solid State Circuits 17, 937–947 (1982).

[6.44] T. N. Nguygen, J. D. Plummer: Physical Mechanisms Responsible for Short-Channel Effects in MOS Devices. Technical Digest IEDM, Washington D. C. (1981), pp. 596–599.
[6.45] V. Marash, R. W. Dutton: Submicron 2-D MOS Modelling. Digest of Technical Papers ICCAD-86. Washington D. C. (1986), pp. 476–479.
[6.46] P. P. Wang: Device Characteristics of Short-Channel, Narrow-Width MOSFETs. IEEE Transactions on Electron Devices ED-25, 779–786 (1978).
[6.47] R. R. Troutman: VLSI Limitations from Drain-Induced Barrier Lowering. IEEE Journal of Solid-State Circuits SC-14, 383–391 (1979).
[6.48] E. Sun: Short-Channel MOS Modelling for CAD. In: Proceedings Asilomar Conference on Circuits and Systems. Pacific Grove, CA (1978), pp. 493–499.
[6.49] T. Skotnicki, W. Marciniak: New Approach to Threshold Voltage Modelling of MOSFETs. Solid-State Electronics 29, 1115–1128 (1986).
[6.50] T. Skotnicki, G. Merckel, T. Pedron: The Voltage-Doping Transformation, A New Approach to Modelling Short-Channel Effects. In: Proceedings ESSDERC 87, Bologna. North-Holland, Amsterdam (1987), pp. 543–546.
[6.51] R. H. Dennard, F. H. Gaensslen, H. N. Yu, V. L. Rideout, E. Bassou, A. R. Leblanc: Design of Ion-Implanted MOSFETs with Very Small Physical Dimensions. IEEE Journal of Solid-State Circuits SC-9, 256–268 (1974).
[6.52] F. M. Klaassen: Design and Performance of Micron-Size Devices. Solid-State Electronics 21, 565–572 (1978).
[6.53] J. R. Brews, W. Fichtner, E. H. Nicollian, S. M. Sze: Generalized Guide for MOSFET Miniaturization. IEEE Electron Device Letters EDL-1, 2–3 (1980).
[6.54] F. M. Klaassen, W. C. J. de Groot: Modelling of Scaled-Down MOS Transistors. Solid-State Electronics 23, 237–242 (1980).
[6.55] G. Merckel: Simple Model of the Threshold Voltage of Short and Narrow Channel IGFETs. Solid-State Electronics 23, 1207–1213 (1980).
[6.56] L. A. Akers, J. J. Sanchez: Threshold Voltage Models of Small Geometry MOSFETs. Solid-State Electronics 25, 621–641 (1982).
[6.57] T. Ando, A. B. Fowler, F. Stern: Electronic Properties of Two-Dimensional Systems. Review of Modern Physics 54, 437–672 (1982).
[6.58] S. R. Hofstein, G. Warfield: Carrier Mobility and Current Saturation in MOS Transistors. IEEE Transactions on Electron Devices ED-12, 129–138 (1965).
[6.59] G. Baum, H. Beneking: Drift Velocity Saturation in MOS Transistors. Solid-State Electronics 13, 789–798 (1970).
[6.60] A. G. Sabnis, J. T. Clemens: Characterization of the Electron Mobility in Inverted Si. Technical Digest IEDM, Washington D. C. (1979), pp. 18–21.
[6.61] G. T. Wright: Physical and CAD Models for the VLSI MOSFET. IEEE Transactions on Electron Devices ED-34, 823–833 (1987).
[6.62] D. M. Caughey, R. E. Thomas: Carrier Mobility in Silicon Empirically Related to Doping and Field. Proc. IEEE 52, 2192–2193 (1967).
[6.63] J. A. Cooper, D. F. Nelson: High-Field Drift Velocity of Electrons at the Si-SiO_2 Interface. Journal of Applied Physics 54, 1445–1456 (1983).
[6.64] G. Merckel, J. Borel, N. Z. Cupcea: Accurate Large-Signal MOS Transistor Model for Use in CAD. IEEE Trans. Electron Devices ED-19, 681–690 (1972).
[6.65] P. Smith, M. Inoue, J. Frey: Electron Velocity in Si and GaAs at Very High Fields. Applied Physics Letters 37, 797–798 (1980).
[6.66] K. K. Thornber: Relation of Drift Velocity to Low-Field Mobility. Journal of Applied Physics 51, 2127–2133 (1980).
[6.67] F. M. Klaassen: Compact MOSFET Modelling. In: Process and Device Modelling. North-Holland, Amsterdam (1986), pp. 393–412.
[6.68] F. M. Klaassen, P. T. J. Biermans, R. M. D. Velghe: The Series Resistance of Submicron MOSFETS and Its Effect on Their Characteristics. ESSDERC 1988, Montpellier. Journal de Physique 257–260 (1988).
[6.69] F. M. Klaassen: (to be published).

References

[6.70] T. Poorter, J. H. Satter, V. A. Satyadharma: Journeé d'Electronique 285–301 (1977)

[6.71] J. E. Schroeder, R. S. Muller: IGFET Analysis Through Numerical Solution of Poisson's Equation. IEEE Trans. Electron Devices *ED-15*, 954–961 (1968).

[6.72] J. A. El-Mansy, A. R. Boothroyd: A Simple 2-D Model for IGFET Operation in Saturation Region. IEEE Trans. Electron Devices *ED-24*, 254–262 (1977).

[6.73] T. Poorter, J. H. Satter: A d-c Model for a MOS Transistor in the Saturation Region. Solid-State Electronics *23*, 765–772 (1980).

[6.74] P. P. Guebels, F. v. d. Wiele: A Small Geometry MOSFET Model for CAD Applications. Solid-State Electronics *26*, 267–273 (1983).

[6.75] P. K. Ko, R. S. Muller, C. Hu: A Unified Model for Hot-Electron Currents in MOSFETs. Technical Digest IEDM *81*, 600–603 (1981).

[6.76] H. Shichman, D. A. Hodges: Modelling and Simulation of Insulated-Gate Field-Effect Transistor Circuits. IEEE Journal of Solid-State Circuits *SC-3*, 285–289 (1968).

[6.77] G. Merckel: CAD Models of MOSFETs. In: Process and Device Modelling for IC Design. Noordhoff, Leyden (1977), pp. 751–764.

[6.78] F. M. Klaassen: A MOS Model for CAD. Philips Research Reports *31*, 71–83 (1976).

[6.79] D. E. Ward, R. W. Dutton: A Charge-Oriented Model for MOS Transistor Capacitance. IEEE Journal of Solid-State Circuits *SC-13*, 703–707 (1978).

[6.80] J. A. Robinson, Y. A. El-Mansy, A. R. Boothroyd: A General Four-Terminal Charging Current Model for the IGFET. Solid-State Electronics *23*, 405–414 (1980).

[6.81] T. J. O'Reilly: Transient Response of IGFETs. Solid-State Electronics *8*, 947–956 (1965).

[6.82] Z. S. Gribnikov, Y. A. Tkhorik: Calculation of Transients in Field Triodes. Radio Engineering and Electronic Physics *11*, 776–781 (1966).

[6.83] A. Möschwitzer: Zum statischen und dynamischen Großsignal-Verhalten des MOS-Transistors. Nachrichten-Technik *20*, 150–154 (1967).

[6.84] K. Goser: Einschaltzeiten und Umladungsvorgänge bei MOS-Transistoren. Archiv Elektrische Übertragung *24*, 21–28 (1970).

[6.85] C. Turchetti, P. Mancini, G. Masetti: A CAD-Oriented Non-Quasi-Static Approach for Transient Analysis of MOS IC's. IEEE Journal of Solid-State Circuits *SC-21*, 827–836 (1986).

[6.86] H. Park, P. Ko, C. Hu: A Non-Quasi Static MOSFET Model for SPICE. Technical Digest IEDM *87*, 652–655 (1987).

[6.87] G. W. Taylor, W. Fichtner, J. G. Simons: Description of MOS Internodal Capacitances for Transient Simulations. IEEE Transactions on CAD *CAD-1*, 150–156 (1982).

[6.88] P. Yang, B. D. Eppler, P. K. Chatterjee: An Investigation of the Charge Conservation Problem for MOSFET Circuit Simulation. IEEE Journal of Solid-State Circuits *SC-18*, 128–138 (1983).

[6.89] B. J. Sheu, D. L. Scharfetter, C. Hu, D. O. Pederson: A Compact IGFET Charge Model. IEEE Transactions on Circuits and Systems *CAS-31*, 745–749 (1984).

[6.90] S. Y. Oh, D. E. Ward, R. W. Dutton: Transient Analysis of MOS Transistors. IEEE Journal of Solid-State Circuits *SC-15*, 636–643 (1980).

[6.91] M. F. Sevat: On the Channel Charge Division in MOSFET Modelling. Digest Technical Papers *ICCAD-87*, 208–210 (1987).

[6.92] J. G. Fossum, H. Jeong, S. Veeragaghavan: Significance of Channel-Charge Partition in the Transient MOSFET Model. IEEE Transactions on Electron Devices *ED-33*, 1621–1623 (1986).

[6.93] K. W. Chai, J. J. Paulos: Comparison of Quasi-Static and Non-Quasi-Static Capacitance Models for the MOSFET. IEEE Electron Device Letters *EDL-8*, 377–379 (1987).

[6.94] P. E. Cottrell, E. M. Buturla: VLSI Wiring Capacitance. IBM Journal of R&D *29*, 277–287 (1985).

[6.95] J. H. H. M. Quint, F. M. Klaassen, R. Petterson: 2-D and 3-D Capacitance Effects in MOS VLSI. In: Proceedings ESSDERC 87, Bologna. North-Holland, Amsterdam (1987), pp. 417–420.
[6.96] H. Murrman, D. Widmann: Current Crowding on Metal Contacts to Planar Devices. IEEE Transactions Electron Devices *ED-16*, 1022–1026 (1969).
[6.97] H. H. Berger: Models for Contacts to Planar Devices. Solid-State Electronics *15*, 145–158 (1972).
[6.98] K. K. Ng, W. T. Lynch: Analysis of the Series Resistance of MOSFETs. IEEE Transactions Electron Devices *ED-33*, 965–972 (1986).
[6.99] G. Baccarani, G. A. Sai-Halasz: Spreading Resistance in Submicron MOSFETs. IEEE Electron Device Letters *EDL-4*, 27–29 (1983).
[6.100] K. K. Ng, R. J. Bayruns, S. C. Fang: The Spreading Resistance of MOSFETs. IEEE Electron Device Letters *EDL-6*, 195–197 (1985).
[6.101] H. El Kamchouchi, A. A. Zaky: A Direct Method for the Calculation of the Edge Capacitance. Journal of Applied Physics *8*, 1365–1371 (1975).
[6.102] E. W Greeneich: An Analytical model for the Gate Capacitance of Small-Geometry MOS Structures. IEEE Transactions on Electron Devices *ED-30*, 1838–1839 (1983).

Models for the Enhancement-Type MOSFET 7

From the viewpoint of application, the enhancement-type transistor, which operates in the off-state mode at zero gate bias, is the most important MOSFET. Usually devices of this class are made on a uniform doped substrate or on a substrate with an implanted channel region. Examples are n-channel transistors with or without a p-type implanted layer in a p-type substrate and p-channel transistors with or without a shallow p-type implanted layer in an n-type substrate. Generally the models concerned have followed the path of progress in processing technology, from devices with structural dimensions longer than 10 µm to present-day devices with possibly submicron dimensions. Therefore we shall first discuss models for long-channel devices. This will be followed by a discussion of models for short-channel devices, in which a number of corresponding effects, like threshold voltage lowering etc., are taken into account. Since MOSFETs are also employed nowadays in analog circuitry, we finally discuss modelling for this more demanding application.

7.1 Long-Channel Models

During the first decade of MOS integrated circuit technology, device dimensions were larger than 3 microns. Consequently 2-D effects were unimportant. In addition the velocity saturation effect could be ignored or at most taken into account as a higher-order effect.
Since the modelling of devices with implanted substrates is more complicated, we start with a model for the drain current of transistors with uniformly doped substrates. This is followed by a current model for implanted structures. Next the modelling of charges and capacitances is discussed. Finally as a transition to the next section on short-channel devices, it is shown how velocity saturation can be taken into account as a second-order correction.

7.1.1 The Drain Current of Transistors in Uniformly Doped Substrates

The drain current expression for the strong-inversion region has already been discussed in principle in section 6.2. Nevertheless two corrections have to be made to Eqs. (6.34) to (6.36). Although the gate oxide is relatively thick in the devices of the process generations considered and the normal field in the channel is consequently rather low, a decrease of mobility at higher gate bias has to be taken into account. This can be achieved [7.1] by taking the first two terms of the denominator in the mobility relation (6.81). Alternatives to this expression have been published too [7.2], but these have led to problems with parameter determination [7.3].

The second correction to be made is designed to allow for the finite drain conductance. In long devices the above effect is small and therefore a form of channel length modulation is usually chosen, which is a simplification of those discussed in section 6.5.2. A fair approach for instance is given by [7.4, 7.5, 7.6, 7.7]

$$\Delta L = \alpha\{(V_p + V_D - V_{DSAT})^{1/2} - V_p^{1/2}\}, \tag{7.1}$$

where α and V_p are parameters.

Since, however, even this form complicates the formulation of the saturation voltage that guarantees continuity of the drain conductance at V_{DSAT}, a more primitive form is often chosen,

$$\Delta L = \alpha(V_D - V_S + \varepsilon)^{1/2}, \tag{7.2}$$

where the constant is chosen to avoid numerical problems at $V_{DS} = 0$ V. In order to avoid too many complications around V_{DSAT}, the channel-length modulation is also applied in the linear mode. Although this is less physical in approach, the above modelling term has proved capable of describing the characteristics rather well [7.8]. Therefore we first give a long-channel model using Eq. (7.2). Taking all voltages with respect to bulk, the associated expressions read:

— threshold voltage

$$V_T(V_S) = V_{T0} + V_S + \gamma\{(V_S + \psi_{sm})^{1/2} - \psi_{sm}^{1/2}\}; \tag{7.3 a}$$

if $V_G \geqslant V_T$ and $V_D > V_S$,

— saturation voltage

$$V_{DSAT} = V_G - V_{T0} + \gamma\psi_{sm}^{1/2}$$
$$+ \frac{1}{2}\gamma^2\left[1 - \left\{1 + \frac{4}{\gamma^2}(V_G - V_{T0} + \psi_{sm} + \gamma\psi_{sm}^{1/2})\right\}^{1/2}\right], \tag{7.3 b}$$

$$g(V_G, V) = (V_G - V_{T0} + \gamma\psi_{sm}^{1/2} - V)^2 + \tfrac{4}{3}\gamma(V + \psi_{sm})^{3/2}, \tag{7.3 c}$$

7.1 Long-Channel Models

— drain current

$$I_{DS} = \frac{\beta}{2} \frac{[g(V_G, V_S) - g(V_G, V_R)]}{[1 + \theta_A(V_G - V_{T0}) + \theta_B V_S]} \frac{1}{1 - (\alpha/L)(V_D - V_S + \varepsilon)^{1/2}}, \tag{7.3 d}$$

where the gain factor

$$\beta = \mu_0 C_{0x} W L^{-1}. \tag{7.3 e}$$

In addition the following conditions apply to V_D: if $V_D < V_{DSAT}$, $V_R = V_D$; if $V_D \geq V_{DSAT}$, $V_R = V_{DSAT}$.
Note that compared with the original equation (6.35), the function g has now been put in a symmetrical form. This has the advantage that for the inverse mode ($V_D < V_S$) Eq. (7.3 d) remains valid, if V_D and V_S are interchanged and

$$V_G \geq V_{TR}(V_D) = V_{T0} + V_D + \gamma\{(V_D + \psi_{sm})^{1/2} - \psi_{sm}^{1/2}\}. \tag{7.3 f}$$

In the above model the following process-related electrical parameters are used: the gain constant $\beta_0 = \mu_0 C_{0x}$, the zero-bias threshold voltage V_{T0}, the body-effect coefficient γ, the mobility correction factors θ_A and θ_B, and the channel length modulation factor α.
As shown in Fig. 7.1, the drain current calculated with the above model equations is in a fair agreement with the measured characteristics [7.8]. The parameters indicated in the figure have been obtained from an optimization of the simulated results (compare chapter 10). The average deviation between simulated and experimental results is 5%. Generally this result holds for p-channel devices with a channel length $L > 3$ μm and for n-channel devices with $L > 10$ μm.
In order to incorporate the more physical relation (7.1) into Eqs. (7.3), the saturation voltage has to be chosen such that continuity of the drain conductance is obtained at the transition point. Calling the new saturation voltage V_{DSAT}^*, it is easily shown from (7.1) and (7.3) that the following condition has to be satisfied:

$$\left.\frac{\partial g(V_G, V_D)}{\partial V_D}\right|_{V_{DSAT}^*} = \frac{\alpha}{2LV_P^{1/2}}[g(V_G, V_S) - g(V_G, V_{DSAT}^*)]. \tag{7.4}$$

Unfortunately the above condition leads to a quartic equation in V_{DSAT}^*. Although a general solution in explicit form can be given, the procedure is very time-consuming. Therefore an approximate solution is desirable. Since α/L is usually small, V_{DSAT}^* is only slightly smaller than the original value. Introducing

$$\Delta V = V_{DSAT}^* - V_{DSAT}, \tag{7.5 a}$$

Eq. (7.4) can be rewritten in terms of a power series in ΔV, disregarding higher-order terms. Keeping only the first-order term, it is easily shown from (7.4) that

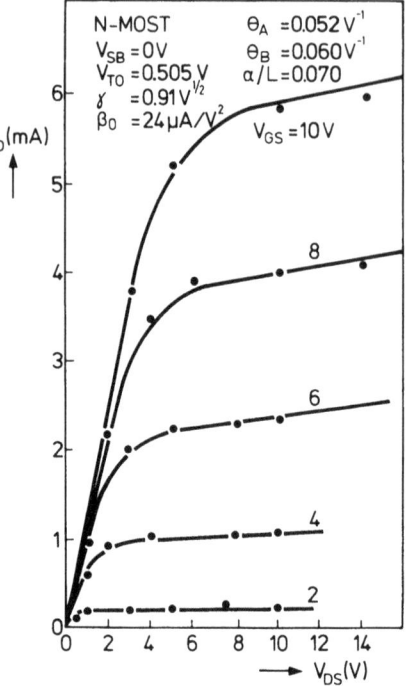

Fig. 7.1 Calculated (fully-drawn lines) and measured characteristics of n-channel MOSFET ($W/L = 100$ μm/10 μm)

$$\Delta V = -\frac{\alpha}{2LV_p^{1/2}} \frac{g(V_G, V_S) - g(V_G, V_{DSAT})}{g''(V_{DSAT})}, \quad (7.5\text{ b})$$

where

$$g''(V) = 2 + \gamma(V + \psi_{sm})^{-1/2}.$$

Naturally, the above approach may introduce a small discontinuity in the drain conductance when α is no longer small. Generally this is the case with poorly scaled short-channel devices. The above problem can be circumvented by extrapolating the ΔL expression (7.1) into the linear region (analogous to the previous model). This can be accomplished with no resulting loss in accuracy by defining

$$\Delta L = \alpha[(V_p + V)^{1/2} - V_p^{1/2}], \quad (7.6\text{ a})$$

where

$$V = \tfrac{1}{2}[(V_D - V_{DSAT}) + \{(V_D - V_{DSAT})^2 + \varepsilon\}^{1/2}] \quad (7.6\text{ b})$$

and ε is a small constant. In this case the ΔV correction is no longer necessary. Finally we remark that in the derivation of (7.1) the effect of free carrier space charge has been neglected when solving Poisson's equation. Attempts to

7.1 Long-Channel Models

incorporate this contribution, via a modification of (7.1), lead to model equations which can only be solved implicitly [7.9, 7.10]. A more successful approach, related to short-channel device modelling, has been introduced in section 6.5.2. Therefore we shall postpone a further evaluation to section 7.3.

In section 6.1.2 it was discussed that in the weak-inversion region the current is determined by drift and diffusion. For this reason it is very difficult to construct a simple yet accurate model. Since an implicit model [7.11], as for instance given by Eqs. (6.29) through (6.32), is not an ideal solution, it is usual for long-channel models to merge a subthreshold current expression like Eq. (6.37) with Eqs. (7.3). This can be achieved as follows [7.12]. Introducing a gate driving voltage

$$V_{Gr} = V_T + RU_T, \qquad (7.7)$$

where R is a fitting constant with a value near 2, the original strong-inversion expressions (7.3) are only maintained for $V_G > V_{Gr}$. However, for $V_G \leq V_{Gr}$ the drain current is defined as

$$I_{DS} = \frac{\beta}{2(1 - \Delta L/L)} \frac{g(V_{Gr}, V_S) - g(V_{Gr}, V_D)}{1 + \theta_A(V_{Gr} - V_{T0}) + \theta_B V_S} \exp\left(\frac{V_G - V_{Gr}}{mU_T}\right). \qquad (7.8)$$

In this approach the drain current is continuous at V_{Gr}, but not the transconductance. For digital applications this shortcoming is not a serious problem. This is shown in Fig. 7.2, where the measured subthreshold char-

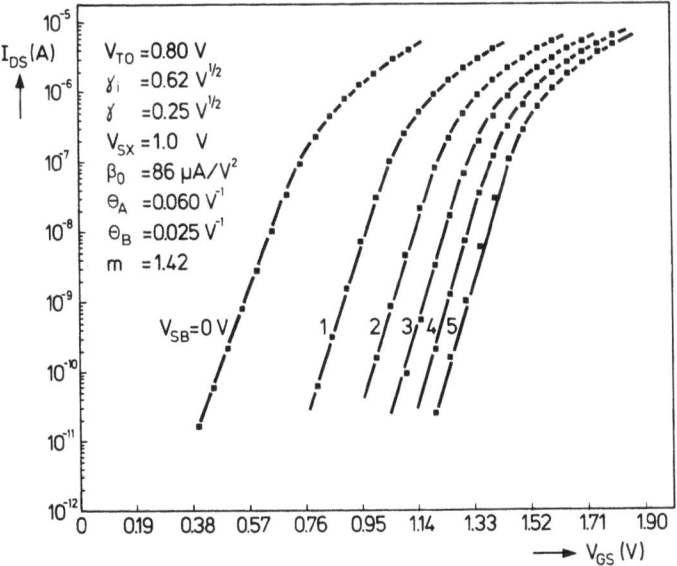

Fig. 7.2 Calculated (fully-drawn lines) and measured subthreshold characteristics of n-channel MOSFET ($W/L = 20$ μm/20 μm)

acteristics are compared with calculated values. However, for analog applications it is a serious drawback. Recently several attempts to correct this deficiency have been discussed [7.13–7.15]. However, some approaches introduce serious physical errors, which become manifest in the drain conductance. A very attractive approach is discussed in sections 7.2 and 7.3, being more directed to the characteristics of short-channel devices.

7.1.2 The Drain Current of Transistors with Threshold Adjustment Implant

For implanted MOSFET structures only the modelling of the *n*-channel transistor has to be reconsidered. Referring to the discussion on *p*-channel devices in section 6.3.3, the threshold voltage implant is opposite to the substrate doping and, owing to the inherently low dose, is always depleted. Consequently the threshold voltage expression (6.57) can be written in the form of Eq. (7.3 a). Only the diffusion potential ψ_{sm} is slightly higher due to the correction term $\Delta\phi$ (compare (6.58)). However, since in practice $\Delta\phi <$ 0.10 Volt, this correction has no practical modelling consequences. Therefore the present approach is also valid for *p*-channel devices with a channel implant [7.16].

In *n*-channel devices the implant is of the same type as the substrate and the implanted layer is not always depleted. For this reason the threshold voltage depends on back-bias in a more complicated manner than appears from (7.3 a). This has been shown in Fig. 6.8. Since the I_D characteristic is more dependent on the integral of the substrate doping than on the actual doping distribution, for modelling purposes [7.17, 7.18, 7.19] a box-type profile is assumed with an average doping N_i and thickness d_i. Introducing similar to 6.3.2 the back-bias value (6.45),

$$V_{SX} = \frac{qN_i d_i^2}{2\varepsilon_s} - \psi_{sm}, \tag{7.9}$$

at which the implanted layer is fully depleted at the onset of channel conduction, three different regions of operation have to be distinguished. When $V_S < V_D < V_{SX}$, the depletion layer remains within the implanted region under the whole channel. In this case the model equations (7.3) remain valid, provided that γ is replaced by $\gamma_i = (2\varepsilon_s q N_i)^{1/2}/C_{ox}$.

When $V_{SX} < V_S < V_D$, the depletion boundary has moved into the lightly doped substrate. Now the depletion charge is no longer given by the last part of Eq. (7.3 a) and therefore has to be recalculated.

When $V_S < V_{SX} < V_D$, the most complicated situation is found. Near the source side the depletion region boundary is located in the implanted region, but at the drain side it is located deeper in the substrate (compare Fig. 7.3). In section 6.3.2 the threshold voltage relations for the various modes have already been given in terms of the physical quantities N_i, d_i and the substrate doping N_B (compare Eqs. (6.40) through (6.46)). Since the latter quantities

7.1 Long-Channel Models

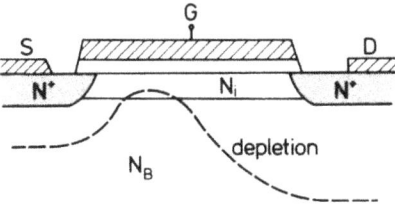

Fig. 7.3 Possible boundary (broken line) of depletion region in a MOSFET with a threshold voltage implant

are not suitable as electrical parameters of the transistor characteristics, we prefer to express them as a function of the related body-effect parameters γ_i and γ, which can be measured directly. The desired relations can be obtained from the condition that the threshold voltage relations (6.40) and (6.46) and their derivatives have to be continuous at the boundary V_{SX}. It is easily verified that the threshold voltage shift can be expressed as

$$\Delta V_T = \frac{q(N_i - N_B)d_i}{C_{ox}} = \left(1 - \frac{\gamma^2}{\gamma_i^2}\right)\gamma_i(V_{SX} + \psi_{sm})^{1/2} \tag{7.10}$$

and the correction voltage term as

$$\Delta \phi = \frac{q(N_i - N_B)d_i^2}{2\varepsilon_s} = \left(1 - \frac{\gamma^2}{\gamma_i^2}\right)(V_{SX} + \psi_{sm}). \tag{7.11}$$

Taking for simplicity the mobility and channel length modulation effect to be implicitly present in the factor β the drain current is now described as follows:

— If $V_S < V_D < V_{SX}$, the threshold voltage reads

$$V_{T1} = V_{T0} + V_S + \gamma_i[(V_S + \psi_{sm})^{1/2} - \psi_{sm}^{1/2}]. \tag{7.12 a}$$

If $V_G > V_{T1}$, the saturation voltage is given by

$$V_{DSAT,1} = V_G - V_{T0} + \gamma_i \psi_{sm}^{1/2}$$
$$+ \frac{\gamma_i^2}{2}\left[1 - \left\{1 + \frac{4}{\gamma_i^2}(V_G - V_{T0} + \psi_{sm} + \gamma_i \psi_{sm}^{1/2})\right\}^{1/2}\right] \tag{7.12 b}$$

and the current is given by

$$I_{DS,1} = \beta[g_1(V_G, V_R) - g_1(V_G, V_S)], \tag{7.12 c}$$

where

$$g_1(V_G, V) = [(V_G - V_{T0} + \gamma_i \psi_{sm}^{1/2})V - \tfrac{1}{2}V^2 - \tfrac{2}{3}\gamma_i(V + \psi_{sm})^{3/2}]. \tag{7.12 d}$$

If $V_D < V_{DSAT,1}$, $V_R = V_D$, but if $V_D \geq V_{DSAT,1}$, $V_R = V_{DSAT,1}$.

— If $V_{SX} \leqslant V_S < V_D$, the threshold voltage is given by (compare (6.46))

$$V_{T2} = V_{T0} + V_S - \gamma_i \psi_{sm}^{1/2} + \Delta V_T + \gamma(V_S + \psi_{sm} - \Delta\phi)^{1/2}. \quad (7.13\text{ a})$$

Since in this case the inversion layer charge is given by

$$Q_{ni}(V) = -C_{0x}[V_G - V_{FB} - \psi_{sm} - V] - Q_B, \quad (7.14)$$

in which the depletion charge is given by (6.45),

$$Q_B = -qN_i d_i - qN_B(y_d - d_i), \quad (7.15\text{ a})$$

and the depletion layer thickness has a value

$$y_d(V) = \frac{\gamma C_{0x}}{qN_B}(V + \psi_{sm} - \Delta\phi)^{1/2}, \quad (7.15\text{ b})$$

the saturation voltage is calculated from the condition $Q_{ni}(V_{DSAT,2}) = 0$. Therefore if $V_G > V_{T2}$,

$$V_{DSAT,2} = V_G - V_{T0} + \gamma_i \psi_{sm}^{1/2} - \Delta V_T$$

$$+ \frac{\gamma^2}{2}\left[1 - \left\{1 + \frac{4}{\gamma^2}(V_G - V_{T0} + \psi_{sm}\right.\right.$$

$$\left.\left. + \gamma_i \psi_{sm}^{1/2} - \Delta V_T - \Delta\phi)\right\}^{1/2}\right]. \quad (7.13\text{ b})$$

Similar to previous cases the drain current is obtained by integrating Eq. (7.14) with respect to V. This procedure results into

$$I_{DS,2} = \beta[g_2(V_G, V_R) - g_2(V_G, V_S)], \quad (7.13\text{ c})$$

where

$$g_2(V_G, V) = [(V_G - V_{T0} + \gamma_i \psi_{sm}^{1/2} - \Delta V_T)V - \tfrac{1}{2}V^2$$

$$- \tfrac{2}{3}\gamma(V + \psi_{sm} - \Delta\phi)^{3/2}]. \quad (7.13\text{ d})$$

To Eq. (7.13 c) the additional condition applies: If $V_D < V_{DSAT,2}$, $V_R = V_D$ etc.

— If $V_S < V_{SX} \leqslant V_D$, the threshold voltage equals the corresponding value of region 1,

$$V_{T3} = V_{T1}. \quad (7.16\text{ a})$$

However, the saturation voltage equals the corresponding value of region 2,

$$V_{DSAT,3} = V_{DSAT,2}. \quad (7.16\text{ b})$$

Finally the current is expressed by

$$I_{DS,3} = I_{DS,1}(V_G, V_{SX}, V_S) + I_{DS,2}(V_G, V_R, V_{SX}). \quad (7.16\text{ c})$$

Compared to the previous models two additional parameters V_{SX} and γ_i are required. Furthermore the model reduces to the previous models by defining $\gamma_i = \gamma$. In Fig. 7.4 the measured characteristics in region 3 are compared with

7.1 Long-Channel Models

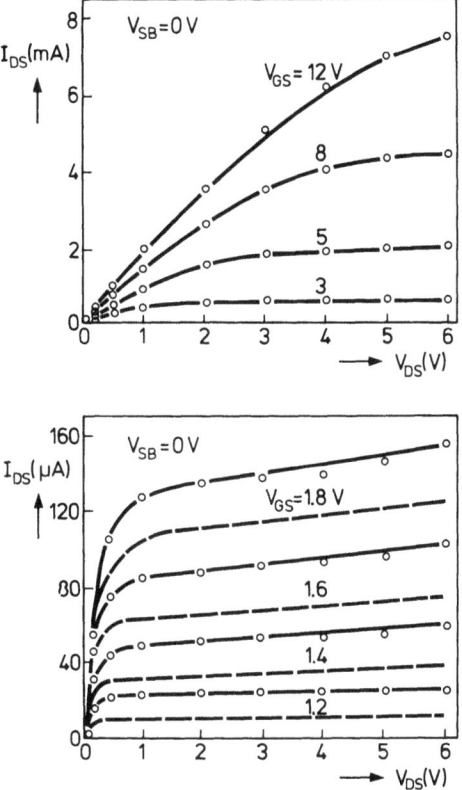

Fig. 7.4 Calculated (fully-drawn lines) and measured characteristics of MOSFET operating according to the situation indicated in Fig. 7.3. Broken lines represent values calculated according to classical model

calculated values [7.20]. Owing to errors in the $V_T - V_{SB}$ relation, the classical model shows marked deviations at gate driving voltages close to the zero back-bias threshold voltage.

7.1.3 Charges and Capacitances

In the cut-off mode ($V_G < V_{FB}$) only an accumulation charge is present, which is given by

$$Q_B = -WLC_{ox}(V_G - V_{FB}). \tag{7.17}$$

In the subthreshold mode ($V_{FB} \leqslant V_G \leqslant V_T$) the bulk charge is made up of a gate-voltage-dependent depletion charge, whose value has already been given in (6.24). Since, however, this charge and its derivative have to be continuous at the two boundaries (note that Q_{depl} becomes independent of V_G above threshold), a modification of (6.24) is required. This is accomplished by defining

$$Q_B = -\gamma WLC_{ox}\left[-\frac{\gamma}{2} + \left\{\frac{\gamma^2}{4} + H(V_G)\right\}^{1/2}\right], \qquad (7.18\ a)$$

where

$$H(V_G) = \tfrac{1}{2}[V_G - V_{FB} + \{(V_T - V_{FB})^2 + \varepsilon^2\}^{1/2}$$
$$- \{(V_G - V_T)^2 + \varepsilon^2\}^{1/2}]. \qquad (7.18\ b)$$

In the above expression ε is a modelling constant, whose value can be estimated as 0.05 Volt.

In addition to the depletion charge $Q_B(V_G = V_T)$, in the active mode the mobile carrier charge dominates. Using the definition for the total charge in the inversion layer (6.92)

$$Q_I = -W \int_0^L qn(x)\,dx,$$

the above integral can be calculated for the strong-inversion condition by making use of the current expression

$$I_D = -Wqn\mu\frac{dV}{dx}.$$

Combining both equations we can rewrite (6.92)

$$Q_I = WL\frac{\int_{V_S}^{V_D} q_i^2\,dV}{\int_{V_S}^{V_D} q_i\,dV}, \qquad (7.19)$$

in which according to (6.33) $q_i = qn(V)$ is given by

$$q_i(V) = -C_{ox}[V_G - V_{FB} - \psi_{sm} - V - \gamma(\psi_{sm} + V)^{1/2}]. \qquad (7.20)$$

Since the above integral results in an expression with a large number of higher-order terms in $(V_D + \psi_{sm})^{1/2}$, and in very lengthy, impractical capacitance expressions [7.21], it has been proposed to approximate the contribution of the depletion charge in (7.20) by the crude simplification

$$\gamma(\psi_{sm} + V)^{1/2} \approx \gamma(\psi_{sm} + V_S)^{1/2}.$$

It can be shown [7.22] that in this case (7.19) yields

$$Q_I = -\frac{2}{3}WLC_{ox}\left(V_1 + V_2 - \frac{V_1 V_2}{V_1 + V_2}\right), \qquad (7.21)$$

in which

$$V_1 = V_G - V_S - V_T \qquad (7.22\ a)$$

and

$$V_2 = V_G - V_D - V_T. \qquad (7.22\ b)$$

Later on it has been shown [7.8] that the effect of the depletion charge is much better represented by Eq. (7.21) if the above definition of V_2 is changed

7.1 Long-Channel Models

into
$$V_2 = V_G - V_D - V_{TR}, \tag{7.23}$$

in which V_{TR} has already been defined in (7.3 f). However, the required split of Q_i into a source and drain part Q_S and Q_D, respectively, (compare section 6.6.2) remains a rather precarious compromise, since not all charge derivatives correspond to physically realistic capacitances. This holds for many alternative approaches too [7.21, 7.22]. However, due to the fact that parasitics contribute equally well to circuit delays in practical MOSFET designs, small errors in Q_S and Q_D are often not harmful.

In fact the best compromise between accuracy and efficiency is obtained by approximating the depletion charge by writing the most right-hand term of Eq. (7.20) in terms of a power series [7.9] in $(V - V_S)$,

$$\gamma(\psi_{sm} + V)^{1/2} \approx \gamma(\psi_{sm} + V_S)^{1/2} + \delta(V - V_S), \tag{7.24}$$

in which
$$\delta \approx \frac{0.3\gamma}{(V_S + \psi_{sm})^{1/2}}. \tag{7.25}$$

In order to avoid an overestimation of the depletion charge, the factor 0.3 in the above equation is chosen instead of 0.5 [7.8]. Alternative approximations with higher accuracy in some subregions have been given in [7.23, 7.24, 7.25].

Using the approximation (7.24), Eq. (7.20) can be rewritten as
$$q_i(V) = -C_{ox}[V_G - V_T - (1 + \delta)(V - V_S)]. \tag{7.26}$$

Substituting the above expression into (7.19) we obtain
$$Q_I = -WLC_{ox}\left[V_{GT} - \frac{1}{3}\frac{3V_{GT} - 2V'_{DS}}{2V_{GT} - V'_{DS}}V'_{DS}\right], \tag{7.27}$$

in which
$$V_{GT} = V_G - V_T = V_{GS} - V_{TS} \tag{7.28}$$

and
$$V'_{DS} = (1 + \delta)V_{DS}. \tag{7.29}$$

Note that for reasons of convenience next the source voltage is taken as a reference in the calculation of terminal charges.

Furthermore in the active region ($V_{GS} \geq V_{TS}$) the depletion charge is made up of a drain-voltage-independent part, which is according to (7.18) given by

$$Q_{B1}(V_{TS})$$

and a drain-voltage-dependent part

$$Q_{B2} = -WC_{ox}\int_0^L \delta(V - V_S)\,dx.$$

Elaborating the latter integral we finally obtain for the total depletion charge

$$Q_B = -\frac{1}{3}\delta WLC_{ox} V_R \frac{3V_{GT} - 2V'_R}{2V_{GT} - V'_R}$$

$$-\gamma WLC_{ox}\left[-\frac{\gamma}{2} + \left\{\frac{\gamma^2}{4} + H(V_{TS})\right\}^{1/2}\right]. \qquad (7.30)$$

Consequently the gate charge is now determined as well,

$$Q_G = -(Q_I + Q_B). \qquad (7.31)$$

Next we have to calculate the source and drain charges. According to (6.99) the drain charge is given by

$$Q_D = -qW \int_0^L n(x)\frac{x}{L}\,dx. \qquad (7.32)$$

In the above relation x can be eliminated by making use of the current expression

$$I_D x = -Wq \int_0^V \mu n(V)\,dV.$$

Substituting the latter result in (7.32) we obtain

$$Q_D = -WLC_{ox}[\tfrac{1}{2}V_{GT} - \tfrac{1}{3}V'_R - F_1 F_2], \qquad (7.33)$$

where

$$F_1 = \frac{(V'_R)^2}{6(2V_{GT} - V'_R)} \qquad (7.34\text{ a})$$

and

$$F_2 = -\frac{5V_{GT} - 2V'_R}{5(2V_{GT} - V'_R)}. \qquad (7.34\text{ b})$$

Similar to the previous procedure, Q_S can be calculated with the result

$$Q_S = -WLC_{ox}[\tfrac{1}{2}V_{GT} - \tfrac{1}{6}V'_R + F_1(1 + F_2)]. \qquad (7.35)$$

Note that according to the approximation (7.24) the saturation voltage has changed. Instead of (7.3 b) we have to use

$$V_{DSAT} = V_S + \frac{V_{GT}}{(1 + \delta)} \qquad (7.36)$$

To get insight into the behaviour of the MOSFET charges in Figs. 7.5 and 7.6 the drain and the bulk charge have been plotted as a function of the terminal voltages.

Since the $V_S = V_D$ line forms the boundary between the forward mode ($V_D > V_S$) and the reverse mode ($V_D < V_S$), actually the left-hand part of Fig. 7.5 shows the Q_S expression. At low values of V_S and V_D the MOSFET operates in the linear mode. At $V_S = V_D = 0$ Volt the maximum value

7.1 Long-Channel Models

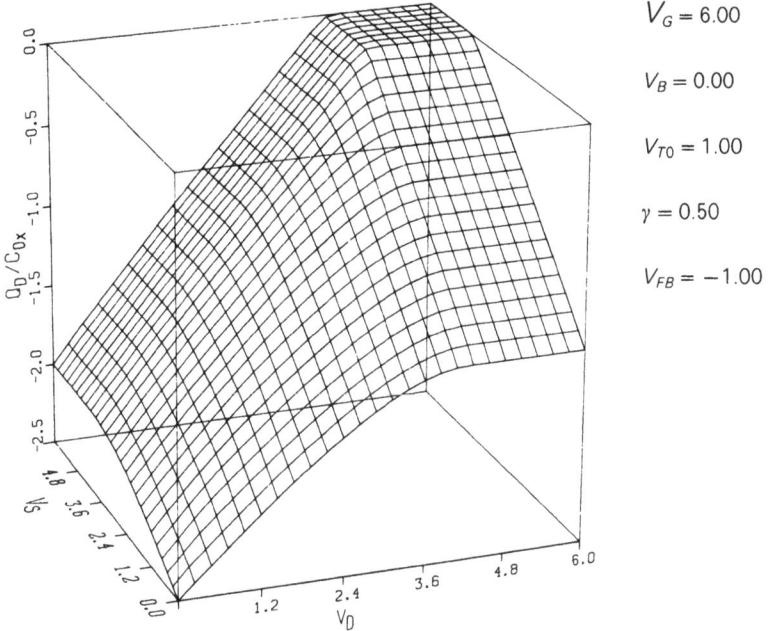

Fig. 7.5 Drain charge of n-channel MOSFET as a function of the source-bulk and drain-bulk voltage at a fixed gate bias of 6 Volts with $V_{TO} = 1.0\,\text{V}$, $\gamma = 0.50\,\text{V}^{1/2}$ and $V_{FB} = -1.0\,\text{V}$

$$Q_D = -\tfrac{1}{2}WLC_{ox}(V_G - V_T)$$

is reached. At low values of V_S and high values of V_D the transistor operates in the forward saturation mode. In this case

$$Q_{DSAT,F} = -\tfrac{4}{15}WLC_{ox}(V_G - V_T).$$

On the other hand, in the reverse mode at large values of V_S the value is slightly changed,

$$Q_{DSAT,R} = -\tfrac{2}{5}WLC_{ox}(V_G - V_T).$$

When both V_S and V_D are large, the device operates in the subthreshold mode and the active charges are virtually zero within the scale of this figure. Actually the charges considered decrease exponentially towards zero, but this fact can be neglected, since the charge behaviour is now dominated by the depletion charge. The latter charge is shown in Fig. 7.6.

In the active mode the change in Q_B is determined more by the absolute value of V_S or V_D (whichever is the larger) than by their difference. Therefore when V_S is increased the depletion layer is expanded and Q_B increases. In contrast to Q_D and Q_S at very high values of V_S (subthreshold mode) this trend is maintained. Note further the continuity of the Q_B hypersurface as a function of gate bias.

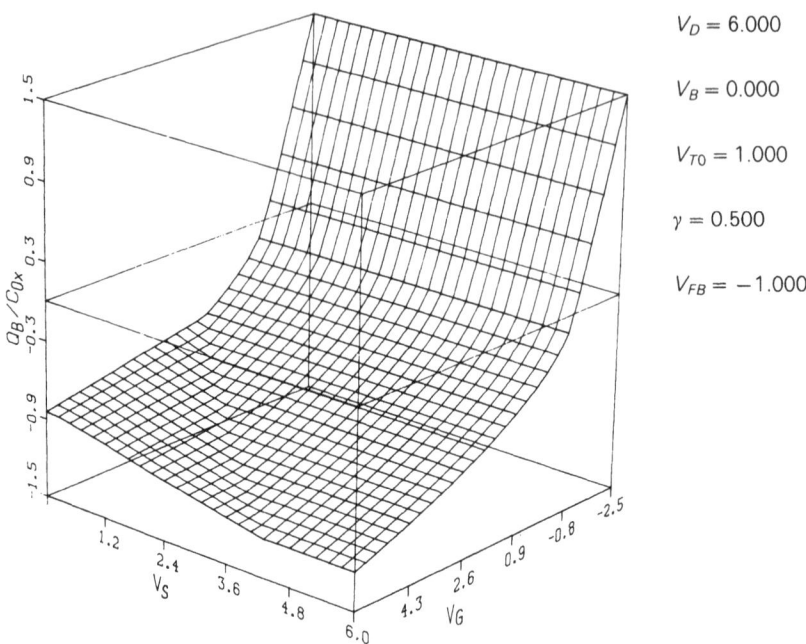

Fig. 7.6 Bulk charge of long-channel MOSFET as a function of gate-bulk and source-bulk voltage at a fixed-drain bias of 6 volts

Although the magnitude of the charges varies considerably in the above figures, it is clear from the small asymmetry shown in Fig. 7.5 that the distribution of the channel charge between both channel ends varies only slowly. In addition the change of Q_D and Q_S with V_G hardly deviates from a planar surface. This is shown in Fig. 7.7, where Q_D has been plotted as a function of V_G and V_S at a fixed value of V_D.

If V_G varies, only the value of dQ_D/dV_G is changed when the device turns from the subthreshold mode via saturation into the linear mode. At large values of V_S the transistor operates in the reverse mode and therefore the right-hand part of Fig. 7.7 shows the behaviour of the Q_S expression.

Since it does not make sense to discuss all possible capacitances [7.26] here, we only give the major ones. From the definition

$$C_{ij} = -\frac{\partial Q_i}{\partial V_j}\bigg|_{i \neq j},$$

successively we obtain

$$C_{GD} = WLC_{0x}\left[\frac{1}{2} - \frac{4V_{GT}V'_R - V'^2_R}{6(2V_{GT} - V'_R)^2}\right], \tag{7.37}$$

$$C_{DG} = WLC_{0x}\left[\frac{1}{2} - \frac{(10V_{GT} - 3V'_R)V'^2_R}{30(2V_{GT} - V'_R)^3}\right], \tag{7.38}$$

7.1 Long-Channel Models

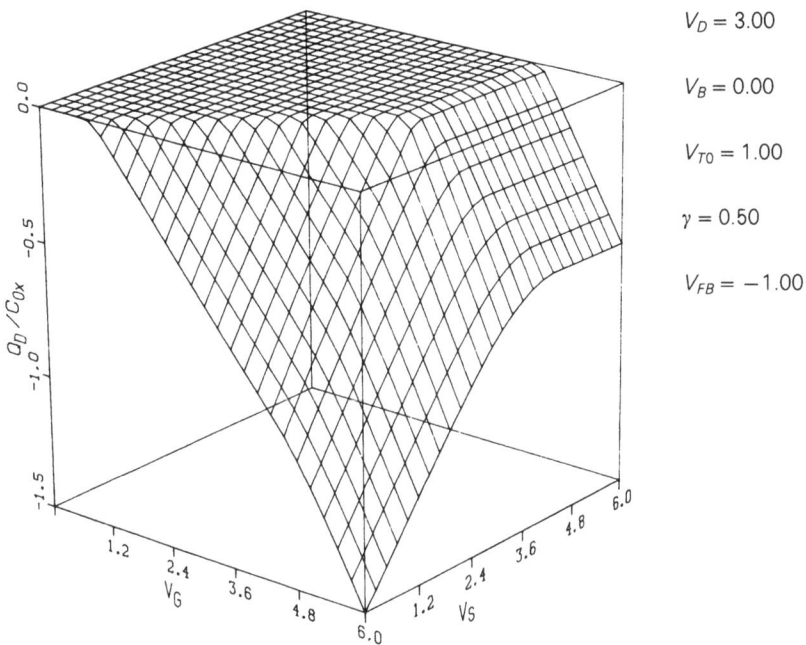

Fig. 7.7 Drain charge of long-channel MOSFET as a function of gate-bulk and source-bulk voltage at a fixed drain bias of 3 Volts

$$C_{SG} = WLC_{0x}\left[\frac{1}{2} - \frac{(10V_{GT} - 7V_R')V_R'^2}{30(2V_{GT} - V_R')^3}\right], \quad (7.39)$$

$$C_{GS} = WLC_{0x}\left[\frac{1}{2} + V_R' \frac{4V_{GT} - 3V_R' + \dfrac{\delta V_{GT} V_R'}{(1+\delta)^2(V_S + \psi_{sm})}}{6(2V_{GT} - V_R')^2}\right]. \quad (7.40)$$

In order to demonstrate their behaviour and the violation of reciprocity, the above capacitance expressions have been plotted in Fig. 7.8 as a function of the ratio $(1 + \delta)V_{DS}/(V_{GS} - V_{TS})$.

In saturation the deviations from reciprocity are particularly striking. The latter result is not contradictory to physics. For instance, when the drain voltage is varied in saturation, the channel is in the pinch-off condition and no additional charge in the channel or gate is required to maintain the drain current. Consequently $C_{GD} = 0$. However, when V_G is varied in saturation, the drain current changes and, since the channel is not completely cut off from the drain, a displacement charge flows into the drain terminal to supply the required charge. Therefore $C_{DG} \neq 0$.

Obviously, the most lengthy expression is found for the capacitance C_{GS}. This is also caused by the fact that Q_G depends indirectly on the variable V_S via V_{TS} and the δ-expression (7.23).

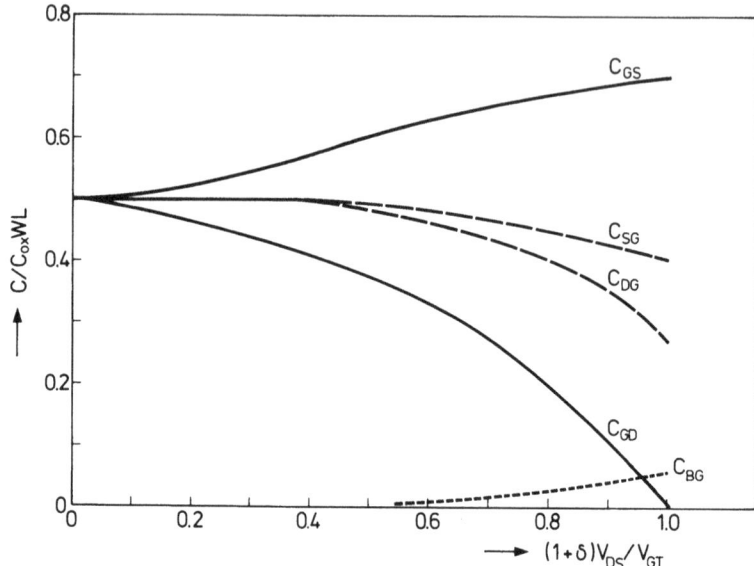

Fig. 7.8 A number of long-channel MOSFET capacitances as a function of normalized drain-source bias

Since the latter is not equal to the derivative of V_{TS}, actually the most correct expression of C_{GS} is more complicated than (7.40). However, it can be shown that (7.40), which has been obtained by taking in the charge expressions

$$\delta = \frac{\partial V_{TS}}{\partial V_S} = \frac{0.5\gamma}{(V_S + \psi_{sm})^{1/2}}, \tag{7.41}$$

leads almost to the same result. In addition, without any loss in accuracy the derivative of δ may be omitted. However, this simplification has not been carried out here. Finally for comparison one of the smaller capacitances is shown in Fig. 7.8 too. We have chosen the capacitance

$$C_{BG} = WLC_{ox}\delta \frac{V_R V_R'}{3(2V_{GT} - V_R')^2}. \tag{7.42}$$

Although the depletion charge is overestimated in the above approach, in particular when (7.41) is used, generally the capacitance values only differ slightly from exact numerical results, based on the charge sheet approximation [7.27]. The deviations are larger (up to 50%) only for the smaller capacitances like C_{GB}, and in contrast to the present approach $C_{GB} < C_{BG}$ [7.27]. However, in practice these errors will not affect transient circuit simulations.

Generally speaking, in the subthreshold region transient behaviour is dominated by the parasitic capacitances discussed in section 6.6.4 and by the intrinsic bulk capacitance

7.1 Long-Channel Models

$$C_{GB} = C_{BG} = WLC_{Ox}\frac{\gamma}{2(\frac{1}{4}\gamma^2 + V_G - V_{FB})^{1/2}}. \tag{7.43}$$

The latter result is easily derived from (7.29).
In order to evaluate the charge model two additional parameters are required:

— the unit area gate oxide capacitance C_{Ox} and
— the flat band voltage V_{FB}.

Strictly, for a complete description the gate-junction overlap capacitances (C_{GSO} and C_{GDO}) have to be added to these parameters.
Although for enhancement devices with a threshold implant, the V_T-model (7.12 a, 7.13 a) applies to the charge modelling as well, it is not practical to modify the charge expressions in this way. Generally this will lead to very complicated expressions, which are very weak functions of the difference between the body coefficients γ_i and γ. Therefore the previous expressions apply rather well in this case if an average value of the body coefficients is taken into account.

7.1.4 Effect of Velocity Saturation on the Drain Current

When the channel length becomes shorter, especially in n-channel devices a field-dependent mobility relation has to be applied. When this effect is not too large (which is the case if $L > 2$ µm) a relation like (6.81) can be used. Substituting the latter expression into the basic equation (6.30) it is easily derived that the current expression (7.3 d) becomes

$$I_{DS} = \frac{\beta}{2(1 - \Delta L/L)} \frac{g(V_G, V_S) - g(V_G, V_R)}{1 + \theta_A(V_G - V_T) + \theta_B V_S + \theta_C V_R}, \tag{7.44}$$

where $\theta_c = (LE_c)^{-1}$ represents the velocity saturation effect.
Naturally, the presence of the additional term in the denominator of (7.44) will affect the value of the saturation voltage. Therefore (7.3 b) is no longer valid. As a first-order approach to obtaining the correct V_{DSAT}, it can be assumed that the saturated drain current is given by the product of the saturation velocity (v_s) and the free carrier density at the pinch-off point [7.8],

$$I_{DSS} = WC_{Ox}v_s[V_G - V_{TO} + \gamma\psi_{sm}^{1/2} - V - \gamma(\psi_{sm} + V)^{1/2}]. \tag{7.45}$$

Equating this value with the value (7.44) at the saturation point yields a quartic equation [7.8] for the saturation voltage, which can be evaluated with some effort. However, if the effect of velocity saturation on V_{DSAT} is only small, it is more practical to apply a direct approximate solution from the condition that the derivative of (7.44) should approach zero at $V_D = V_{DSAT}^*$,

$$\left.\frac{dI_{DS}}{dV_D}\right|_{V_{DSAT}^*} = 0. \tag{7.46}$$

Defining $\Delta V_{SAT} = V_{DSAT} - V^*_{DSAT}$, where the original V_{DSAT} satisfies the condition (compare (6.33))

$$g'(V_{DSAT}) = \left.\frac{dg}{dV_D}\right|_{V_{DSAT}} = 0,$$

the condition (7.46) leads to

$$\frac{\theta_C[g(V^*_{DSAT}) - g(V_S)]}{1 + \theta_A V_{GT} + \theta_B V_S + \theta_C(V^*_{DSAT} - V_S)} + g'(V^*_{DSAT}) = 0.$$

Since ΔV_{SAT} may be considered as only a correction to V_{DSAT}, the g-function in the above equation can be expanded and we obtain [7.8]

$$\Delta V_{SAT} \approx \frac{\theta_C[g(V_{DSAT}) - g(V_S)]}{[1 + \theta_A V_{GT} + \theta_B V_S + \theta_C(V_{DSAT} - V_S)]g''(V_G)}, \qquad (7.47\text{ a})$$

where

$$g''(V_G) = -2 - \gamma\left[-\frac{\gamma}{2} + \left\{\frac{\gamma^2}{4} + V_G - V_{TO} + \gamma\psi_{sm}^{1/2} + \psi_{sm}\right\}^{1/2}\right]^{-1}.$$

(7.47 b)

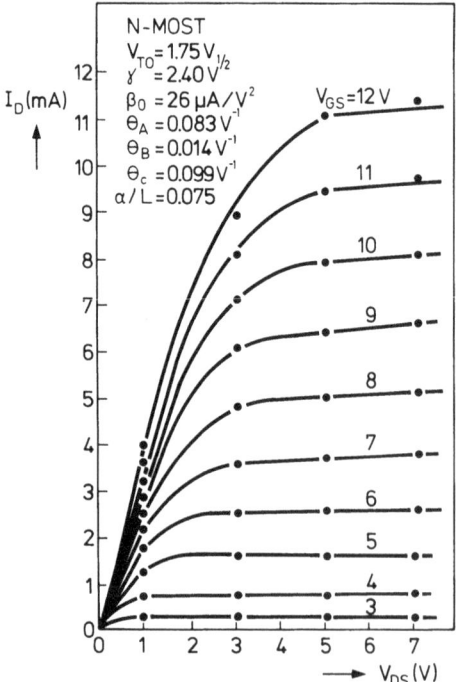

Fig. 7.9 Calculated (fully-drawn lines) and measured characteristics of n-channel MOSFET ($W/L = 100$ µm/3 µm)

7.2 Small Transistor Models

Now the corrected saturation voltage V_{DSAT}^* is given by Eqs. (7.3 b) and (7.47).

Fig. 7.9 gives a comparison of the measured characteristics of an *n*-channel MOSFET (with $L_{eff} = 3.5$ μm) and the I_D-values calculated according to Eqs. (7.44), (7.3 b) and (7.47). Note that at larger values of V_{GS} the saturated current values become equidistant. This is a direct consequence of the velocity saturation effect.

Although the above comparison shows fair agreement, it is clear that for shorter devices several shortcomings appear, which are caused by a larger change of V_{DSAT}, misfits from the ΔL-modelling and other short-channel effects (compare section 6.3.5). Therefore this section forms only a transition to the next section, in which short-channel effects are discussed in depth.

7.2 Small Transistor Models

Taking into account a number of short-channel and narrow-width effects and the effect of additional process steps required for producing smaller devices, we shall first discuss the modelling of the drain current in small MOSFETS. In order to avoid an impractical solution, several useful simplifications are introduced. Owing to this and to a few other shortcomings the resulting model is more suitable for digital than analog applications. This discussion will be followed by the derivation of a charge model compatible with the drain current expressions. In this case too, several simplifications are made. In submicron devices the source and drain series resistances have a major effect on the characteristics, and therefore it is shown how these quantities can be taken into account in the intrinsic model equations. Owing to the higher electric fields occurring in short-channel devices, the associated hot carriers generate a substrate current, which in turn is a first-order indication of detrimental effects. Finally, therefore a model for the substrate current is given.

7.2.1 The Drain Current in Small MOSFETS

7.2.1.1 The Threshold Voltage

Since the threshold voltage is a key parameter, the small-size effects on this quantity will first be taken into account. In section 6.3.5 the threshold voltage lowering in short-channel devices has been discussed in physical terms as, for instance, charge sharing between gate and junctions (see Eq. (6.64)) or a reduction of the effective substrate doping by field lines originating from the junctions according to (see Eq. (6.69))

$$N_{eff} = N_B - \frac{2\varepsilon_s V_{DS}^*}{qL^2}, \tag{7.48}$$

where V_{DS}^*, as is shown next, differs slightly from V_{DS}. Since the resulting threshold voltage expressions (6.70) and (6.71) are given in physical quantities, we have to transform them into a form more suitable for parameter extraction. This can be achieved by expanding the square-root terms in N_{eff}, since in the case of well-scaled devices the last term of (7.48) can be considered as a higher order correction. Considering that [7.29]

$$V_{DS}^* \approx V_{DS} + \psi_{sm},$$

we obtain from (6.70)

$$V_T = V_{T0}^\infty - \gamma_i \psi_{sm}^{1/2} + (\gamma_i^\infty - \Delta\gamma_{iL} - sV_{DS})(\psi_{sm} + V_S)^{1/2},$$

where V_{T0}^∞ is the zero-bias threshold voltage of a long-channel device, γ_i^∞ is the body-effect coefficient in long channels, $\Delta\gamma_{iL}$ is the change of the latter coefficient at zero drain bias when an actual channel length L is considered, and s is the threshold voltage shift with an increase of drain bias.

The above expression can be extended for small devices by taking into account the narrow-width effect. Defining according to (6.74) a threshold voltage change

$$\Delta V_T = \Delta\gamma_{iW}(\psi_{sm} + V_S)^{1/2},$$

we finally obtain for the threshold voltage

$$V_T = V_{TS} - s(L)(\psi_{sm} + V_S)^{1/2} V_{DS}, \tag{7.49}$$

where

$$V_{TS} = V_{T0} + \gamma_i\{(\psi_{sm} + V_S)^{1/2} - \psi_{sm}^{1/2}\} \tag{7.3 a}$$

and

$$\gamma_i = \gamma_i^\infty - \Delta\gamma_{iL} + \Delta\gamma_{iW}. \tag{7.50}$$

For well-scaled processes, in which 2-D threshold lowering effects have been minimized, one usually finds according to (6.64) and (6.74)

$$\Delta\gamma_{iL} \approx \frac{\gamma_{iL}}{L} \tag{7.51 a}$$

and

$$\Delta\gamma_{iW} \approx \frac{\gamma_{iW}}{W}. \tag{7.51 b}$$

Furthermore according to (7.48)

$$s \approx \frac{s_L}{L^2}. \tag{7.51 c}$$

It is shown in chapter 11 that the above relations are confirmed experimentally [7.28–7.31].

In the case of a substrate with threshold-voltage adjustment implant the more complicated relation (6.71) has to be elaborated. Unfortunately for a

7.2 Small Transistor Models

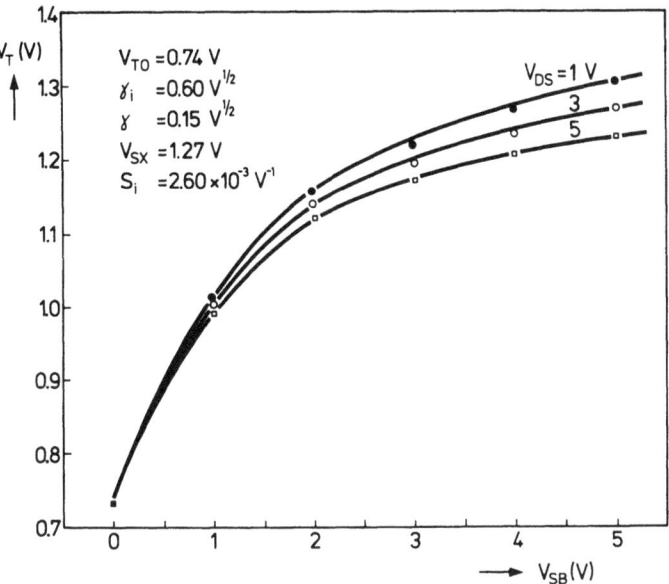

Fig. 7.10 Threshold voltage of short-channel MOSFET as a function of source-bulk and drain-source bias. Fully drawn lines are modelled results; dots are measured values

process on a lightly doped substrate, in which the values of γ_i and γ may differ considerably, it is observed that the considered relation cannot satisfactorily describe the threshold voltage shift at increased drain and substrate back-bias. Owing to additional 2-D effects occurring deeper in the substrate a rather complicated V_T behaviour is observed. An example is shown in Fig. 7.10, where V_T has been plotted as a function of V_{SB} at several values of V_{DS}.

In order to obtain a more accurate description of the subthreshold characteristics for the drain-voltage-dependent shift term an empirical relation has to be chosen. Usually a good fit is observed when the threshold voltage for $V_S > V_{SX}$ is defined,

$$V_T = V_{TS} - s_i(\psi_{sm} + V_S)V_{DS}, \tag{7.52}$$

where

$$V_{TS} = V_{TO} - \gamma_i \psi_{sm}^{1/2} + \left(1 - \frac{\gamma^2}{\gamma_i^2}\right)\gamma_i(V_{SX} + \psi_{sm})^{1/2}$$

$$+ \gamma\left[(V_S + \psi_{sm}) - \left(1 - \frac{\gamma^2}{\gamma_i^2}\right)(V_{SX} + \psi_{sm})\right]^{1/2} \tag{7.53}$$

and

$$\gamma = \gamma^\infty - \frac{\gamma_L}{L} + \frac{\gamma_W}{W}. \tag{7.54}$$

The most right-hand term of (7.52) is an empirical relation. Fig. 7.10 shows a good fit with experimental values of a 2-μm n-channel LDD-type MOSFET.

7.2.1.2 The Substrate Effect

In order to represent the effects of normal and lateral field on the mobility in short-channel devices, an expression like those given in (6.78) and (6.79) has to be used:

$$\mu_s = \frac{\mu_{s0}}{[1 + \theta_A(V_{GS} - V_{TS}) + \theta_B\{(V_{SB} + \psi_{sm})^{1/2} - \psi_{sm}^{1/2}\}](1 + \theta_C LE_x)}. \quad (7.55)$$

In section 7.1.4 we have already seen that the substitution of (7.55) in drain current expressions like (7.3) or (7.12) yields a quartic or even higher-order equation, from which the saturation voltage has to be solved. For a short-channel model the same approach is therefore chosen as in the modelling of the inversion layer charge. Taking the approximation (7.22) for the depletion charge distribution across the channel, it is easily shown that the drain current in the linear mode of operation can be written as

$$I_{DS} = \beta \frac{[V_{GT} - \frac{1}{2}(1 + \delta)V_{DS}]V_{DS}}{[1 + \theta_A(V_{GS} - V_{TS}) + \theta_B\{(V_{SB} + \psi_{sm})^{1/2} - \psi_{sm}^{1/2}\}](1 + \theta_C V_{DS})}, \quad (7.56)$$

where the effective gate driving voltage is defined as

$$V_{GT} = V_{GS} - V_T \quad (7.57)$$

and where

$$\delta \approx \frac{0.3\gamma}{(V_{SB} + \psi_{sm})^{1/2}} \quad (7.23)$$

represents the effect of increased charge in the depletion layer when the drain bias is increased. Note that for practical reasons all voltages in this case refer to the source terminal.

In addition to the approximation (7.23) several alternatives have been proposed [7.23, 7.24, 7.25], from which we mention here only

$$\delta \approx \frac{0.5\gamma}{(1 + V_{SB} + \psi_{sm})^{1/2}} \quad (7.58)$$

and

$$\delta \approx \frac{e\gamma}{2(V_{SB} + \psi_{sm})^{1/2}},$$

7.2 Small Transistor Models

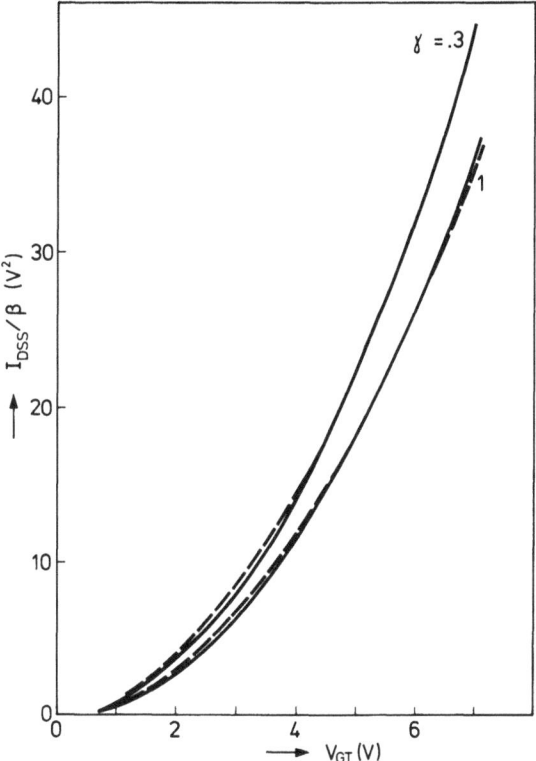

Fig. 7.11 Saturated drain current versus gate bias calculated according to original model (fully-drawn line) and using several approximations for the bulk effect (broken lines)

where

$$e = 1 - [1.7 + 0.85(V_{SB} + \psi_{sm})]^{-1}. \tag{7.59}$$

Since the effect of the above approximation on the current is most critical for long-channel devices at zero back-bias, in Fig. 7.11 a comparison is made between the saturated current I_{DSS} calculated according to (7.3 d) (fully drawn lines) and according to (7.56) (broken lines) using the above three δ expressions. The two γ values refer to extreme process conditions. Naturally in this comparison the long-channel drain saturation voltage is needed. Although the latter quantity is the subject of the next section, we can already refer to Eq. (7.36). From Fig. 7.11 it is observed that the difference between the various approximations is negligible. In addition the agreement with the classical approach is good. Only at medium gate drive do the δ-approximations produce an underestimation of the depletion layer charge. Since we have to extend the considered approximation to the case of implanted substrates, we prefer to continue here on the basis of the simplest expression (7.23). Other approaches using less simplifications lead to implicit relations for the drain current [7.36].

As in the discussion given in section 7.1.2 for the implanted substrate, we have to distinguish three different modes of operation.

- When $V_S < V_D < V_{SX}$, the previous approach applies, except that γ has to be replaced by the high value of the body coefficient γ_i.
- When $V_{SX} < V_S < V_D$, the expansion of the depletion layer occurs in the lower doped substrate, and therefore in this region (7.23) remains valid if the lower value of the body coefficient is chosen for γ.
- When $V_S < V_{SX} < V_D$, the most complicated situation occurs, since in this case the depletion layer boundary near the source end is located in the implanted layer, but near the drain end it lies deeper in the substrate. When the simple approximation (7.23) is chosen, taking the larger γ_i-value, the saturation voltage in this case is estimated from (7.36) too low at high values of gate bias. Therefore a modelling relation for δ is required, which decreases at larger values of gate drive too. In practice the following relation has been found to satisfy all required conditions

$$\delta = \frac{0.3\gamma}{(V_{SB} + \psi_{sm})^{1/2}} \left[1 + \frac{\left(\frac{\gamma_i}{\gamma} - 1\right)}{1 + \left\{\frac{V_{SB} + cV_{GT}}{V_{SX}}\right\}^2} \right], \quad (7.60)$$

where c is a modelling constant.

When $V_{SB} = 0$ and V_{GS} has a low value, δ is given by (7.23) taking the high value γ_i; when V_{SB} has a high value, again δ is given by (7.23) taking the low value γ. For all other values of V_{GS} and V_{SB}, δ decreases monotonically from the first mentioned to the second extreme value.

Fig. 7.12 gives a plot of the saturation voltage as a function of the gate driving voltage for the case of a process in which the two γ parameters differ considerably. The fully drawn line has been calculated according to Eqs. (7.12 b) and (7.15 b) and the broken lines according to Eqs. (7.36) and (7.60), taking c as a parameter. Although the result of Fig. 7.12 suggest choosing a value $c > 0.2$ for the above modelling constant, from the analysis of measured characteristics as a function of gate and back-bias rather a value $c = 0.1$ is found. Fig. 7.13 gives a comparison between the measured characteristics and the calculated current values (fully drawn lines) using the approximation (7.60) with $c = 0.10$. Some relevant parameter values are given in the figure. While this approach gives satisfactory agreement, this is not the case if only the simple formula (7.23) is used (broken line).

7.2.1.3 The Drain Saturation Voltage

For long-channel devices the saturation voltage can be calculated in good approximation from the zero value of the derivative of the current expression (in the linear mode) with respect to the drain voltage. For short-channel devices this procedure is less self-evident, since the saturated drain conduc-

7.2 Small Transistor Models

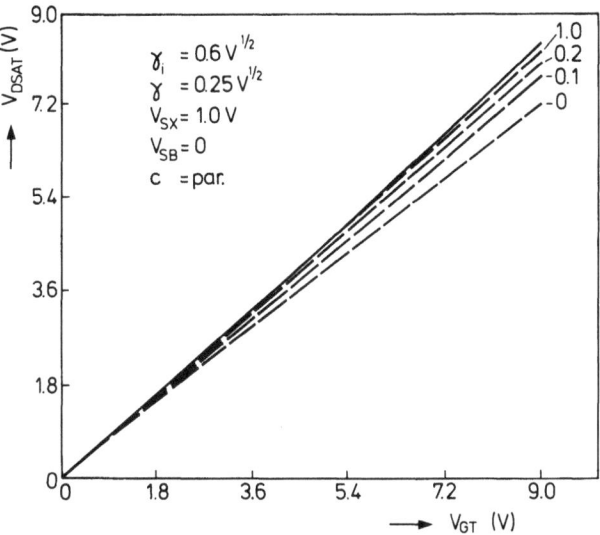

Fig. 7.12 Saturation voltage for implanted MOSFET as a function of gate driving voltage. Fully-drawn line represents values calculated to original, complicated model and broken lines represent values using an approximation for the bulk effect

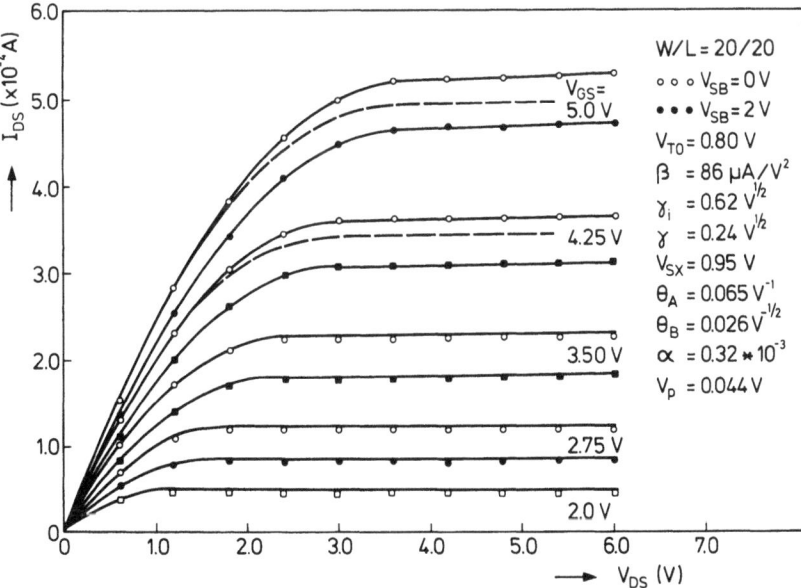

Fig. 7.13 Modelled (fully-drawn lines) and measured characteristics of long-channel, implanted MOSFET at $V_{SB} = 0$ and 2.0 Volts. Broken lines are calculated using a too simple approximation for the bulk effect

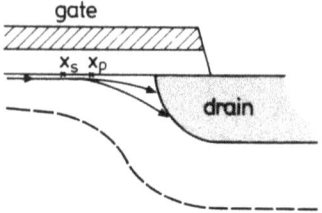

Fig. 7.14 Sketch of current path for MOSFET operating in the saturation mode

tance may have a considerable value. Usually the saturation voltage is calculated by defining the saturated drain current to be equal to the product of the free carrier concentration at the drain end and the saturated drift velocity [7.7, 7.9, 7.30, 7.32]. Fig. 7.14 gives an illustration of the situation. When a high drain voltage is applied, the normal field in the gate insulator is inverted near the drain end of the channel, and beyond a point x_p mobile carriers are pushed into the bulk and move towards the drain with a saturated drift velocity. In fact, x_p corresponds to the classical pinch-off situation. Numerical calculations [7.30] show that the carriers have already reached almost velocity saturation at a point $x_s < x_p$. Assuming that the normal field at x_s is still sufficiently high to guarantee the validity of the so-called gradual channel approximation [7.33], and defining the potential and the lateral field at x_s to be equal to V_{DSS} and E_{xs}, respectively, the saturated drain current at the right-hand side of x_s can be expressed by [7.7, 7.30]

$$I_{DSS} = \beta[V_{GT} - (1 + \delta)V_{DSS})]\frac{LE_{xs}}{1 + E_{xs}/E_c}. \tag{7.61}$$

Since I_{DSS} at the left-hand side of x_s is still given by Eq. (7.56), we obtain by equating (7.56) and (7.61) an expression for the drain saturation voltage,

$$V_{DSS} = V_c\left[\left\{1 + \frac{2xLE_cV_{GT}}{(2x-1)(1+\delta)V_c^2}\right\}^{1/2} - 1\right], \tag{7.62}$$

where

$$x = [1 + \theta_A V_{GT} + \theta_B\{(V_S + \psi_{sm})^{1/2} - \psi_{sm}^{1/2}\}]\frac{E_{xs}/E_c}{1 + E_{xs}/E_c}$$

and

$$V_c = \left[\frac{xLE_c}{2x-1} + \frac{(1-x)V_{GT}}{(1+\delta)(2x-1)}\right].$$

Since the field E_{xc} is not known exactly, it has to be considered as a parameter. However, in order to avoid an additional parameter, a specific

7.2 Small Transistor Models

value is usually chosen for the ratio E_{xs}/E_c. This value varies from $E_{xs}/E_c = 1$ [7.33], $E_{xs}/E_c = 2$ [7.32] to $E_{xs}/E_c > 4$ [7.30]. The first choice yields a very simple expression for the saturation voltage,

$$V_{DSS} \approx \frac{V_{GT}}{1 + \delta + \theta_c V_{GT}}, \tag{7.63}$$

where $\theta_c = (LE_c)^{-1}$.

However, owing to the assumption used, one should realize that in this case the carriers move through the saturation region at only one half of the theoretical saturated velocity. Note that in addition the term $(1 + \theta_A V_{GT} +$ etc.) has disappeared. In fact this manipulation is only allowed, if the value of θ_A is small. Usually, this is the case. However, in submicron devices, where this parameter can reach a high value owing to the effect of series resistances, the neglect of the above factor does introduce an error (see section 7.2.3).

If on the other hand $E_{xs} \gg E_c$, E_{xs} can be eliminated from (7.62) and the most frequently used expression

$$V_{DSS} = \frac{2V_{GT}}{1+\delta}\left[1 + \left\{1 + \frac{2\theta_c V_{GT}}{1+\delta}\right\}^{1/2}\right]^{-1} \tag{7.64}$$

is obtained. This result has the additional advantage that, for $V_{DS} = V_{DSS}$, the intrinsic drain conductance as derived from (7.56) becomes zero. This makes the further addition of static feedback and channel modulation effects to the current expressions less of a numerical problem. The foregoing argument also applies to the continuity of short-channel capacitance expressions.

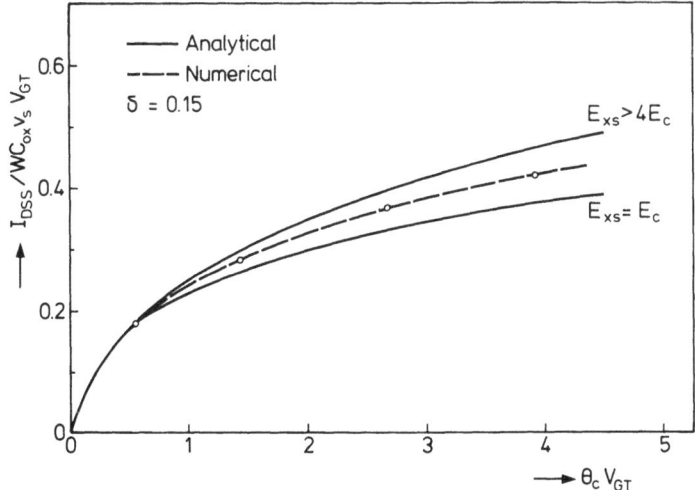

Fig. 7.15 Modelled value of saturated drain current as a function of normalized gate driving voltage at several values of short-channel parameter E_{xs}/E_c. The broken line represents values calculated numerically using a 2-D device simulator

In order to clarify the uncertainty in the ratio E_{xs}/E_c, we have plotted in Fig. 7.15 the saturated drain current as a function of the normalized gate driving voltage V_{GT}/LE_c with the ratio E_{xs}/E_c as a parameter. As expected the latter value only has an effect on the value of the drain current when $\theta_c V_{GT} > 1$. However, the observed differences are less of a problem in practice, when θ_c is used as a model parameter and usually $\theta_c V_{GT} < 2$. In this case the difference in I_{DSS} can virtually be eliminated by adjusting the θ_c-parameter value. In the same figure we have also added the result of a 2-D device simulator [7.34], in which the considered mobility expression (7.55) can be implemented. From the simulated values of the saturation current we conclude that $E_{xs}/E_c \approx 2$. Therefore having regard to all previous facts, we follow in this chapter the approximation leading to (7.64).

For drain voltages in excess of V_{DSS} the drain current is in the saturation mode and can increase only by a static feedback effect and a slight shift of x_s towards the source. This will be the subject of the next section.

7.2.1.4 Static Feedback and Channel Length Modulation

The static feedback effect was introduced in section 6.5.1. It was shown that this effect can be interpreted as an apparent increase of the effective gate driving voltage (compare Eq. 6.82)

$$V_{GT} = V_{GS} - V_{TS} + \eta V_{DS}.$$

Unfortunately it has been observed that the value of η in the strong-inversion region differs from the threshold voltage shift with applied drain bias, which is crucial for the subthreshold behaviour. Therefore, in order to achieve a smooth transition from the subthreshold model to the strong inversion model, it is necessary to decouple the η term from the s_i terms given in (7.49) and (7.52). This can be done by modifying the effective gate driving voltage, defined in (7.57),

$$V_{GT} = V_{GS} - V_{TS} + f V_{DS}, \tag{7.65}$$

where

$$f = \sigma + (\eta - \sigma)\frac{(V_{GS} - V_{TS})^2}{V_0^2 + (V_{GS} - V_{TS})^2}$$

and

$$\sigma = s(V_S + \psi_{sm})^{1/2} + s_i(V_S + \psi_{sm}).$$

In the above equation V_0 is an empirical voltage, whose value can be chosen as 0.25 Volt.

In addition to the above effect the saturated drain current may increase with V_{DS} by a slight shift of the saturation point x_s towards the source.

In section 6.5.2 two approximations for this channel length modulation effect have been given [7.33, 7.35]. However, when for instance according

7.2 Small Transistor Models

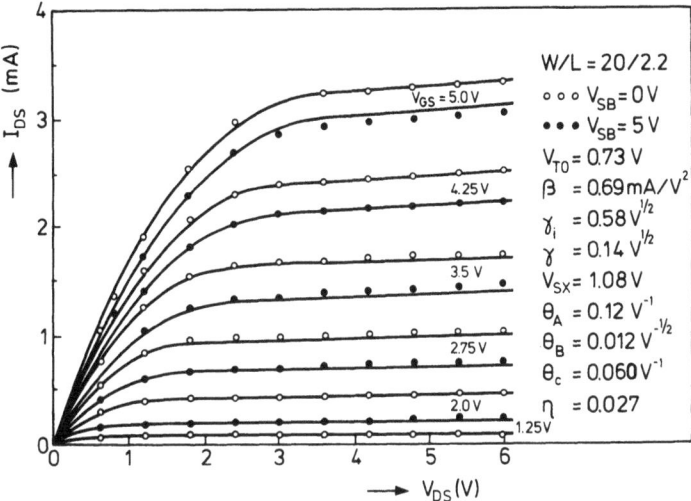

Fig. 7.16 Calculated (fully-drawn lines) and measured characteristics of short-channel MOSFET at several values of gate-source and source-bulk bias

to (6.87) the saturated drain current is multiplied by a term

$$1 + \alpha \ln\left(1 + \frac{V_{DS} - V_{DSS}}{\alpha V_p}\right),$$

it is observed, for devices with a channel length <5 μm, that the latter effect is an order of magnitude smaller than the static-feedback effect [7.30].
In addition when the channel length modulation effect for larger devices is taken into account, only a minor improvement of the simulated characteristics is observed. In fact this improvement does not overcome the increased problem which is to correct the saturation voltage in such a way that continuity of the drain conductance is preserved at V_{DSS}. Therefore for digital applications it is sufficient to model the saturated drain conductance using the drain-voltage-dependent gate drive (7.65).
In Fig. 7.16 it is shown that this approach leads to a satisfactory agreement between measured and calculated characteristics. For the calculations use has been made of Eqs. (7.52), (7.53), (7.56), (7.60), (7.64) and (7.65). One might argue that the sample used here is not very short. However, except for the effects of series resistance, its characteristics are troubled with all other short-channel effects. For submicron devices, in which the resistance effects become relatively large, a further modification of the above model is necessary. This will be discussed in section 7.2.3.

7.2.1.5 The Subthreshold Mode

An expression for the drain current in the weak inversion regime was derived in section 6.2. However, for reasons of simplicity this result was obtained for the zero back-bias condition and using a reference voltage V_{Gr}, located halfway along the subthreshold slope. Nevertheless the results given in (6.27), (6.37) and (6.38) can be easily extended to account for an applied back-bias, using the drain-voltage-dependent threshold voltage (7.49) or (7.52) as a reference voltage. Then we obtain (if $V_{DS} \gg U_T$)

$$I_{DSUB} = \frac{I_0}{(1 + V_S/\psi_{sm})^{1/2}} * \exp\left(\frac{V_{GT}}{MU_T}\right), \qquad (7.66)$$

where

$$I_0 = 2\beta \psi_{sm}^{-1/2}(MU_T)^2 \qquad (7.67)$$

and

$$M = 1 + \frac{m}{(V_S + \psi_{sm})^{1/2}}. \qquad (7.68)$$

Although I_0 is expressed in terms of known physical quantities or parameters defined already, for flexibility its value usually is considered as a parameter. Since the value of the slope factor m may depend on the number of surface states, this quantity too is usually considered as a parameter. Owing to the fact that M is inversely proportional to the square root of the back-bias voltage, the subthreshold slope becomes steeper at higher values of V_{SB}. Obviously the subthreshold current varies strongly with temperature

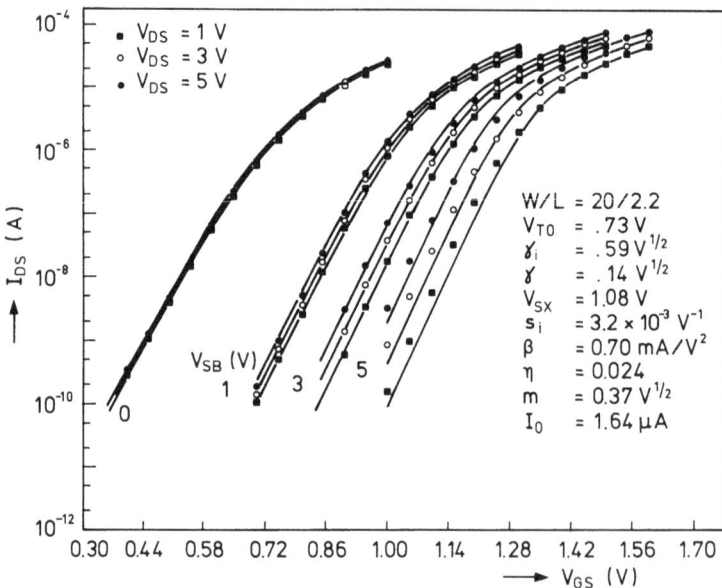

Fig. 7.17 Calculated (fully-drawn lines) and measured subthreshold characteristics of short-channel MOSFET at several values of source-bulk and drain-source bias

7.2 Small Transistor Models

owing to the exponential factor. This causes the current to vary at lower temperatures with a steeper slope too.

In Fig. 6.6 it has been shown that the total current is the sum of the drift component and the diffusion component. Furthermore the latter reaches a saturation value at gate voltages in excess of the threshold voltage.

In line with this exact (implicit) result it has been proposed [7.35, 7.25, 7.37] that the drain current be calculated by adding Eq. (7.66) to equations like (7.56) to (7.61). In order for the subthreshold current to reach a saturation value, actually a current

$$I_{DS2} = \frac{I_{DSUB}[1 - \exp(-V_{DS}/U_T)]}{1 + zI_{DSUB}/I_0}, \qquad (7.69)$$

in which z is a saturation parameter, has to be used for the considered summation. Usually z has a value of approximately 0.5.

Fig. 7.17 gives a comparison between the measured and calculated drain current in the subthreshold and the weak inversion regions. Using the parameters given in the inset, a good agreement is obtained.

7.2.2 Charges

Until now all published special models for the charges in short-channel MOSFETs [7.25, 7.38, 7.39, 7.40] make use of approximate expressions. This implies that either these charges are estimated from an integration of some of the major capacitances or from a division of the channel charge based on physical intuition. However, in practice these approaches result in errors for the minor capacitances or discontinuities between some capacitances at the boundary between different modes of operation. Since these problems are negligible for the charge partitioning scheme discussed in 7.1.3, the charge model for long channels is often adopted for short-channel devices too [7.41, 7.42, 7.43]. Naturally this extension leads to errors for the charges in the saturation region, as the saturation voltage of short-channel devices is generally lower than for long-channel devices. In addition the velocity saturation effect changes the potential distribution and consequently changes the distribution of the charges across the channel. Therefore, for instance, the basic equation (7.32) for the drain charge has to be recalculated by taking into account the velocity saturation effect.

7.2.2.1 Strong-Inversion Region

As in the long-channel case, the integral (7.32) can be calculated by changing the integration with respect to the coordinate x to the channel potential V. This can be done by making use of the basic current expression for the linear region

$$I(x) = \frac{\beta L[V_{GT} - (1+\delta)V]E_x}{1 + E_x/E_c}.$$

Rewriting this expression to obtain the field explicitly yields

$$dx = \left[\frac{\beta L[V_{GT} - (1+\delta)V]}{I_D} - \frac{1}{E_c}\right]dV.$$

By substituting into the latter expression the drain current relation (7.56) we finally obtain

$$E_c \, dx = \frac{V_{GT}V_c + (1+\delta)(\tfrac{1}{2}V_{DS}^2 - VV_c - VV_{DS})}{V_{GT}V_{DS} - \tfrac{1}{2}(1+\delta)V_{DS}^2} \, dV, \tag{7.70}$$

where $V_c = LE_c = \theta_c^{-1}$.
Using the latter result the integral (7.32) can be recalculated. After lengthy algebra we finally obtain [7.44]

$$Q_D = -WLC_{0x}[\tfrac{1}{2}V_{GT} - \tfrac{1}{3}(1+\delta)V_R - F_1 F_2], \tag{7.71}$$

where

$$F_1 = \frac{(1 + \theta_c V_R)(1+\delta)^2 V_R^2}{6[2V_{GT} - (1+\delta)V_R]}, \tag{7.72}$$

$$F_2 = \frac{1}{2}\theta_c V_R - \frac{[5V_{GT} - 2(1+\delta)V_R](1 + \theta_c V_R)}{5[2V_{GT} - (1+\delta)V_R]} \tag{7.73}$$

and $V_R = V_{DS}$, etc.
Like the evaluation of Q_D the charge at the source terminal can be evaluated from (6.97), yielding

$$Q_S = -WLC_{0x}[\tfrac{1}{2}V_{GT} - \tfrac{1}{6}(1+\delta)V_R + F_1(1 + F_2)]. \tag{7.74}$$

Furthermore the drain-source voltage-dependent part of the depletion layer charge can be recalculated. From its definition (7.28) it follows that

$$Q_B = -\delta WLC_{0x}V_R \frac{6V_{GT} - 4(1+\delta)V_R - (1+\delta)\theta_c V_R^2}{6[2V_{GT} - (1+\delta)V_R]}. \tag{7.75}$$

To the above relations the following conditions for the drain voltage apply:
If $V_{DS} < V_{DSS}$, $V_R = V_{DS}$; but if $V_{DS} \geqslant V_{DSS}$, $V_R = V_{DSS}$.
Although formally the definition (7.65) has to be used for the effective gate driving voltage V_{GT}, the inclusion of the feedback factor means that a large number of negligible capacitance contributions are calculated. Therefore to gain computational speed, this effect can be omitted. In fact the same arguments apply to the substrate coefficient δ. Without loss of accuracy, (7.60) can be replaced by the simple relation (7.25), in which an average value for the factor δ may be used. Finally we remark that for the same reason the

7.2 Small Transistor Models

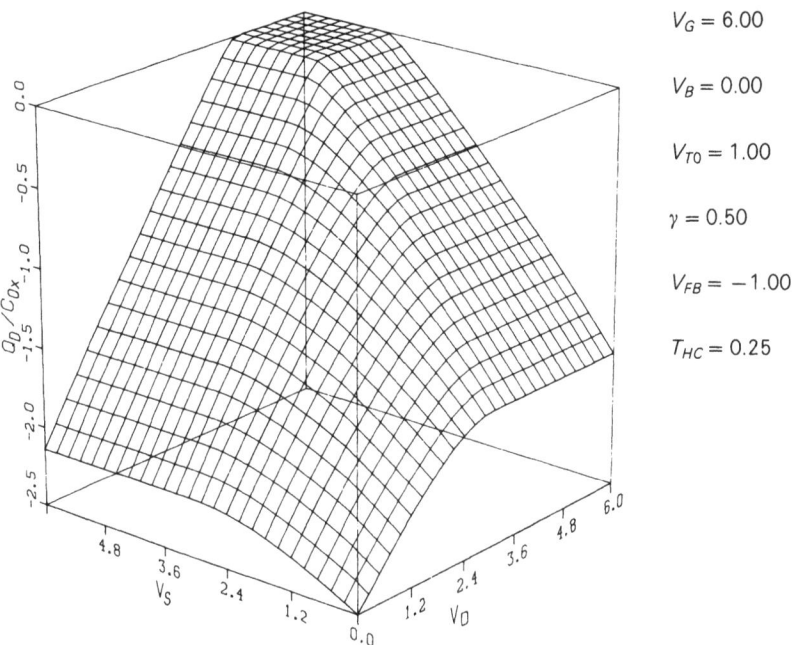

Fig. 7.18 Drain charge of short-channel MOSFET as a function of the source-bulk and drain-bulk voltage at a fixed gate bias of 6 Volts, with $V_{TO} = 1$ V, $\gamma = 0.50$ V$^{1/2}$, $\theta_C = 0.25$ V^{-1} and $V_{FB} = -1.0$ V

term $(1 + \theta_A V_{GT})$ from the general current expression has already been neglected in (7.70). Of course the price of the above simplifications is that the charge expressions need a slightly different expression for the saturation voltage V_{DSS}. However, as already explained, in reality its numerical value differs little from the general expression (7.62).

Although the above charge expressions are more complicated than their long-channel counterparts, their numerical value behaves qualitatively in the same way. This is demonstrated in Fig. 7.18, where the drain charge is plotted as a function of the source and drain potential for a submicron MOSFET with a typical value $\theta_C = 0.25$ V^{-1}. Although the charges have been expressed in voltages referring to the source, in this plot all voltages refer to bulk. This has the advantage that the charges pass through more modes of operation. Similar to Fig. 7.5 the right-hand side of the figure shows the behaviour of the Q_D expression, while the left-hand side gives a view of the Q_S expression. Comparing both figures it is observed that the overall behaviour is the same. In addition, at the boundary $V_S = V_D$ between forward and inverse mode both expressions and their derivatives are continuous. A more detailed inspection shows that small differences are present, which are entirely due to the lower saturation voltage of short-channel devices.

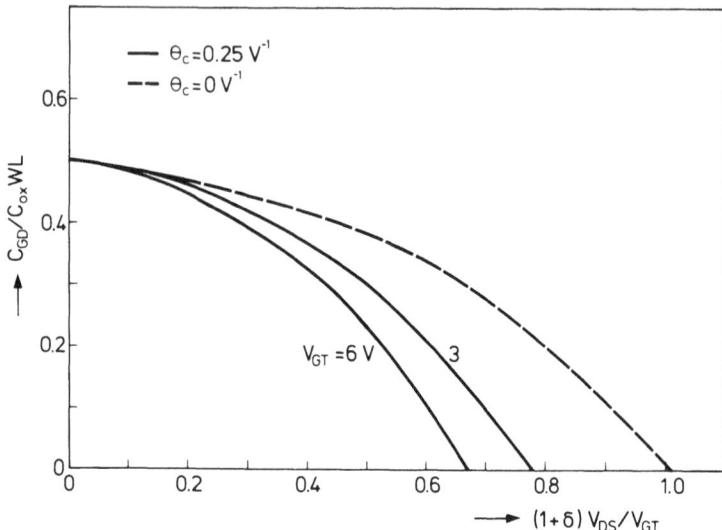

Fig. 7.19 Comparison of long-channel (broken line) and short-channel gate-drain capacitance behaviour

7.2.2.2 Capacitances

While the difference between the charges of short-channel devices and long-channel devices is relatively small, it becomes more explicit when considering the small signal capacitances. As an example we consider here the capacitance

$$C_{GD} = -\frac{\partial Q_G}{\partial V_{DS}},$$

where Q_G is the sum of (7.71), (7.74) and (7.75). Carrying out the differentiation we obtain

$$C_{GD} = WLC_{0x}\frac{[3V_{GT} - (1+\delta)V_R]}{3[2V_{GT} - (1+\delta)V_R]^2}$$
$$* [2V_{GT} - 2V_R(1+\delta) - \theta_c(1+\delta)V_R^2]. \qquad (7.76)$$

In Fig. 7.19 the above capacitance together with its long-channel counterpart (broken line) has been plotted as a function of $(1+\delta)V_{DS}/V_{GT}$. Owing to the value of the parameter $\theta_c = 0.25$ V^{-1}, the short-channel capacitance already becomes zero at $V_{DS} = V_{DSS}$, which value is considerably smaller than for the long-channel expression.

This fact is important for analog applications, where the use of capacitance expressions based on the long-channel charge relations would lead to major errors in short-channel applications. For digital applications this deficiency is probably less harmful.

7.2 Small Transistor Models

7.2.2.3 Charge in the Subthreshold Region

In the subthreshold region the major charge component is the charge present in the depletion layer. Here we give a derivation of this quantity for the case of a substrate with threshold voltage adjustment implant.

When $V_S < V_{SX}$ (compare (7.9)) the situation is equivalent to the case with uniform substrate doping, and the depletion charge is given by (7.18 a), provided that γ_i is used instead of γ.

When $V_S > V_{SX}$ the situation is more complicated. For a certain gate voltage V_G^* the implanted layer becomes fully depleted. In this case it follows from (6.24) that

$$-\frac{\gamma_i}{2} + \left\{\frac{\gamma_i^2}{4} + V_G^* - V_{FB}\right\}^{1/2} = (V_{SX} + \psi_{sm})^{1/2}.$$

Solving the latter result for V_G^* yields

$$V_G^* = V_{FB} + V_{SX} + \psi_{sm} + \gamma_i(V_{SX} + \psi_{sm})^{1/2}. \tag{7.77}$$

Again for $V_G < V_G^*$ the situation is equivalent to the previous case and we have

$$Q_B = -WLC_{ox}\frac{\gamma_i^2}{2}\left[\left\{1 + \frac{4(V_G - V_{FB})}{\gamma_i^2}\right\}^{1/2} - 1\right]. \tag{7.78}$$

However, when $V_G^* < V_G < V_{T2}$, where V_{T2} is given by (7.13), the depletion charge is given by

$$Q_B = -qN_id_i - qN_B(y_d - d_i),$$

where y_d is expressed in terms of the surface potential

$$y_d = \gamma C_{ox}(\psi_s - \Delta\phi)^{1/2}(qN_B)^{-1}.$$

In the above relation ψ_s is given by

$$V_G = V_{FB} + \psi_s + \Delta V_T + \gamma(\psi_s - \Delta\phi)^{1/2}.$$

Combining the latter three expressions we finally obtain

$$Q_B = -WLC_{ox}\left[\Delta V_T + \frac{\gamma^2}{2}\left[\left\{1 + \frac{4}{\gamma^2}(V_G - V_{FB} - \Delta V_T - \Delta\phi)\right\}^{1/2} - 1\right]\right]. \tag{7.79}$$

Eqs. (7.78) and (7.79) together with the definition (7.77) completely describe the depletion charge in the subthreshold region. Naturally, when $\gamma \ll \gamma_i$, the effect of the last term of (7.79) on the increase of Q_B with V_G becomes very small and the maximum value of Q_B is simply given by the first term of (7.79).

7.2.3 Effect of Series Resistance on the Drain Current

In section 6.7.1 we have seen that the contact barrier, the junction sheet resistance and crowding effects at the channel side give rise to a resistance

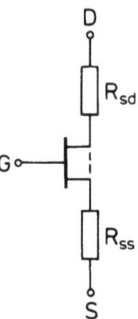

Fig. 7.20 Equivalent circuit for graded source/drain-junction MOSFET

in series with source and drain. In conventional devices, made in a well-scaled process, these resistances have only a minor effect on the characteristics, but this is no longer true for MOSFETS with graded junctions, like for instance LDD transistors. In these devices the value of the series resistance becomes so large that both the current value and the saturation voltage are changed considerably. When both resistances are known, their effect could be taken into account by using the equivalent circuit given in Fig. 7.20. However, from the viewpoint of efficiency this is not a good solution, since two additional nodes per transistor have to be used in this case for circuit simulation. In addition the problem arises that these resistances are not constant, but a (weak) function of the terminal voltages [7.45]. Finally their value can only be measured from the device characteristics. Therefore it is much more practical to include these resistances directly in the terminal device equations.

Defining a source resistance R_{ss}, a drain resistance R_{sd} and a total series resistance $R_T = R_{ss} + R_{sd}$, from the Kirchhoff relations associated with Fig. 7.20 the drain current can be written as

$$I_{DS} = \beta \frac{(V_{GT} - I_{DS}R_{ss})(V_{DS} - I_{DS}R_T) - \frac{1}{2}(1+\delta)(V_{DS} - I_{DS}R_T)^2}{1 + \theta_A(V_{GT} - I_{DS}R_{ss}) + \theta_B V_{SB} + \theta_C(V_{DS} - I_{DS}R_T) - \frac{1}{2}\theta_A V_{DS}}. \tag{7.80}$$

For reasons of simplicity and didactics the mobility expression (6.81) is used in the above equation. In principle the following calculation can be carried out too using (6.79) and (6.80) [7.46], but the final answer is more complicated and less clear. In addition the inclusion of the last term in the denominator corrects for the fact that the normal field in the channel is not constant, but decreases at the increase of the channel potential [7.9].

Solving Eq. (7.80) for I_{DS} a square-law relation is obtained with the solution

$$I_{DS} = \frac{a_2 - (a_2^2 - 4a_1 a_3)^{1/2}}{2a_1}, \tag{7.81}$$

where

7.2 Small Transistor Models

$$a_1 = \beta(R_{ss}R_T - \tfrac{1}{2}R_T^2) + \theta_A R_{ss} + \theta_C R_T,$$
$$a_2 = 1 + (\theta_A + \beta R_T)V_{GT} + \theta_B V_{SB} + (\theta_C - \tfrac{1}{2}\theta_A - \beta R_{sd})V_{DS},$$
$$a_3 = \beta[V_{GT}V_{DS} - \tfrac{1}{2}(1+\delta)V_{DS}^2].$$

Since the term a_1 is much smaller than the others, in practice one often meets the condition

$$\frac{a_1 a_3}{a_2^2} < 0.10,$$

so that the square-root term (7.81) can be expanded. Taking into account the value of the original parameters θ_A, θ_B and θ_C and the common bias voltages the condition used above can be rewritten in the more practical form

$$I_{DS}(R_{ss} + R_{sd}) < 0.50 \text{ Volt}.$$

In this case a very useful result is obtained,

$$I_{DS} = \beta \frac{[V_{GT} - \tfrac{1}{2}(1+\delta)V_{DS}]V_{DS}}{1 + \theta'_A V_{GT} + \theta'_B V_{SB} + \theta'_C V_{DS}}, \tag{7.82}$$

where

$$\theta'_A = \theta_A + \beta(R_{ss} + R_{sd}) \tag{7.83 a}$$

and

$$\theta'_C = \theta_C - \tfrac{1}{2}\theta_A - \beta R_{sd}. \tag{7.83 b}$$

Comparing (7.82) with the short-channel current expression with no series resistances, we conclude that the form is exactly the same. In fact the inclusion of the series resistances leads to two generalized mobility parameters given in (7.83). Obviously the presence of both resistances reduces the current. On the other hand the reduction is less than the effect of the total series resistance R_T, since the negative term in θ'_C provides some compensation.

Applying the previous procedure to the saturated drain current yields

$$I_{DSS} = \beta \frac{(V_{GT} - I_{DSS}R_{ss}) - (1+\delta)(V_{DSS} - I_{DSS}R_T)}{\theta_C}, \tag{7.84}$$

where V_{DSS} is the external saturation voltage. Finally by equating the expressions (7.82) and (7.84) the following relation for V_{DSS} is obtained,

$$V_{DSS} = \frac{2V_{GT}}{1+\delta} * \left[\left\{1 + \frac{2\theta'_C V_{GT}}{(1+\delta)(1 + \theta'_A V_{GT} + \theta_B V_{SB})}\right\}^{1/2} + 1\right]^{-1}. \tag{7.85}$$

Owing to the fact that the drain resistance reduces the parameter value θ'_C, the effect of velocity saturation is partly reduced and the external saturation voltage is increased. This latter fact also causes less reduction of the drain current.

Comparing the result (7.85) with the common expression for short-channel devices (7.64), in this case an additional term

$$(1 + \theta'_A V_{GT} + \theta_B V_{SB})$$

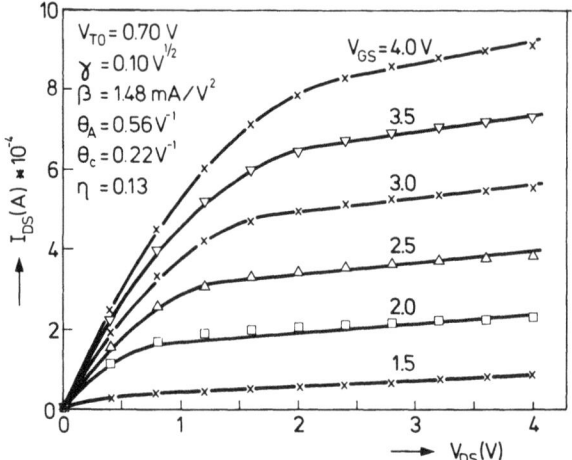

Fig. 7.21 Comparison of calculated (fully-drawn lines) and measured (dots) characteristics of submicron LDD-type MOSFET ($W/L = 4$ μm/0.5 μm)

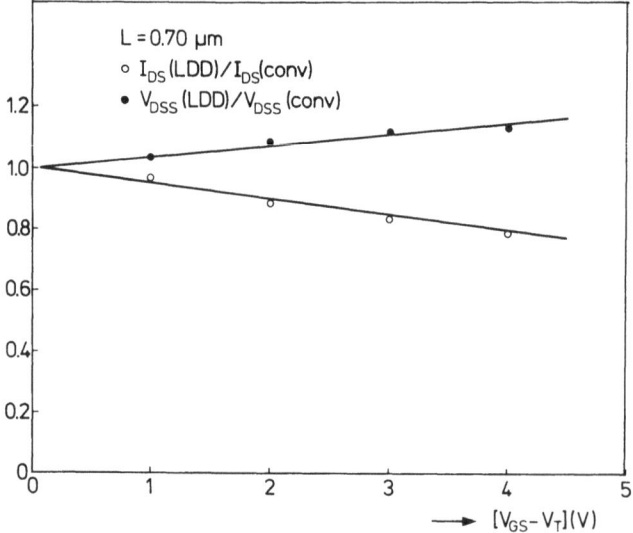

Fig. 7.22 Drain current and saturation voltage of 0.7 μm LDD MOSFET (normalized to similar values of a non-LDD type) as a function of gate drive

7.2 Small Transistor Models

is present in the square-root part. Since in practice $\theta'_A V_{GT}$ may reach a considerable value, this term can no longer be neglected.

Since both series resistances present themselves differently in the two parameters θ'_A and θ'_C, careful measurement of both parameter values provides in principle a means of determining both resistances separately. This can be achieved by measuring θ'_A and θ'_C for MOSFETS of different gate length and plotting both parameters as a function of β. The resistances then follow from the slope value [7.47]. However, in this case the original parameter θ_C has to be known exactly.

In Fig. 7.21 it is shown that the above modification of the short-channel model can represent well the characteristics of an LDD transistor. Several relevant parameter values for this submicron MOSFET ($L_{\text{eff}} = 0.50\ \mu\text{m}$) are given in the subscripts. Finally Fig. 7.22 gives a comparison of the saturated drain current and the saturation voltage between an LDD transistor and a conventional device of the same channel length. Note that according to the compensation effects mentioned earlier, the reduction of the current is less than the direct change of the parameter values θ_A and θ_C.

7.2.4 The Substrate Current

In order to obtain useful characteristics the short-channel MOSFET has to be designed in accordance with scaling rules [7.48]. This implies usually that the gate insulator thickness is reduced, the channel implant is enhanced and the junction depth is reduced. Consequently at high drain voltage operation the high field region near the drain becomes restricted to a smaller depleted area. Therefore at comparable drain bias the peak field increases in a shorter channel and causes low-level avalanche multiplication, resulting in significant substrate current [7.49] (compare Fig. 7.23). Owing to a higher ionization rate the problem is greatest in n-channel MOSFETS, where the substrate current is made up of holes. In fact the above substrate current gives rise to several constraints in VLSI design.

The use of negative substrate bias is a common technique to minimize junction capacitance, sensitivity to body effect and punch-through. However, since this bias is generated on-chip, the value of the substrate current must be considered in the design of the back-bias generator right from the

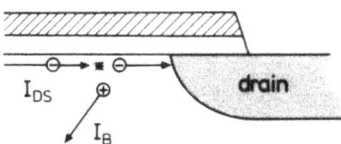

Fig. 7.23 Illustration of low-level avalanche multiplication in a MOSFET, resulting in the generation of substrate current

very beginning. In fact a too high value of substrate current may cause malfunctioning of this generator.

Secondly, because of the high resistivity of substrate material, the considered current will cause a significant voltage drop across the substrate resistance. Consequently if the substrate current is large enough, a source-substrate junction may be forward-biased and turn on a parasitic bipolar transistor [7.50].

Thirdly the secondary electrons, which are proportional to the substrate current, may be injected into the substrate and hence discharge dynamic nodes [7.51].

Finally in the high field region primary electrons can be injected into the insulator and damage the material. This phenomenon is usually observed as a change-in time of the device characteristics [7.52, 7.53] and plays a major role in the device lifetime. Although the substrate current is not always uniquely correlated to this hot carrier effect, in practice its value can be considered as a first-order measure of this detrimental effect.

In view of the above facts and in order to have a good understanding of the problems related to the substrate current, a simple but fairly accurate model with easily obtainable parameters is required. Several models have been published [7.54, 7.55, 7.56], but unfortunately several of them are too complex to provide the desired accuracy. This is not so surprising, since the current in question depends very critically on the peak field, which is difficult to model with satisfactory precision.

The ionization rate has already been given in principle in Eq. (2.77) of section 2.7. Following this approach the substrate current is easily obtained,

$$I_B = aI_{DS} \exp(-b/E_{xd}), \tag{7.86}$$

where a and b are parameters and E_{xd} is the peak value of the lateral drain field. Strictly speaking, the above equation is only valid for a uniform field and therefore the substrate current should be obtained from an integration procedure. Furthermore, in the case of narrow field peaks even a dead-space correction has to be applied [7.57]. In practice, however, in particular for LDD structures the field distribution is sufficiently broadened to allow all the above corrections to be dispensed with. Eq. (7.86) can then be considered as a fair approximation. On the other hand, owing to the 2-D field distribution, any analytical solution of the field distribution is an approximation too. Unfortunately even a small error in E_{xd} will introduce large errors via the exponential term in (7.86).

Taking for instance one of the better analytical approaches given in Eq. (6.86), we observe that E_{xd} should be proportional to $(V_{DS} - V_{DSS})$. Given this fact, it is easily explained from (7.86) that, at constant V_{DS}, I_B first increases with V_{GS} owing to an increase of the primary current I_{DS}, but I_B decreases at larger values of V_{GS} owing to the increase of V_{DSS}, which causes the exponential term to decrease. This is shown in Fig. 7.24. However, when the modelled characteristics are compared in detail with measurements, the calculated values are seen to decrease too fast at high gate bias.

7.3 Models for Analog Applications

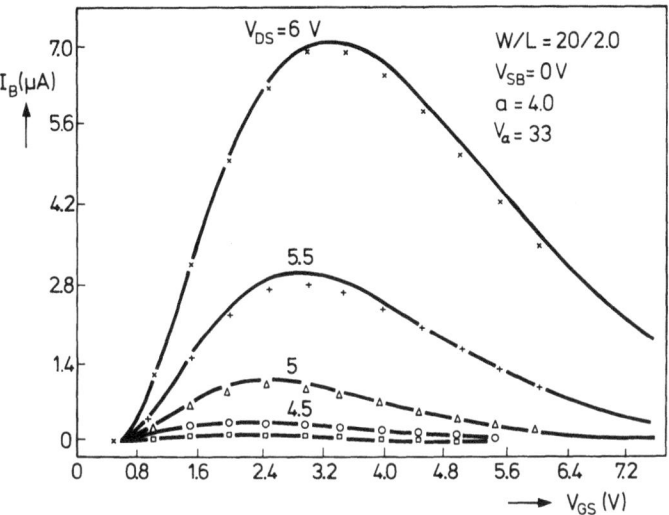

Fig. 7.24 Modelled (fully-drawn lines) and measured substrate current versus gate bias for several values of drain-source voltage

This deviation is already an indication that the calculation presented in (6.86) results in an underestimation of the peak field. In practice this deficiency can be corrected by modifying the voltage dependence of E_{xd} with an empirical term $(V_{DS} - 0.8 * V_{DSS})$ instead of with the physical term $(V_{DS} - V_{DSS})$. For completeness we present this modelling result:

$$I_B = aI_{DS} \exp\left(\frac{-V_a}{V_{DS} - 0.8 V_{DSS}}\right). \tag{7.87}$$

Owing to its empirical character, this result is less suitable for predicting the substrate current in an unknown process. However for an existing process, where the parameters a and V_a can be easily determined, the calculated values are close to experimental data in a relatively large range of current and bias voltages. This is shown in Fig. 7.24 for a short-channel structure. As an alternative to the exponential term the ionization rate had been presented in terms of a power function in E_{xd}^m [7.58]. Although obviously any relation to the physical background is lost in this manner, in practice satisfactory results are achieved. Experimentally m is found to have a value of approximately 5.

7.3 Models for Analog Applications

7.3.1 Review of Existing Models

Since their introduction in 1960, MOSFET ICs have been realized with an exponential growth [7.59] in the size of memories and digital systems.

Owing to several shortcomings like higher differential off-set voltages, lower transconductance for a given current and higher $1/f$ noise, the MOSFET is often less suitable for analog applications than the bipolar transistor. Requirements for integrated systems that combine analog and digital functions, however, have made MOSFET technology mandatory for analog circuits [7.60].

Due to the mostly digital history of the MOSFET, in transistor modelling more attention has been paid to current behaviour of even smaller and more complex devices than to small-signal properties. While the transconductance is reasonably described, if the absolute value of the current is modelled with a fair accuracy, the latter result is usually no guarantee for a satisfactory description of the drain conductance. In this case most (digital) models produce errors in excess of 100%. To some extent the same situation exists for the transconductance of small transistors in the weak inversion region. For the a.c. capacitances the situation is slightly better, but often several mutual capacitances are omitted.

With a view to improving this situation, several approaches have been published [7.24, 7.33, 7.61, 7.62, 7.15]. The first cited two are based on a more accurate modelling of the MOSFET saturation current for the strong inversion region. This has been achieved by taking into account static feedback, velocity saturation and channel length modulation. However, the weak inversion region is totally neglected. In addition several deficiencies exist. In [7.24] an empirical relation for channel length modulation is used, which lacks a physical basis. Although on account of this an extensive analysis is given of the behaviour of transconductance and drain conductance as a function of gate bias, a comparison with experimental results has been avoided. Furthermore it can be shown that the dependence of the above quantities on drain bias is incorrect. In [7.33] the modelled drain conductance agrees rather well with experimental data, but the model has an error in the saturation voltage, which may cause convergence problems in a simulator.

The next cited two approaches attempt to describe the current in weak and strong inversion with a smooth transition between both modes. This is achieved by making use of the implicit relation for the surface potential ψ_s (compare Eqs. (6.16) and (6.18)), taking into account a short-channel barrier-lowering correction factor for the depletion charge. Furthermore by taking into account the basic mobility equation (6.75), expressing the normal field in terms of the inversion and depletion charge (compare (2.62)), the drain current is expressed as a quartic equation in the value of ψ_s at the drain and the source. As an additional result the saturation voltage has been obtained from another implicit relation. Despite this rigorous, time-consuming approach and a refined description of the channel length modulation effect, when compared with experimental data, the modelling results are no more accurate than those presented in section 7.2.1. To some extent this is caused by the mixture of a crude barrier lowering approximation with the implicit calculation of ψ_s, which may introduce significant errors in

7.3 Models for Analog Applications

the subthreshold current values and in the drain conductance at strong inversion.

Finally, in the last cited result [7.15] a smooth transition beween both modes of operation has been forced by introducting an effective gate driving voltage in the strong-inversion current equation according to

$$V_{GT}^* = 2MU_T \ln\left[1 + \exp\left(\frac{V_{GT}}{2MU_T}\right)\right]. \tag{7.88}$$

In this equation the original gate driving voltage V_{GT} has already been defined in (7.57) and the subthreshold slope factor M in (7.68). When the gate voltage is a few (kT/q) Volts above threshold, V_{GT}^* reduces to (the linear relation) V_{GT}. When the gate voltage is a few (kT/q) below threshold, the effective gate voltage becomes

$$V_{GT}^* \approx 2MU_T \exp\left(\frac{V_{GT}}{2MU_T}\right)$$

and the saturation voltage becomes

$$V_{DSS} = \frac{V_{GT}^*}{1 + \delta}.$$

Thus the saturated drain current reduces to the form

$$I_{DS} = I_0 \exp(V_{GT}/MU_T),$$

where

$$I_0 \approx 2\beta(MU_T)^2.$$

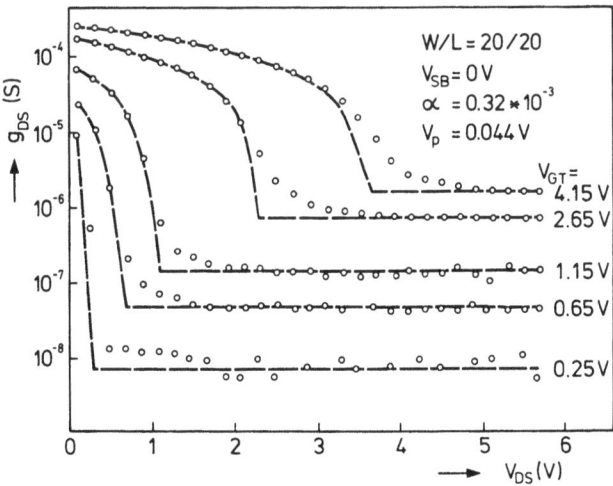

Fig. 7.25 Illustration of the failure of most models to describe the measured drain conductance (dots) around the saturation voltage

Obviously the above saturation voltage for the subthreshold current has not the correct value, which is only a few times (kT/q) Volts. Therefore the total current is multiplied by a correction term which provides the correct saturation voltage for the subthreshold region (see the next section). In spite of this major improvement, the poor description of effects of substrate implant on threshold voltage and channel length modulation causes this model to behave satisfactorily only in digital applications (although not better than the model given in section 7.2.1). However, for analog applications the above deficiencies give problems.

Finally it should be remarked that most models have problems to describe the rather smooth transition of the drain conductance between the linear and the saturation region, in particular at high values of gate drive (compare Fig. 7.25). This is due to the fact that the second derivative $\partial^2 I_{DS}/\partial V_{DS}^2$ is mostly discontinuous at the saturation voltage $V_{DS} = V_{DSS}$.

7.3.2 Improved Description of the Drain Current

Since in fact all published models have shortcomings, we attempt in this section to describe the current characteristics by combining strong points of the previous models.

The nucleus of the model is formed by the expressions for the strong inversion region already defined in 7.2.1, taking into account the generalized form of the gate driving voltage. Thus the drain current is given by

$$I_D = \beta \frac{[V_{GT}^* - \frac{1}{2}(1+\delta)V_{DS}]V_{DS}}{[1 + \theta_A(V_{GS} - V_{TS}) + \theta_B\{(V_{SB} + \psi_{sm})^{1/2} - \psi_{sm}^{1/2}\}](1 + \theta_C V_{DS})}, \quad (7.56)$$

where V_{GT}^* is given in (7.88), V_{GT} is given in (7.65) and V_{TS} is given in (7.53). In addition to (7.56) the expression (7.64) for the saturation voltage applies. Next the saturated drain current is multiplied by a channel length modulation factor (6.87), which has a more physical basis [7.33] than (7.1). For completeness sake this relation is repeated here,

$$F_L = 1 + \alpha \ln\left(1 + \frac{V_{DS} - V_{DSS}}{\alpha V_p}\right). \quad (7.89)$$

In order to guarantee a smooth transition of the first and second derivative of the current with respect to V_{DS}, which otherwise would cause a sharp bend of the drain conductance at V_{DSS}, a numerical manipulation on V_{DS} is required [7.64].

Finally a last correction term is needed to provide the correct saturation voltage in the subthreshold region. Therefore the drain current in (7.56) is multiplied by a factor [7.15]

$$F_T = \frac{t[1 - \exp(-V_{DS}/U_T)] + F_L \exp(V_{GT}^*/2MU_T)}{(1/t) + \exp(V_{GT}^*/2MU_T)} \quad (7.90)$$

7.3 Models for Analog Applications

In weak inversion this has the form

$$F_T = t^2[1 - \exp(-V_{DS}/U_T)],$$

where t is a parameter used to adjust the magnitude of the weak-inversion current and so to correct for the various approximations involved in the derivation. Above threshold, however, this weak-inversion mechanism becomes inoperative and $F_T = F_L$.

Compared to the model presented in 7.2.1. the above one does not provide a more accurate description in the weak inversion regime. However, it has the advantage that a smooth transition to the strong-inversion regime is automatically guaranteed. In the other case this is only true, if the parameters have been determined carefully. For a comparison of the calculated current characteristics to experimental data we therefore refer to Figs. 7.16 and 7.17.

Using the drain current expression given by (7.56) and (7.88), the transconductance is easily derived by differentiation with respect to V_{GS}.

For the linear region we obtain

$$g_m \approx \frac{\beta V_{DS}[1 + \exp(-V_{GT}/2MU_T)]}{[1 + \theta_A V_{GT} + \theta_B\{(V_{SB} + \psi_{sm})^{1/2} - \psi_{sm}^{1/2}\}](1 + \theta_C V_{DS})}. \quad (7.91)$$

On the other hand for the saturation region the following relation applies

$$g_{ms} \approx \frac{\beta V_{DSS}[1 + \exp(-V_{GT}/2MU_T)]}{[1 + \theta_A V_{GT} + \theta_B\{(V_{SB} + \psi_{sm})^{1/2} - \psi_{sm}^{1/2}\}]\left\{1 + \frac{2\theta_C V_{GT}}{1 + \delta}\right\}^{1/2}}. \quad (7.92)$$

Using the definition of the saturation voltage (7.64) it can be proved that continuity is preserved between both regions. In Fig. 7.26 the calculated transconductance data are compared with measured values as a function of gate bias for devices operating in the saturation region. For this comparison the same device as used for Fig. 7.16 was taken.

Generally the expression for the drain conductance becomes rather complicated. Therefore we only give here the result for the most important region. Differentiating with respect to V_{DS} the saturated drain conductance in the strong inversion region is given by

$$g_{dss} \approx \frac{\eta \beta V_{DSS}}{[1 + \theta_A V_{GT} + \theta_B V_{SB}](1 + \theta_C V_{DSS})} + \frac{\alpha I_{DSS}}{\alpha V_p + (V_{DS} - V_{DSS})}. \quad (7.93)$$

Since in long-channel devices the second term dominates, in this case $g_{dss} \sim V_{GT}^2$. However, owing to the increase of the θ parameters and η, in short-channel devices $g_{dss} \sim V_{GT}$.

In order to show the model's additional capability a comparison is given in Fig. 7.27 between the modelling result (calculated numerically) and

240   7 Models for the Enhancement-Type MOSFET

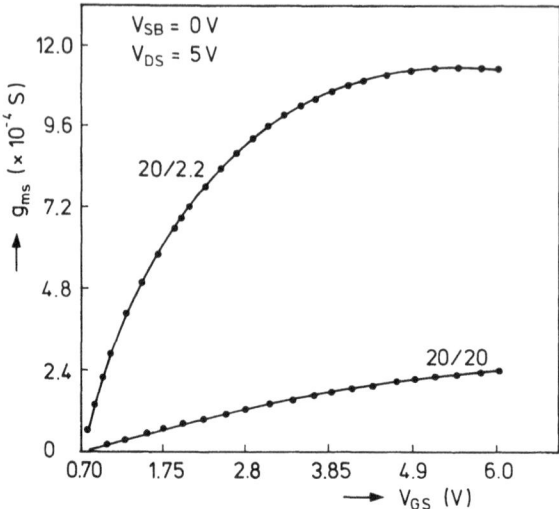

Fig. 7.26 Comparison of modelled (fully-drawn lines) and measured saturation transconductance as a function of gate bias for the MOSFET of Fig. 7.16

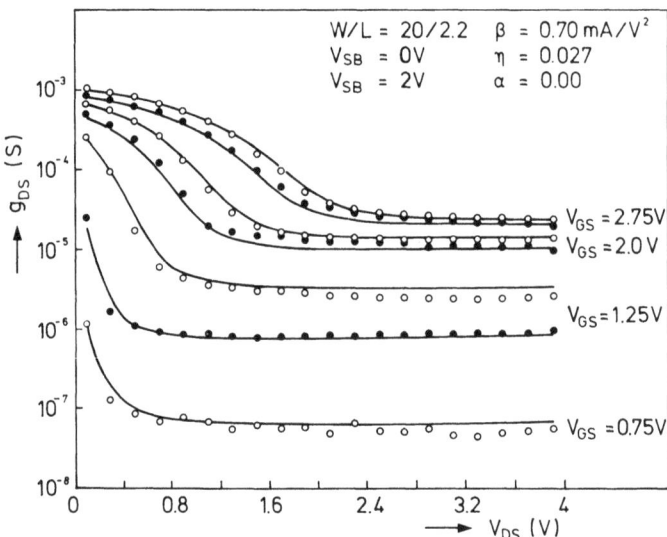

Fig. 7.27 Comparison of modelled (fully-drawn lines) and measured drain conductance versus drain-source voltage at several values of gate bias

measured values of the drain conductance in both the linear and saturation region.
In fact the calculated values in the saturation region, shown in Fig. 7.25 for a long-channel device, were already based on (7.93).

7.3.3 Capacitances

On the basis of the intrinsic charges defined in section 7.2.2, a complete list of possible small-signal capacitances can be obtained. Depending on the reference voltage chosen, a wide variety of capacitance schemes exists. Here we only present a scheme based on the source voltage as a reference. Taking $V_S = 0$ it easily follows from (6.100) that [7.26]

$$i_D = C_{dd}\frac{dV_{DS}}{dt} - C_{dg}\frac{dV_{GS}}{dt} - C_{db}\frac{dV_{BS}}{dt}, \qquad (7.94\text{ a})$$

$$i_G = -C_{gd}\frac{dV_{DS}}{dt} + C_{gg}\frac{dV_{GS}}{dt} - C_{gb}\frac{dV_{BS}}{dt}, \qquad (7.94\text{ b})$$

$$i_B = -C_{bd}\frac{dV_{DS}}{dt} - C_{bg}\frac{dV_{GS}}{dt} + C_{bb}\frac{dV_{BS}}{dt}. \qquad (7.94\text{ c})$$

Like all other possible cases, the scheme contains nine independent capacitances. Together with small-signal quantities like the transconductance g_m and the drain conductance g_{ds}, these capacitances have been represented in an equivalent circuit given in Fig. 7.28. For completeness several parasitic capacitances have been added as dotted figures. Their value has already been discussed in section 6.6.4. Note that out of all intrinsic capacitances only three can be recognized as capacitances in the usual sense (being controlled

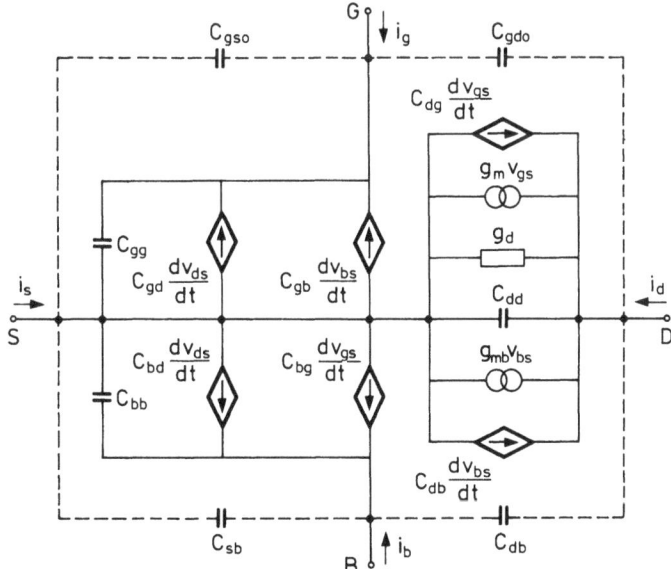

Fig. 7.28 Small-signal equivalent circuit of a MOSFET, showing all intrinsic capacitances (with fully drawn interconnections) and parasitic capacitances (with interrupted interconnections)

by the associated terminal voltages). All others may be called transcapacitances, since they are controlled by a third terminal as well. Therefore in Fig. 7.28 a different symbol is used.

For long-channel devices expressions for the major capacitances have already been discussed in section 7.1.3. Owing to the approach chosen these expressions have to be considered as applicable to quasi-static situations only. Therefore their value is limited up to one half of the cut-off frequency f_T [7.65, 7.66]. For practical n-channel devices (with a channel length shorter than 3 µm) and micron-size p-channel devices no analytical, physics-based results have yet been published, apart from a few empirical expressions for some capacitances [7.36, 7.38, 7.39]. This is not surprising. Owing to the back-gate effect via V_{SB} and the partitioning of channel charge according to section 6.6.2, most capacitance expressions become very complicated in the case of short-channel transistors. Only in a few cases is the algebraic evaluation worth the effort. Examples are the capacitances C_{gd} given in Eq. (7.76) and C_{bd}. From the charge expression (7.75) it is easily derived that the latter quantity equals

$$C_{bd} = WLC_{ox}\delta \frac{[3V_{GT} - (1+\delta)V_R][2V_{GT} - 2(1+\delta)V_R - \theta_c(1+\delta)V_R^2]}{3[2V_{GT} - (1+\delta)V_R]^2}.$$

For other capacitances approximate results can be derived (neglecting the effect of V_{SB} upon δ or taking simplifications for the saturation voltage), but these apply more to design estimation than to accurate simulation. Therefore Fig. 7.29 gives results for other capacitances, which have been calculated by numerical differentiation of the charge equations (7.71)–(7.75).

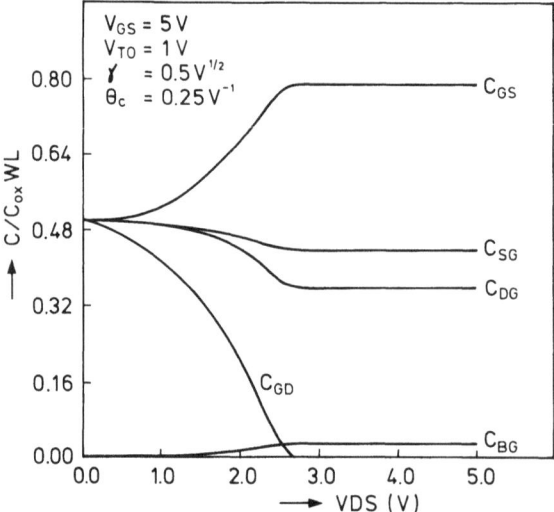

Fig. 7.29 Intrinsic short-channel MOSFET capacitances versus normalized drain-source voltage calculated numerically from the charge expressions

7.3 Models for Analog Applications

Qualitatively these results agree with those presented in Fig. 7.8, but quantitatively they are different owing to the velocity saturation effect.
Similar to the long-channel results the above ones assume quasi-static behaviour and therefore apply only to frequencies below the cut-off frequency. However, calculated values of the frequency response of an inverter are close to non-quasistatic results [7.66]. In principle capacitance expressions applicable to a much higher frequency range can be derived by a direct small-signal series expansion of the continuity equations (2.50), but so far only results for long-channel devices have been given [7.67, 7.63, 7.65].

7.3.4 Noise

7.3.4.1 Thermal Noise

Since the MOSFET channel can be considered as a nonlinear resistor, the channel current is subject to thermal noise. Applying small-signal series expansion directly to the continuity equation (2.50), and assuming a thermal noise source to be present at each infinitesimal part of the channel it has been shown that the noise spectral density is given by a generalized Nyquist relation [7.68],

$$S_{I_D}(f) = \frac{4kT}{L^2} \int_0^L g(x)\,dx, \tag{7.95}$$

where $g(x)$ is the local specific channel conductance. Using the transform appearing above Eq. (7.70), the above result can be rewritten as an integral with respect to V,

$$S_{I_D}(f) = \frac{4kT}{I_D L^2} \int_0^{V_{DS}} g^2(V)\,dV, \tag{7.96}$$

where

$$g(V) = \beta L[V_{GT} - (1+\delta)V] - \theta_c L I_{DS}.$$

Note that, owing to the velocity saturation effect, the local channel conductance is reduced at the drain side.
Solving the latter integral (7.96), we finally obtain

$$S_{I_D}(f) = 4kT \left[\frac{\beta_0 W[V_{GT} - (1+\delta)H V_{DS}](1 + \theta_c V_{DS})}{L[1 + \theta_A V_{GT} + \theta_B\{(V_{SB} + \psi_{sm})^{1/2} - \psi_{sm}^{1/2}\}]} \right.$$
$$\left. - 2\theta_c I_{DS}\left(1 + \frac{1}{2}\theta_c V_{DS}\right) \right], \tag{7.97}$$

where the function H reads

$$H = \frac{1}{2} - \frac{(1+\delta)V_{DS}}{6[2V_{GT} - (1+\delta)V_{DS}]}.$$

In the above form the spectral density has never been published. However, in the case of long-channel devices operating in the strong-inversion saturation mode, H is reduced to a value of one third and Eq. (7.97) can be simplified to the form

$$S_{I_D}(f) = \frac{4kT * \frac{2}{3}\beta_0 WL^{-1}V_{GT}}{1 + \theta_A V_{GT} + \theta_B\{(V_{SB} + \psi_{sm})^{1/2} - \psi_{sm}^{1/2}\}}.$$

Keeping in mind that the transconductance is given by Eq. (7.92), the above result can be put into the well-known form

$$S_{I_D}(f) = 4kT * \{\tfrac{2}{3}(1 + \delta)g_{ms}\}. \tag{7.98}$$

Except for the factor $(1 + \delta)$ this expression has been obtained along different lines [7.68, 7.69, 7.70].

The reason for presenting thermal noise in the form of Eq. (7.97) is that the latter result applies to short-channel devices as well, at all modes of operation. However, in order to demonstrate that the result is applicable to the weak-inversion region, we have to present (7.97) in a slightly different form for gate voltages close to threshold voltage. Since in this case the effect of velocity saturation can be neglected for the saturation mode, $H \approx \tfrac{1}{3}$. Furthermore, considering that in this case the saturated drain current is given by

$$I_{DSS} = \frac{\tfrac{1}{2}\beta V_{GT}^2}{1 + \theta_A V_{GT} + \theta_B\{(V_{SB} + \psi_{sm})^{1/2} - \psi_{sm}^{1/2}\}},$$

we can now rewrite (7.97) in the form

$$S_{I_D}(f) \approx 4kT * \frac{4I_{DSS}}{3V_{GT}}.$$

Defining that, for $V_{GT} < V_{GTR} = \tfrac{8}{3}U_T$, the latter quantity is kept to the value V_{GTR} in the noise expression, the thermal noise in the weak-inversion mode reduces to the form

$$S_{I_D}(f) \approx 2qI_{DSS}, \tag{7.99}$$

which is the general result for diffusion noise. In fact Eq. (7.99) has been confirmed experimentally for the subthreshold mode [7.71, 7.72].

Owing to capacitive coupling between gate and channel, the fluctuating channel current induces noise in the gate terminal at high frequencies. Unfortunately the calculation of the latter component is too complicated [7.73] to provide a result applicable to all modes of operation, which is desirable for implementation in a circuit simulator. Therefore it is more practical to derive the desired result from a high-frequency equivalent circuit presentation given in Fig. 7.30 a. Owing to the earlier mentioned capacitive coupling, part of the channel conductance is present as a resistance in series with the gate input capacitance. From a direct calculation of the input admittance it follows that [7.74] this resistance is given by

7.3 Models for Analog Applications

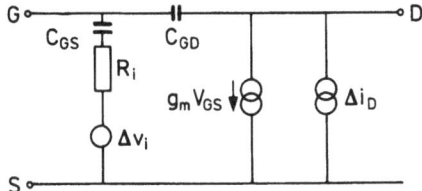

Fig. 7.30 a MOSFET equivalent circuit with internal noise sources

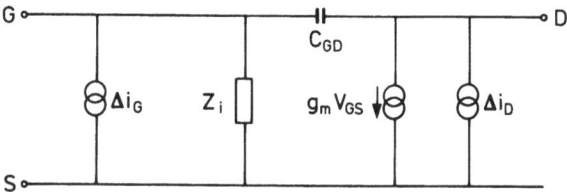

Fig. 7.30 b MOSFET equivalent circuit with noise sources transferred to the terminals

$$R_i = (3g_m)^{-1}.$$

Since this resistor produces thermal noise too, two noise sources have been added to the scheme of Fig. 7.30 a. In addition to the channel noise current Δi_D an input noise voltage source ΔV_i is present, whose value is simply given by Nyquist's theorem. In practice it is more convenient to use a scheme with a noise-free MOSFET where the noise sources are present at the terminals only. This scheme is given in Fig. 7.30 b. Now ΔV_i is easily transformed into an input noise current Δi_G.
Since

$$\langle \Delta i_G^2 \rangle = \langle \Delta V_i^2 \rangle / Z_i^2,$$

where Z_i is the input impedance made up of C_{GS} and R_i, we finally obtain

$$S_{I_G}(f) \approx 4kT \frac{(\omega C_{GS})^2}{3g_m}. \tag{7.100}$$

This result is identical with a noise calculation based on a solution of the transport equations for long-channel devices [7.75]. In practice it is advisable to add a default value to the denominator. This has the advantage of avoiding numerical problems in a circuit simulator, when g_m may become zero in the linear region of the characteristics.
Generally, it follows from the direct calculation that Δi_G and Δi_D are correlated. For the saturation mode this correlation is a purely imaginary number and given by [7.76]

$$\langle \Delta i_G^* \, \Delta i_D \rangle = 0.4j [\langle \Delta i_G^2 \rangle \cdot \langle \Delta i_D^2 \rangle]^{1/2}, \tag{7.101}$$

but this value may be used for the linear mode too.

Experimentally it is found, in particular for short-channel devices, that the channel noise is somewhat larger than the value given by (7.98). This increase may be caused by carrier heating [7.77]. Therefore it is advisable to multiply Eq. (7.97) by an empirical noise parameter A_N.

7.3.4.2 Flicker Noise

Usually thermal or white noise is not the only source of noise present. At low frequencies flicker (or $1/f$) noise becomes dominant in MOSFETs. Despite the efforts taken, its physical origin is not yet clear. During the first decade of MOST developments the noise level was rather high and a close correlation with surface state density could be established [7.78, 7.79]. Although some facts remained unexplained, McWorther's model of spontaneous trapping of free carriers via a tunnelling mechanism could explain most phenomena observed. Owing to improvements in processing technology the magnitude of the noise has decreased and at present no clear correlation with surface-states density can be observed. However, it has to be admitted that the latter quantity too is hard to measure. Only in short-channel devices, subjected to stress experiments, is the noise increased by hot-carrier-induced defects to such a level that the considered correlation again becomes visible [7.80]. As an alternative explanation the observed $1/f$ noise has been interpreted in terms of mobility fluctuations [7.81]. However, in MOSFETs the characteristic model constants [7.82] are much lower than for passive material.

Owing to this unsatisfactory situation, a rather empirical approach has to be taken for circuit modelling. This implies that only a well established

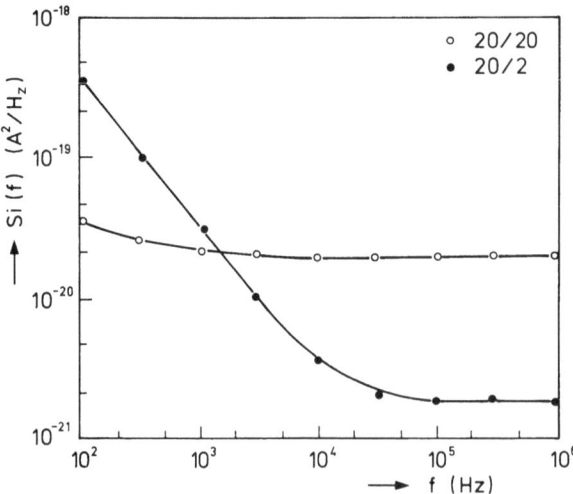

Fig. 7.31 Noise spectral density versus frequency for long- and short-channel n-type MOSFET

dependence on process and bias parameters will be taken into account. For instance, it has been observed that, for the saturation mode, the $1/f$ noise of n-channel devices is considerably higher than for p-channel devices and virtually independent of bias condition. In fact the main factor is the dependence of the noise on the gate area, which is only a consequence of Gaussian statistics [7.78]. Consequently the $1/f$ noise in MOSFETs is modelled by the following empirical relation,

$$S_{I_D}(f) = \frac{B_N g_m^2}{WLf}. \tag{7.102}$$

With present processes the noise parameter B_n has a value between 10^{-25} A^2 cm^2 Hz and 4×10^{-25} A^2 cm^2 Hz for n-channel devices and between 10^{-26} A^2 cm^2 Hz and 4×10^{-26} A^2 cm^2 Hz for p-channel devices. Figure 7.31 gives a typical noise spectrum of a MOSFET operating in saturation. Although generation-recombination noise spectra have been reported too, this type of noise is mainly present owing to inappropriate anneal steps. In an established process this kind of noise is always absent.

References

[7.1] F. M. Klaasen: A MOS Model for Computer-Aided Design. Philips Research Reports *31*, 71–83 (1976).
[7.2] L. W. Nagel: SPICE-2: A Computer Program to Simulate Semiconductor Circuits. Univ. of California, Berkeley, Memo ERL-M520 (1975).
[7.3] D. E. Ward, K. Doganis: Optimized Extraction of MOS Model Parameters. IEEE Transactions on CAD of Integrated Circuits and Systems *CAD-1*, 163–168 (1982).
[7.4] S. R. Hofstein, G. Warfield: Carrier Mobility and Current Saturation in the MOS Transistor. IEEE Transactions Electron Devices *ED-12*, 129–138 (1965).
[7.5] H. Schichman, D. A. Hodges: Modelling and Simulation of Insulated-Gate Field-Effect Transistor Switching Circuits. IEEE Journal of Solid-State Circuits *SC-3*, 285–289 (1968).
[7.6] D. Frohman-Bentchkowsky, A. S. Grove: Conductance of MOS Transistors in Saturation. IEEE Transactions on Electron Devices *ED-16*, 108–113 (1969).
[7.7] G. Baum, H. Beneking: Drift-Geschwindigkeitssättigung bei MOS-Feldeffekt-Transistoren. Solid State Electronics *13*, 789–798 (1970).
[7.8] F. M. Klaassen: Review of Physical Models for MOS Transistors. In: Process and Device Modelling for Integrated Circuit Design. Noordhoff, Leyden (1977).
[7.9] G. Merckel, J. Borel, N. Z. Cupcea: An Accurate Large-Signal MOS Model for Use in CAD. IEEE Transactions on Electron Devices *ED-19*, 681–690 (1972).
[7.10] P. Rossel, H. Martinot, G. Vassilieff: Accurate Two-Section Model for MOS Transistors in Saturation. Solid-State Electronics *19*, 51–56 (1976).
[7.11] F. v. d. Wiele: A Long-Channel MOSFET Model. Solid-State Electronics *22*, 991–997 (1979).
[7.12] A. Vladimirescu, S. Liu: The Simulation of MOS Integrated Circuits. Memo, University of California, Berkeley, ERL M80/7 (1980).
[7.13] R. F. Vogel: Analytical MOSFET Model with Easily Extracted Parameters. IEEE Transactions on CAD *CAD-4*, 127–134 (1985).
[7.14] G. G. de Jong, M. M. Abu-Zeid: MOSFET Model Continuous from Weak to Strong Inversion. Electronics Letters *23*, 1299–1300 (1987).

[7.15] G. T. Wright: Physical and CAD Models for the VLSI MOSFET. IEEE Transactions on Electron Devices *ED-34*, 823–833 (1987).
[7.16] F. M. Klaassen, W. Hes: Compensated MOSFET Devices. Solid-State Electronics *28*, 359–373 (1985).
[7.17] V. L. Rideout, F. H. Gaensslen, A. LeBlanc: Device Design Considerations for Ion-Implanted n-Channel MOSFETs. IBM Journal of Research and Development *19*, 50–59 (1975).
[7.18] E. Demoulin, F. van de Wiele: Ion Implanted MOS Transistors. In: Process and Device Modelling for Integrated Circuit Design. Noordhoff, Leyden (1977).
[7.19] N. Herr, B. Garbs, J. Barnes: A Statistical Modelling Approach for Simulation of MOS Circuit Designs. IEDM Technical Digest (1982) pp. 290–293.
[7.20] D. M. Rogers, J. D. Hayden, D. D. Rinerson: Model for the Channel-Implanted Enhancement-Mode IGFET. IEEE Transactions on Electron Devices *ED-33*, 955–964 (1986).
[7.21] R. S. C. Cobbold: Theory and Applications of Field-Effect Transistors. Wiley-Interscience, New York (1970).
[7.22] J. E. Meyer: MOS Models and Circuit Simulation. RCA Review *32*, 42–63 (1971).
[7.23] H. C. Poon: V_T and Beyond. Presented at Workshop on Device Modelling for VLSI, Burlingame, CA (1979).
[7.24] S. Liu, L. W. Nagel: Small Signal MOSFET Modelling for Analog Circuit Design. IEEE Journal of Solid-State Circuits *SC-17*, 983–988 (1982).
[7.25] F. M. Klaassen: In: Process and Device Modelling for CAD. Elsevier, Amsterdam (1986), chapt. 12.
[7.26] Y. P. Tsividis: Operation and Modelling of the MOS Transistor. McGraw-Hill, New York (1987), chapt. 8.
[7.27] C. Turchetti, G. Masetti, Y. P. Tsividis: On the Small-Signal Behaviour of the MOS Transistor in Quasi-Static Operation. Solid-State Electronics *26*, 941–949 (1983).
[7.28] P. P. Wang: Device Characteristics of Short-Channel and Narrow Width MOSFETs. Transactions on Electron Devices *ED-25*, 779–786 (1978).
[7.29] G. Merckel: A Simple Model of the Threshold Voltage in Short and Narrow Channel IGFETS. Solid-State Electronics *23*, 1207–1213 (1980).
[7.30] F. M. Klaassen, W. C. J. de Groot: Modelling of Scaled Down MOS Transistors. Solid-State Electronics *23*, 237–242 (1980).
[7.31] T. Skotnicki, G. Merckel, T. Pedron: The Voltage-Doping Transformation, A New Approach to Modelling Short-Channel Effects. In: Proceedings ESSDERC 1987, Bologna. North-Holland, Amsterdam (1987), pp. 543–546.
[7.32] B. Hoeflinger, H. Sibbert, G. Zimmer: Model and Performance of Hot-Electron MOS Transistor for VLSI. IEEE Transactions on Electron Devices *ED-26*, 513–520 (1979).
[7.33] T. Poorter, and J. H. Satter: A d-c Model for a MOS Transistor in the Saturation Region. Solid-State Electronics *23*, 765–772 (1980).
[7.34] CURRY, Proprietary Philips 2-D Device Simulation Routine.
[7.35] G. Merckel: CAD Models for MOSFETs. In: Process and Device Modelling for IC Design. Noordhoff, Leyden (1977), pp. 751–764.
[7.36] L. Lauwers, K. de Meyer: Novel Calculations in the Field of Accurate MOS Transistor Model. ESSDERC 1988. Journal de Physique *C4*, 249–252 (1988).
[7.37] B. J. Sheu, D. L. Scharfetter, P. K. Ko, M. C. Jeng: BSIM, Berkeley Short-Channel IGFET Model. IEEE Journal of Solid State Circuits *SC-22*, 558–566 (1987).
[7.38] S. Liu: A Unified CAD Model for MOSFETs (Memo ERL-M81/31). Electron Research Labs, University of California, Berkeley (1981).
[7.39] P. Yang, B. D. Eppler, P. K. Chatterjee: An Investigation of the Charge Conservation Problem for MOSFET Circuit Simulation. IEEE Journal of Solid-State Circuits *SC-18*, 128–138 (1983).
[7.40] G. W. Taylor, W. Fichtner, J. G. Simmons: A Description of MOS Internodal Capacitances for Transient Simulations. IEEE Transactions on Computer-Aided Design *CAD-1*, 150–156 (1983).

[7.41] D. E. Ward, R. W. Dutton: A Charge-Oriented Model for MOS Transistor Capacitances. IEEE Journal of Solid-State Circuits *SC-13*, 703–707 (1978).
[7.42] Y. P. Tsividis: Operation and Modelling of the MOS Transistor. McGraw-Hill, New York (1987).
[7.43] D. E. Ward: Charge-Based Modelling of Capacitance in MOS Transistors (Technical Report G201-11). Integrated Circuits Laboratory. Stanford University, California (1981).
[7.44] T. Smedes: (to be published).
[7.45] J. Y. Sun, M. R. Wordeman, S. E. Laux: On the Accuracy of Channel Length Characterization of LDD MOSFETS. IEEE Transactions on Electron Devices *ED-33*, 1556–1562 (1986).
[7.46] P. T. J. Biermans: (to be published).
[7.47] F. M. Klaassen, P. T. J. Biermans, R. M. D. Velghe: The Series Resistance of Submicron MOSFETs and Its Effect on Their Characteristics. ESSDERC 1988. Journal de Physique *C4*, 257–260 (1988).
[7.48] R. H. Dennard, F. H. Gaensslen, H. N. Yu, V. L. Rideout, A. R. LeBlanc: Design of Ion-Implanted MOSFETs with Small Physical Dimensions. IEEE Journal of Solid-State Circuits *SC-9*, 256–268 (1974).
[7.49] T. Toyabe, K. Yamaguchi, S. Asai, M. S. Mock: A Two-Dimensional Avalanche Breakdown Model of Submicron MOSFETs. Technical Digest IEDM (1977), pp. 432–435.
[7.50] E. Sun, B. Alders, L. Forbes: The Effect of Electron Trapping on the Performance of Short-Channel MOS Transistors. IEEE Transactions on Electron Devices *ED-26*, 1849 (1979).
[7.51] P. K. Chatterjee: VLSI Dynamic nMOS Design Constraints Due to Drain Induced Primary and Secondary Impact Ionization. Technical Digest IEDM (1979), pp. 14–17.
[7.52] C. Hu, S. C. Tam, F. C. Hsu, P. K. Ko, T. Y. Chan, K. W. Terrill: Hot Electron-Induced MOSFET Degradation—Model, Monitor and Improvement. IEEE Transactions on Electron Devices *ED-32*, 375–385 (1985).
[7.53] P. K. Ko, R. S. Muller, C. Hu: A Unified Model for Hot-Electron Currents in MOSFETs. Technical Digest IEDM (1980), pp. 600–603.
[7.54] J. Mar, S. S. Li, S. Y. Yu: Substrate Current Modelling for Circuit Simulation. IEEE Transactions on Computer-Aided Design of Integrated Circuits *CAD-1*, 183–186 (1982).
[7.55] F. Hsu, P. K. Ko, S. Tam, C. Hu, R. S. Muller: Hot-Electron-Induced Excess Current in n-Channel MOSFETs. IEEE Transactions on Electron Devices *ED-29*, 1735–1740 (1982).
[7.56] W. Müller, L. Risch, A. Schütz: Short-Channel MOS Transistors in the Avalanche-Multiplication Regime. IEEE Transactions on Electron Devices *ED-29*, 1778–1784 (1982).
[7.57] C. Werner, R. Kuhnert, L. Risch: Optimization of Lightly Doped Drain MOSFETs Using a New Quasiballistic Simulation Tool. Technical Digest IEDM (1984), pp. 770–773.
[7.58] J. W. Sing, B. Sudlow: Modelling and VLSI Design Constraints of Substrate Current. Technical Digest IEDM (1980), pp. 732–735.
[7.59] G. Moore: VLSI: Some Fundamental Challenges? IEEE Spectrum *16*, 30–37 (1979).
[7.60] D. A. Hodges, P. R. Gray, R. W. Brodersen: Potential of MOS Technologies for Analog IC's. IEEE J. Solid-State Circuits *SC-13*, 285–294 (1978).
[7.61] P. P. Guebels, F.v.d. Wiele: A Small-Geometry MOSFET Model for CAD Applications. Solid-State Electronics *26*, 267–273 (1983).
[7.62] H. Ogney, S. Cserveny: Modele du Transistor MOS Valable Dans un Grand Domaine de Courants. Reprint from Bulletin des SEV/VSE *73*, 113–116 (1982).
[7.63] J. J. Paulos, D. A. Antoniadis: Limitations of Quasi-Static Capacitance Models for the MOS Transistor. IEEE Electron Device Letters *EDL-4*, 221–224 (1983).

[7.64] F. M. Klaassen, R. Velghe: (to be published).
[7.65] M. Bagheri, Y. P. Tsividis: A Small-Signal d.c.-to-High-Frequency Non-Quasi-Static Model for the Four-Terminal MOSFET. IEEE Transactions on Electron Devices *ED-32*, 2383–2391 (1985).
[7.66] H. J. Park, P. K. Ko, C. Hu: A Non-Quasistatic MOSFET Model for SPICE. Technical Digest IEDM (1987), pp. 652–655.
[7.67] J. A. van Nielen: A Simple and Accurate Approximation to the h.f. Characteristics of IGFETs. Solid-State Electronics *12*, 826–829 (1969).
[7.68] F. M. Klaassen, J. Prins: Thermal Noise in MOS Transistors. Philips Research Reports *22*, 505–514 (1967).
[7.69] A.v.d. Ziel: Thermal Noise in Field-Effect Transistors. Proceedings of the IRE *50*, 1808–1812 (1962).
[7.70] R. Paul: Thermisches Rauschen von MOS-Transistoren. Nachrichtentechnik *17*, 458–466 (1967).
[7.71] J. Fellrath: Short Noise Behaviour of Subthreshold MOS Transistors. Revue de Physique Appliqué *13*, 719–723 (1978).
[7.72] G. Reimbold, P. Gentil: White Noise of MOS Transistors Operating in Weak Inversion. IEEE Transactions on Electron Devices *ED-29*, 1722–1725 (1982).
[7.73] F. M. Klaassen: A Computation of the h.f. Noise Quantities of a MOSFET. Philips Research Reports *24*, 559–571 (1969).
[7.74] F. M. Klaassen, J. Prins: Noise of Field-Effect Transistors at Very High Frequencies. IEEE Transactions on Electron Devices *ED-16*, 952–957 (1969).
[7.75] M. Shoji: Analysis of High-Frequency Thermal Noise of MOS Field Transistors. IEEE Transactions on Electron Devices *ED-13*, 520–524 (1966).
[7.76] A.v.d. Ziel: Gate Noise in Field Effect Transistors at High Frequencies. Proceedings IEEE *51*, 461–467 (1963).
[7.77] A.v.d. Ziel: Noise in Solid-State Devices. Advances in Electronics and Electron Physics *46*, 313–383 (1978).
[7.78] F. M. Klaassen: Characterization of Low $1/f$ Noise in MOS Transistors. IEEE Transactions on Electron Devices *ED-18*, 887–891 (1971).
[7.79] G. Abowitz, E. Arnold, E. A. Leventhal: Surface States and $1/f$ Noise in MOS Transistors. IEEE Transactions on Electron Devices *ED-14*, 775–777 (1967).
[7.80] Z. H. Fang, H. Haddara, S. Cristoloveanu, G. Ghibaudo, A. Chovet: Aging Characterisation of Short Channel MOSFET by $1/f$ Noise (Europhysics Conference Abstracts). Digest ESSDERC (1985), pp. 75–76.
[7.81] F. N. Hooge, T. G. Kleinpenning, L. K. J. Vandamme: Experimental Studies on $1/f$ Noise. Reports on Progress in Physics *44*, 479–532 (1981).
[7.82] L. K. J. Vandamme, H. M. M. de Werd: $1/f$ Noise Model for MOST's Biased in Nonohmic Region. Solid-State Electronics *23*, 325–329 (1980).

Models for the Depletion-Type MOSFET 8

In this chapter we discuss models for MOS transistors with a channel implant region opposite to the substrate doping, which causes the transistor to operate in the on state at zero gate bias [8.1, 8.2]. Usually this class of devices is called depletion-type MOSFET. In practice mainly n-channel devices are applied. When the device is only used as a load, the characteristics can be satisfactorily described by an enhancement model with an appropriate threshold shift. We shall return to this later. However, a wider utilization [8.3] exploiting the depletion-mode operation, requires a separate model.

Generally four different modes of operation may occur and have to be distinguished. When the gate is biased just above threshold (compare section 6.3.3), a conducting channel is formed in a part of the buried n-type region, which is located away from the surface. Thus a buried channel is formed which is relatively weakly controlled by the gate electrode. When the channel implant is fairly strong, a hole inversion layer at the surface may occur, which prevents further depletion of the buried channel on a decrease in gate bias. Since this is an unwanted situation for practical application, we omit here modelling of this second mode of operation. As the gate bias is increased, electrons are also attracted at the surface, as in the enhancement device, and an additional surface channel is formed. Once this surface channel is established, the buried channel is shielded from the effects of the gate and will thus be very weakly controlled by it, whereas the surface channel conduction will increase steadily as the gate drive increases.

However, when the drain bias is increased too, a part of the surface channel at the drain side will disappear owing to reversal of the gate insulator field. This is the most complicated situation. At the source side conduction occurs by the buried channel and the surface channel; at the drain side by the partly depleted buried region only. Fig. 8.1 gives a summary of the three possible modes.

First we discuss a model for long-channel devices, in which the buried channel region is approximated by a shallow sheet with uniform doping [8.1, 8.4–8.6]. Although this model represents well the characteristics of devices with a channel length longer than 3 µm, the lack of practical parameters constitutes a drawback. This is not a problem in empirical

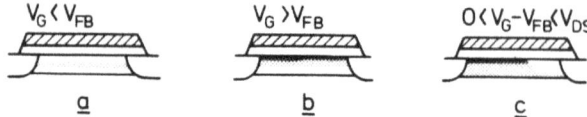

Fig. 8.1 Modes of operation of a buried-channel depletion MOSFET

models like that given in [8.2] and others to be discussed next [8.7, 8.8], which additionally take into account the effects of a short, nonuniform buried channel. However, in this case a more empirical gate-bias dependence of the channel conductance has to be utilized. This chapter closes with several remarks on charge modelling.

8.1 Long-Channel Model

8.1.1 Mobile Charge Density

Fig. 8.2 gives a cross-section of a buried n-channel device (in a direction perpendicular to current flow) together with the boundaries of space-charge regions, which modulate the conducting channel. As mentioned previously, the latter channel is assumed to be doped uniformly (doping N_i, depth d_i). In principle the mobile charge density Q_m is given by [8.4]

$$Q_m = -Q_i + Q_j + Q_s, \tag{8.1}$$

where the various components (per unit area) are defined as follows.

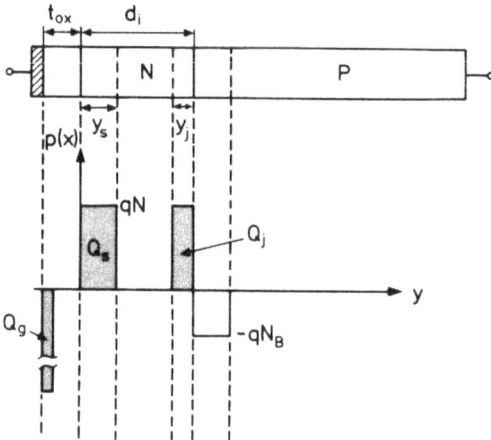

Fig. 8.2 Elementary cross-section of the buried-channel at a position along the channel where the surface is depleted. Also shown is the charge distribution

8.1 Long-Channel Model

The implanted layer charge density is given by

$$Q_i = qN_id_i. \tag{8.2}$$

The p-n junction space-charge density Q_j is given by

$$Q_j = qN_iy_j = k_j[V']^{1/2}, \tag{8.3}$$

where

$$k_j = \left[\frac{2\varepsilon_s qN_iN_B}{N_i + N_B}\right]^{1/2}, \tag{8.4}$$

and

$$V' = V + \phi_j. \tag{8.5}$$

V is the channel potential with respect to bulk at any point between source and drain and ϕ_j is the built-in potential.

The component Q_s takes the following values depending on the gate bias:

— when $V_G - V_{FB} > V$, at the surface accumulation occurs and

$$Q_s = -C_{ox}(V'_G - V'), \tag{8.6}$$

where

$$V'_G = V_G - V_{FB} + \phi_j; \tag{8.7}$$

— when $V_G - V_{FB} \leqslant V$, at the surface a depletion layer is formed and

$$Q_s = qN_iy_s = k_i\{(V' - V_G + V_i)^{1/2} - V_i^{1/2}\}, \tag{8.8}$$

where

$$k_i = (2\varepsilon_s qN_i)^{1/2} \tag{8.9}$$

and

$$V_i = \frac{q\varepsilon_s N_i}{2C_{ox}^2}. \tag{8.10}$$

As mentioned previously, a possible surface inversion is excluded here for practical devices. The set of Eqs. (8.1) to (8.10) is valid at any point along the channel between the source and the drain. Whether all conditions mentioned actually occur will depend on the channel potential.

8.1.2 Threshold and Saturation Voltages

Taking the source potential V_S as the reference voltage, the threshold voltage is obtained by equating Q_m at the source end to zero.
The result has already been obtained in section 6.3.3 via another manner, but will be repeated here for the sake of completeness;

$$V_T = V_{FB} - \frac{N_B}{N_i + N_B}(\phi_j + V_S) - \frac{Q_i(1 + C_{ox}/2C_i)}{C_{ox}}$$
$$+ \frac{k_i(\phi_j + V_S)^{1/2}}{C_{ox}}(1 + C_{ox}/C_i), \qquad (8.11)$$

where

$$C_i = \varepsilon_s/d_i.$$

Note that in contrast to (6.55) no use has been made of a γ symbol. This is due to the fact that in this case the current cannot be expressed in terms of γ. In fact, in this model only physical quantities like N_i, N_B and d_i have to be considered as parameters. Their value cannot, however, be extracted from the characteristics in a simple manner. For instance, the slope of V_T versus $(V_S + \phi_j)^{1/2}$ provides no direct information on the three parameters mentioned. Only in the limiting case $(d_i \to 0, N_i \gg N_B)$ is the situation more simple. In this case (8.11) approaches the relation for the enhancement device threshold

$$V_T = V_{FB} - Q_i/C_{ox} + \gamma(V_S + \phi_j)^{1/2}$$

and γ can be used as a parameter.

Similar to the enhancement case for a fixed gate voltage, the saturation voltage is defined as the drain voltage at which Q_m approaches zero value at the drain end of the channel. Using the expression for Q_m in (8.1) and replacing V' by V'_{DSAT}, we obtain

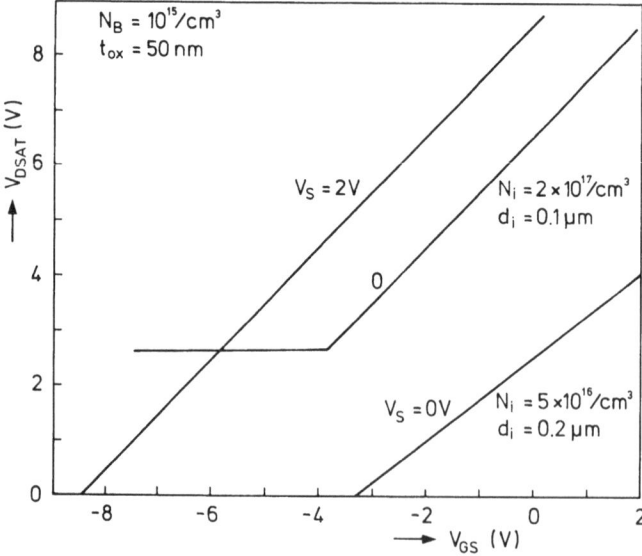

Fig. 8.3 Drain saturation voltage as a function of gate drive. Relevant parameters are given in the figure

8.1 Long-Channel Model

$$V_{DSAT} = -\phi_j + 2B^2 + C - 2B(B^2 + C)^{1/2}, \tag{8.12}$$

where

$$B = \frac{Q_i k_j - k_i k_j V_i^{1/2}}{k_i^2 - k_j^2} \tag{8.12a}$$

and

$$C = \frac{Q_i^2 + k_i^2 V_G' + 2Q_i k_i V_i^{1/2}}{k_i^2 - k_j^2}. \tag{8.12b}$$

Despite the complicated relation above, for low-doped substrates V_{DSAT} is almost a linear function of $(V_G - V_S)$. This is shown in Fig. 8.3. Such a relation is valid as long as the surface region at the drain is depleted. If the region is accumulated, the device cannot be pinched off.

8.1.3 Channel Current

Neglecting diffusion current, the channel current can be written as

$$I_D(x) = -\mu_n W Q_m \frac{dV'}{dx}.$$

Integrating this expression between source and drain

$$I_D = -\frac{\mu_n W}{L} \int_{V_S'}^{V_D'} Q_m \, dV',$$

where W and L are the width and length of the device, respectively. Substituting for Q_m from (8.1) in the latter integral, we obtain

$$I_D = \frac{\mu_n W}{L} \left[Q_i (V_D' - V_S') - \frac{2}{3} k_j (V_D'^{3/2} - V_S'^{3/2}) - F_s \right], \tag{8.13}$$

where k_j has been defined in (8.4) and F_s is the contribution of the surface space-charge region to the current defined as

$$F_s = \int_{V_S'}^{V_D'} Q_s \, dV'. \tag{8.14}$$

The function F_s takes different values depending on the conditions existing at the surface. We now evaluate this function for the different situations ranging from depletion to accumulation (along the surface).

a. *Depletion along the entire surface* ($V_G' < V_S'$). The charge Q_s is now given by (8.8) and the function F_s becomes

$$F_s = \tfrac{2}{3} k_i \{ (V_D' - V_G' + V_i)^{3/2} - (V_S' - V_G' + V_i)^{3/2} - \tfrac{3}{2} V_i^{1/2} (V_D' - V_S') \}. \tag{8.15}$$

Obviously for large values of the drain voltage in the above expression V'_D has to be replaced by V'_{DSAT}.

b. *Accumulation along the entire surface* $(V'_G > V'_D > V'_S)$. The charge Q_s is then given by (8.6) and F_s is

$$F_s = -C_{ox}\{(V'_G - V'_S)(V'_D - V'_S) - \tfrac{1}{2}(V'_D - V'_S)^2\}. \tag{8.16}$$

Owing to the above voltage conditions, saturation of the drain current does not occur. Furthermore, it should be noted that for terms arising from accumulation of electrons at the surface (Eq. (8.16)) a surface mobility μ_{ns} should be used in place of the bulk mobility. As will be shown later, the surface mobility required to fit the measured values is considerably smaller than the bulk mobility. In practice C_{ox} can formally be multiplied by the empirical relation

$$\frac{\mu_{ns}/\mu_n}{1 + \theta_A V'_G}. \tag{8.17}$$

c. *Accumulation at source, depletion at drain* $(V'_S < V'_G < V'_D)$. In this case Q_s is defined as follows:

— If $V'_S < V' < V'_G$,

$$Q_s = C_{ox}(V' - V'_G). \tag{8.6}$$

— If $V'_G < V' < V'_D$,

$$Q_s = k_i\{(V' - V'_G + V_i)^{1/2} - V_i^{1/2}\}. \tag{8.8}$$

Note that, owing to the definition of V_i in (8.10), Q_s and its derivative are continuous at the boundary $V' = V'_G$.
Integrating the above expression of Q_s (8.14) yields

$$F_s = -\tfrac{1}{2}C_{ox}(V'_G - V'_S)^2 + \tfrac{2}{3}k_i\{(V'_D - V'_G + V_i)^{3/2} - V_i^{3/2} + \tfrac{3}{2}V_i^{1/2}(V'_D - V'_G)\}. \tag{8.18}$$

Again for this situation, at high values of the drain voltage, the previous procedure for V_{DSAT} has to be applied.

In addition to N_i, d_i and N_B, the other parameters are V_{FB}, C_{ox}, μ_n, μ_{ns} and θ_A. Since some of these parameters do not appear directly in the current expression above, their extraction from measured data is not an easy procedure. This complication and the fact that such a major quantity like the threshold voltage is only present implicitly in the current expressions, constitute a serious drawback.

Nevertheless, if the physical parameters are known, the above model equations represent well the measured characteristics of long-channel devices. This is shown in Fig. 8.4 and Fig. 8.5 [8.4]. Note from the first figure that the measured data in the accumulation region do indeed require an additional surface mobility. Furthermore, the sample shown in Fig. 8.5 has a relatively

8.1 Long-Channel Model

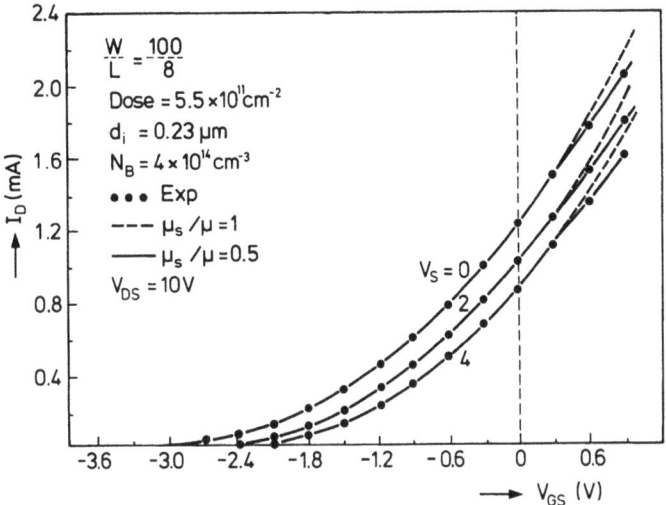

Fig. 8.4 The saturated drain current as a function of gate drive for a lightly implanted buried-channel device. The need for using a surface mobility is illustrated

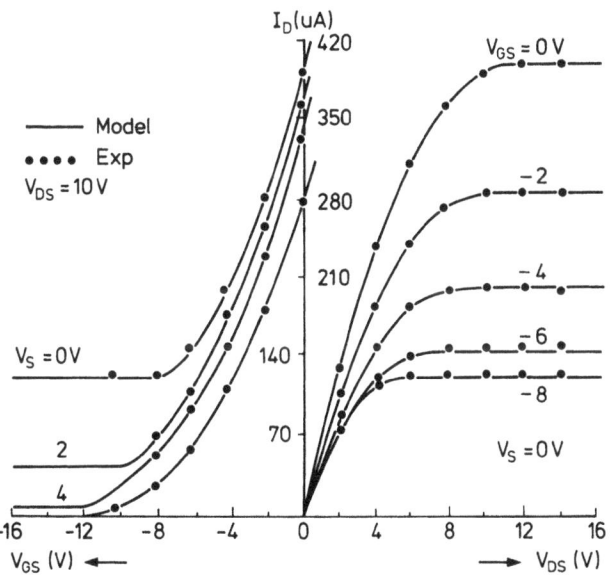

Fig. 8.5 The drain current as a function of gate and drain bias for a heavily implanted long-channel device

high channel implant. This causes the device not to be cut off completely, not even at high negative values of gate bias.

Naturally, the above model only applies for devices with not too short channels. However, several shortcomings are also present for long-channel devices. For instance, an extension for the finite drain conductance and the subthreshold region is lacking. For the majority of applications this is not a problem. A general long-channel model valid in all regions of operation (analogous to the charge-sheet model of section 6.2) has been proposed [8.9] showing natural extension to pinched-off and sub-threshold modes of operation. However, such a model contains a series of implicit relations, which are difficult to handle in a general-purpose circuit simulator.

8.2 Short-Channel Model

8.2.1 Specific Problems

In addition to the shortcomings already mentioned, the previous model causes more problems when compared to measured characteristics, in particular in the case of short-channel depletion transistors. This is caused by physical mechanisms typical for small, implanted buried-channel devices.
Similar to enhancement transistors, at higher values of drain bias the mobile carrier transport becomes impeded by velocity saturation. Naturally, it is always possible to include this effect in the drain current expression. However, even when a simple modelling expression for the high field mobility like Eq. (6.79) is substituted in Eq. (8.13), the usual saturation condition generally leads to a quartic equation for the saturation voltage. In order to obtain a more practical solution, therefore, an approximation of (8.13) is required [8.7]. Furthermore Eq. (8.13) has to be changed as in fact the auxiliary function F_s (compare Eq. (8.15)) does not satisfactorily describe the depletion of the implanted channel by the gate bias. In a modern device the implanted doping distribution deviates considerably from the box-type profile as assumed in the previous model. In fact, at larger depth the channel dope decreases linearly towards the substrate level [8.10]. Consequently, the depletion layer expands more rapidly with gate bias than according to Eq. (8.8) underlying the result (8.15). In the next section we shall return to this point.
An additional shortcoming of the previous model is the fact that a short-channel device cannot be described by a single threshold [8.11]. Usually for the saturation mode the threshold voltage is lower than for the linear mode. This is caused by the drain-bias-induced barrier lowering effect. When the device is biased below threshold at a low value of drain bias, a residual current flows, which is determined by the potential barrier between the undepleted source and the depleted channel area. However, the minimum of this potential barrier is in this case located relatively deep into the bulk (compare section 6.3.3). Owing to the weakening of the screening by the gate, an increase of drain bias causes this minimum to be lowered considerably. Consequently, the residual current is much increased, leading to an apparent change in the threshold voltage [8.12]. Similar to enhancement devices, this effect can be taken into account by introducing a term like Eq. (7.65).

8.2 Short-Channel Model

However, the latter effect is not the only cause of non-zero drain conductance. When the device operates in saturation, the implanted layer is pinched off close to the drain side. Again, for the buried-channel device the pinched-off region is no longer located at the interface, but deep into the bulk. As shown by numerical simulation [8.13], a dipole is formed. At the source side of the pinch-off region mobile carriers are accumulated owing to current impeding, but at the drain side depletion of mobile carriers occurs. At larger values of drain bias, the dipole slowly moves towards the source. This change can be interpreted as channel length modulation. However, up till now no analytical modelling results for this effect have been published.

8.2.2 Depletion-Mode Channel Conductance for a Linear Doping Profile

Since in a practical depletion-type MOSFET the buried-channel doping distribution has a shape in between a box-type and a linearly graded profile, we discuss here an approximation for the depletion-mode drain conductance, which is very useful for modelling short-channel devices [8.8]. Assuming a doping profile according to Fig. 8.6, we have

$$N(y) = \frac{N_{i0}}{d_i}(d_i - y). \tag{8.19}$$

Substituting this equation into Poisson's equation and solving for the thickness of the surface depletion layer (y_s), we obtain for the potential drop across this layer

$$\Delta V_S = \frac{qN_{i0}}{2\varepsilon_s d_i}[-\tfrac{1}{3}d_i^3 - \tfrac{2}{3}y_c^3 + y_c^2 d_i], \tag{8.20}$$

where

$$y_c = d_i - y_s.$$

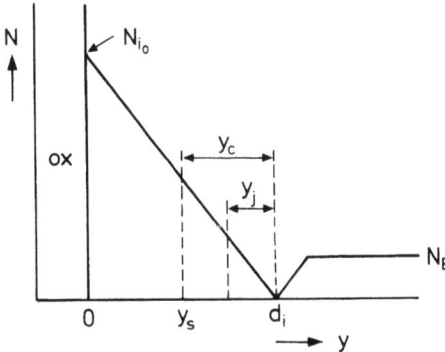

Fig. 8.6 Definition of symbols used to describe a linearly graded buried-channel device

Furthermore, the charge in this layer amounts to

$$Q_s = \frac{qN_{i0}}{2d_i}[d_i^2 - y_c^2]. \qquad (8.21)$$

Since according to Gauss' law

$$V_{GS} = V_{FB} + \Delta V_S - Q_s/C_{ox}, \qquad (8.22)$$

we then obtain

$$V_{GS} = V_{FB} - \frac{qN_{i0}}{2\varepsilon_s d_i}\left[\frac{d_i^3}{3} + \frac{2}{3}y_c^3 - y_c^2 d_i\right] - \frac{qN_{i0}t_{ox}}{2\varepsilon_{0x}d_i}(d_i^2 - y_c^2). \qquad (8.23)$$

Considering that at threshold condition $y_c = y_j$, where y_j is the p-n junction depletion layer thickness (compare Fig. 8.6), we have for the threshold voltage

$$V_T = V_{GS}(y_c = y_j). \qquad (8.24)$$

Consequently, the effective gate drive ($V_{GT} = V_{GS} - V_T$) can be written as follows,

$$V_{GT} = -\frac{qN_{i0}}{3\varepsilon_s d_i}(y_c^3 - y_j^3) + \frac{qN_{i0}}{2\varepsilon_s d_i}\left(d_i + \frac{\varepsilon_s}{\varepsilon_{0x}}t_{ox}\right)(y_c^2 - y_j^2). \qquad (8.25)$$

Solving Poisson's equation for the p-n junction depletion layer we obtain

$$\phi_j = \frac{q}{\varepsilon_s}\left[\frac{N_{i0}y_j^3}{3d_i} + \frac{N_{i0}^2 y_j^4}{4N_B d_i^2}\right], \qquad (8.26)$$

where ϕ_j is the built-in junction potential. From the given doping profile and the boundary values y_c and y_j, we finally obtain for the conductance at location $x(y_c)$ in the channel

$$g_c = \frac{q\mu_n N_{i0} W}{2d_i L}(y_c^2 - y_j^2). \qquad (8.27)$$

Via Eqs. (8.25), (8.26) and (8.27), the channel conductance is indirectly related to the external gate drive. In Fig. 8.7 a plot is given of the drain conductance g_d versus V_{GS}, using a number of relevant data given in the figure. In almost the entire depletion region the considered relation is linear. Only near $V_{GS} = V_T$ does a deviation occur, which is due to the fact that the depletion boundary in this case has moved far away from the gate control. The linear result of Fig. 8.7 can be directly obtained from the corresponding equations by approximating the cubic term in (8.25) according to

$$y^3 = \alpha d_i y^2,$$

where

$$\alpha \approx \tfrac{1}{2}.$$

Using this approximation we obtain from Eqs. (8.25), (8.26) and (8.27)

8.2 Short-Channel Model

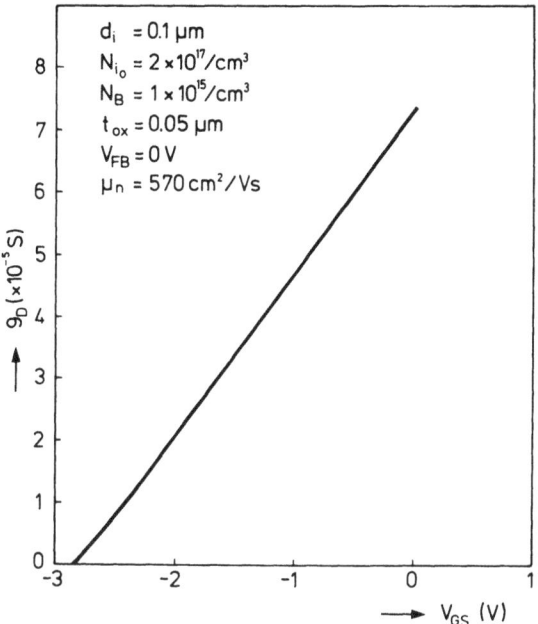

Fig. 8.7 Drain conductance as a function of gate-source voltage for a linearly graded buried-channel device in the depletion mode

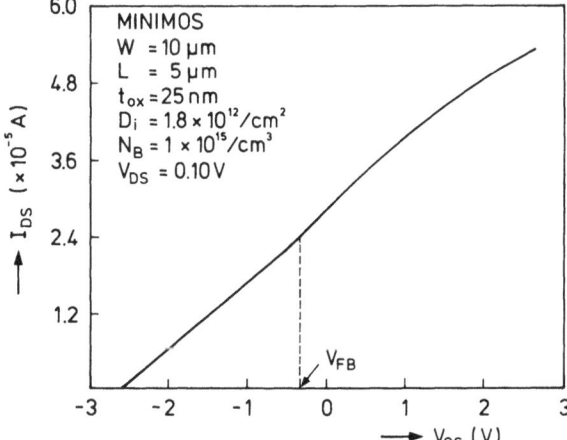

Fig. 8.8 Simulated drain conductance of an As-implanted, buried-channel MOSFET. Relevant parameters have been indicated

$$g_d = \mu_n C_i \frac{W}{L}(V_{GS} - V_T), \tag{8.28}$$

where the effective gate capacitance C_i is given by

$$C_i \approx \frac{\varepsilon_s}{\frac{2}{3}d_i + \varepsilon_s t_{ox}/\varepsilon_{ox}}. \tag{8.29}$$

The above result corrects for the fact that the centre of mobile carrier density is no longer located at the interface [8.2].
The linear relation (8.28) is further confirmed by numerical device simulation [8.14]. This is shown in Fig. 8.8, which gives the drain conductance of an As-implanted, buried-channel MOSFET.
Naturally the above approximation only holds for a gate bias which is not too close to the flat-band voltage V_{FB}. Although for the latter region a correction term has to be applied to (8.28), the above approximation has the advantage that the characteristics of the depletion MOSFET can be modelled in terms of an enhancement device and therefore velocity saturation effects are much simpler to implement. This is discussed in the next section.

8.2.3 The Drain Current of a Short-Channel Depletion MOSFET

In addition to the channel conductance in the depletion mode, the behaviour of this quantity in the accumulation mode is also important for modelling the drain current. Since in this case the mobile carriers are constrained to the interface, a charge-sheet modelling approach applies.
Therefore for $V_{GS} > V_{FB}$,

$$g_d = \mu_n C_i \frac{W}{L}[V_{FB} - V_T + r(V_{GS} - V_{FB})], \tag{8.30}$$

where

$$r = \frac{r_0}{1 + \theta_A(V_{GS} - V_{FB})}. \tag{8.31}$$

Here the first term of (8.30) represents the conductance due to the undepleted implanted layer and the second part that of the accumulation sheet.
The correction factor r takes into account that for the accumulation mode the effective gate capacitance is equal to C_{ox} rather than C_i, and that the mobility is not only lower than the bulk value, but also dependent on the normal field. Usually $1 < r_0 < 2$, owing to the smaller effective gate capacitance in the depletion mode.
Introducing an apparent threshold voltage

$$V_T' = \frac{V_T + (r_0 - 1)V_{FB}}{r_0}, \tag{8.32}$$

8.2 Short-Channel Model

Eq. (8.30) can be easily rewritten into the useful form

$$g_d = \mu_n C_i r \frac{W}{L}(V_{GS} - V_T'). \tag{8.33}$$

Note that the form of Eq. (8.33) makes it possible to model the characteristics of the accumulation mode as a pure enhancement device with an apparent threshold voltage V_T'. This is often done for depletion load applications. However, the most important point is that in the more general cases Eqs. (8.28) and (8.33) are still similar to the basic equation of an enhancement transistor and therefore effects of velocity saturation, p-n junction space charge and drain-induced barrier lowering can be easily adapted. Of course, following this approach it is necessary to correct in the above equations for the difference in slope of the conductance at the boundary V_{FB}. In this way the deviation of the depletion-mode conductance near V_{FB} can be corrected as well. The above aims can be achieved by introducing a smoothing function for the effective gate drive according to

$$\begin{aligned} V_{GT} = \frac{1}{2}\bigg[& V_{GS} - 2V_T + V_{FB} + r(V_{GS} - V_{FB}) \\ & + \frac{r-1}{r}\frac{V_0^2}{V_{FB} - V_T} + \{r^2(V_{GS} - V_{FB})^2 + V_0^2\}^{1/2} \\ & - \{(V_{GS} - V_{FB})^2 + V_0^2\}^{1/2} \bigg]. \end{aligned} \tag{8.34}$$

The factor V_0 is a modelling constant, which has a value of approximately 0.25 Volt in practice.

Using the above gate drive the drain current of a short-channel depletion MOSFET can be easily derived from the enhancement model. For the sake of completeness we repeat here a complete set of equations, by which the characteristics can be described.

— The threshold voltage

$$V_T = V_{T0} + \gamma[(V_{SB} + \phi_j)^{1/2} - \phi_j^{1/2}] - \eta V_{DS}, \tag{8.35}$$

where γ represents the back-bias control from the p-n junction and η represents the drain-induced barrier lowering effect.

— The drain current

$$I_D = \frac{W}{L}\beta_0 \frac{[V_{GT} - \frac{1}{2}(1 + \delta)V_R]V_R}{1 + \theta_C V_R}, \tag{8.36}$$

where

$$\delta \approx \frac{0.3\gamma}{(V_{SB} + \phi_j)^{1/2}} \tag{8.37}$$

represents the effect of bulk depletion charge and θ_C represents the velocity saturation effect.

— The saturation voltage

$$V_{DSS} = \frac{2V_{GT}}{1+\delta}\left[\left\{1 + \frac{2\theta_C V_{GT}}{1+\delta}\right\}^{1/2} + 1\right]^{-1}. \tag{8.38}$$

When $V_{DS} < V_{DSS}$, then in (8.36) $V_R = V_{DS}$, but when $V_{DS} \geqslant V_{DSS}$, then $V_R = V_{DSS}$. Since usually the drain-induced barrier lowering effect is more important than channel length modulation, we have neglected the latter effect here. Note that compared to the pure enhancement transistor two extra parameters are needed: V_{FB} and r_0. Note further that the above approach is rather similar to a piecewise linear approach [8.7], which takes into account the three different modes separately. However, the cited approach still suffers from discontinuities of the channel conductance at the various boundaries. When comparing the above model equations to the measured characteristics of depletion MOSFETS a fair agreement is obtained. This is shown in Fig. 8.9, where the measured data for a short-channel depletion transistor is compared to calculated values.

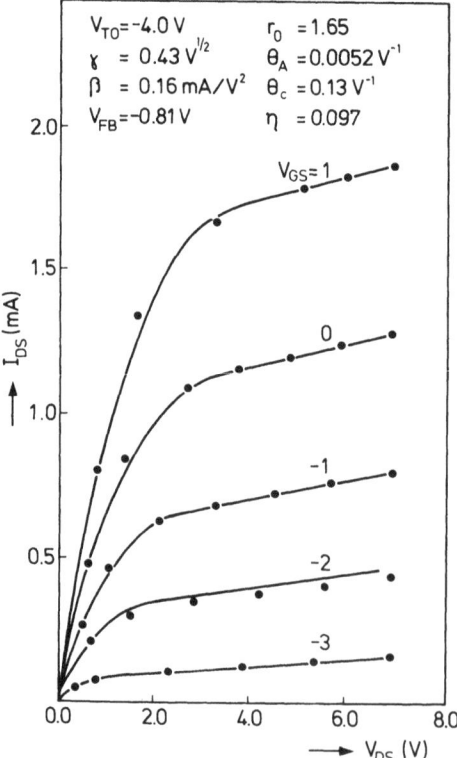

Fig. 8.9 Characteristics of a short buried-channel MOSFET in all modes of operation ($L_{eff} \approx 2$ µm). Dots are measured values and fully drawn lines represent modelled results

Several relevant parameter values have been given as an inset in the figure. Not only is the average deviation very low, but most important is that some primary parameters like V_{T0} and β_0 can be easily obtained and that all others have a physical meaning.

8.3 Charges and Charge Distribution

In principle the terminal charges can be calculated making use of the auxiliary functions for the drain current. For instance, making use of the fact that the unit-area gate charge Q_g equals Q_s, the gate charge is given by [8.4]

$$Q_G = W \int_0^L Q_g \, dx = \frac{\mu_n W^2}{I_D} \int_{V_S'}^{V_D'} Q_s Q_m \, dV', \qquad (8.39)$$

where the charges Q_s and Q_m have been defined already in Eqs. (8.1) through (8.8). The same procedure can be applied to the bulk charge

$$Q_B = \frac{\mu_n W^2}{I_D} \int_{V_S'}^{V_D'} Q_j Q_m \, dV'. \qquad (8.40)$$

The real problem, however, lies in the calculation of the source and drain charge. Although the total channel charge can be easily obtained from (8.39) and (8.40), the required procedure to split the channel charge into a source and drain part (compare section 6.6.2) leads to cumbersome calculations. This is due to the numerous square-root terms in the various equations for Q_s and Q_j. In fact up till now no complete and correct charge model corresponding to the long-channel current model has been published.
In practice therefore it is more useful to apply the charge model for the enhancement device, which has been discussed in 7.2.2. By defining a gate driving voltage according to (8.34), in a manner similar to the calculation of current, all expressions from 7.2.2 can in fact be adapted for the depletion MOSFET. In practice this approach is sufficiently accurate.

References

[8.1] J. R. Edwards, G. Mar: Depletion-Mode IGFET Made by Deep Ion Implantation. IEEE Transactions on Electron Devices *ED-20*, 283–289 (1973).
[8.2] J. S. T. Huang: Characteristics of a Depletion-Mode IGFET. IEEE Transactions on Electron Devices *ED-20*, 513–515 (1973).
[8.3] R. W. Knepper: Dynamic Depletion Mode: An E/D MOSFET Circuit-Method for Improved Performance. IEEE Journal of Solid-State Circuits *SC-13*, 542–548 (1978).
[8.4] J. A. El-Mansy: Analysis and Characterization of the Depletion-Mode IGFET. IEEE Journal of Solid-State Circuits *SC-15*, 331–340 (1980).
[8.5] P. E. Schmidt, M. B. Das: D. C. and High-Frequency Characteristics of Built-In Channel MOSFETS. Solid-State Electronics *21*, 495–505 (1978).

[8.6] R. A. Haken: Analysis of the Deep-Depletion MOSFET and the Use of the DC Characteristics for Determining Bulk-Channel Charge Coupled Device Parameters. Solid-State Electronics *21*, 753–761 (1978).

[8.7] G. Merckel: Ion Implanted Depletion-Mode MOS Transistors. In: Process and Device Modelling for IC Design. Noordhoff, Leiden (1977), pp. 617–676.

[8.8] F. M. Klaassen, W. C. J. de Groot: Modelling of the Depletion-Mode IGFET. Europhysics Conference Abstracts, ESSDERC 80 (1980), pp. 107–108.

[8.9] C. Turchetti, G. Masetti: Analysis of the Depletion-Mode MOSFET Including Diffusion and Drift Currents. IEEE Transactions on Electron Devices *ED-32*, 773–782 (1985).

[8.10] G. R. Mohan Rao: An Accurate Model for a Depletion Mode IGFET Used as a Load Device. Solid-State Electronics *21*, 711–714 (1978).

[8.11] M. Wordeman: Characterization of Depletion-Mode MOSFETS. Digest IEDM, Washington DC (1979), pp. 26–29.

[8.12] N. Ballay, B. Baylac: Analytical Modelling of the Depletion-Mode MOSFET with Short- and Narrow-Channel Effects. IEE Proceedings *127*, 225–230 (1981).

[8.13] D. P. Kennedy, R. R. O'Brien: Electric Current Saturation in a Junction Field-Effect Transistor. Solid-State Electronics *12*, 829–836 (1969).

[8.14] MINIMOS, A 2-D Device Simulator for MOSFET Devices. Technical University of Vienna.

Models for the JFET and the MESFET 9

Although from a viewpoint of manufacturing and application the differences between the junction-gate field-effect transistor (JFET) and the metal-gate field-effect transistor (MESFET) are considerable, their physical operation is almost identical. The modelling of these devices is therefore discussed in one chapter. In both cases transistor operation is achieved by depleting an already existing channel region via a gate-controlled p-n junction or a Schottky diode. The channel region can be realized as an n-type epitaxial layer in a p-type substrate (as for instance for the discrete JFET), as an implanted p-type layer in an n-well (bipolar IC-compatible JFET) or as an epitaxial layer on a semi-insulating substrate (GaAs MESFET). In the latter case a channel implant is often added for achieving better process control. Figs. 9.1a and 9.1b give a cross-section of the JFET and the MESFET, respectively.
For analog applications both devices are operating in the depletion mode and therefore their characteristics are qualitatively similar to that of depletion MOSFETs. However, for logic applications normally-off type, n-channel GaAs MESFETs have been developed. In these devices the implanted channel is sufficiently thin to become completely depleted at zero gate bias. By operating the Schottky gate lightly in forward bias, the depletion of the channel area is diminished and transistor action occurs. However, due to the increase in gate current, the applicable range of gate voltages is rather narrow.
Usually a circuit model for the JFET is identical to the description [9.1] found in most textbooks, and which is based on the depletion of a uniformly doped channel via an abrupt junction. After a short discussion of this long-channel model, it is further refined for short-channel devices by taking into account the effects of series resistance and velocity saturation [9.2, 9.3]. Although the latter model seems suitable to describe the MESFET characteristics as well, it has been reported [9.4] that for modern sub-micron-gate GaAs devices the saturation of the characteristics is modelled unsatisfactorily. This has been attributed to the effects of non-uniform channel doping and velocity overshoot, which occurs in GaAs already at rather low fields [9.5]. In particular the latter effect may cause the current to saturate

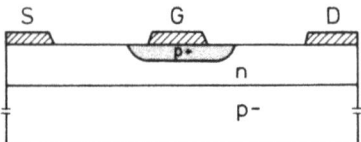

Fig. 9.1 a Cross-section of a JFET

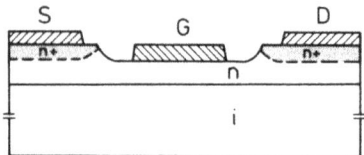

Fig. 9.1 b Cross-section of a MESFET

at much lower drain voltages than according to a Si-based model. Since incorporation of a velocity overshoot model [9.6, 9.7] in the basic current equations leads to cumbersome, implicit relations for the drain current expression [9.8], only empirical models have been proposed for circuit simulation [9.4, 9.9, 9.10, 9.11, 9.12]. Unfortunately, after an evaluation of the above models, none of them appears to be more accurate in describing measured characteristics than a simple short-channel MOSFET model. Therefore it is discussed how an empirical model can be improved to enhance the accuracy [9.13]. This chapter is ended with some proposals for charge modelling.

9.1 The Drain Current of the Junction-Gate FET

9.1.1 The Classical Description

When an *n*-channel JFET, made on a high-resistivity or semi-insulating substrate, is operating at negative gate bias V_{GS} and positive drain bias V_{DS} (taking all voltages with respect to the source), the conducting channel has a configuration as given in Fig. 9.2.

For each position x located in the channel, the current is given by

$$I(x) = WqN_D\mu_n(a - h(x))\frac{dV(x)}{dx}. \qquad (9.1)$$

where N_D is the doping of the channel, a is the thickness and $h(x)$ is the depleted part of that region. Assuming an abrupt junction between gate and channel, the depletion thickness is given by

9.1 The Drain Current of the Junction-Gate FET

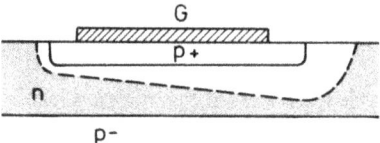

Fig. 9.2. View of channel and depletion width of a JFET under biased conditions ($V_G < 0$, $V_D > 0$)

$$h(x) = a \frac{[V(x) - V_{GS} + \phi_j]^{1/2}}{V_p^{1/2}}, \tag{9.2}$$

where ϕ_j is the built-in potential difference and

$$V_p = \frac{qN_D a^2}{2\varepsilon_s} \tag{9.3}$$

is the pinch-off voltage, that is the total voltage ($V_{DS} - V_{GS} + \phi_j$) at which the channel is entirely depleted ($h(L) = a$).

Since the gate reverse current is orders of magnitude smaller than the channel current, $I(x)$ is a constant and equals the drain current I_{DS}. Substituting Eq. (9.2) into Eq. (9.1) and integrating the latter expression from $x = 0$ to $x = L$ yields

$$I_{DS} = I_p \left[\frac{3V_{DS}}{V_p} - 2 \left\{ \left(\frac{V_{DS} - V_{GS} + \phi_j}{V_p} \right)^{3/2} - \left(\frac{-V_{GS} + \phi_j}{V_p} \right)^{3/2} \right\} \right], \tag{9.4}$$

in which the current constant

$$I_p = \frac{W\mu_n q^2 N_D^2 a^3}{6\varepsilon_s L}. \tag{9.5}$$

The drain current in the saturation region is obtained from Eq. (9.4) by setting ($h(L) = a$)

$$V_p = V_{DS} - V_{GS} + \phi_j.$$

Hence I_{DSS} is given by

$$I_{DSS} = I_p \left[1 - 3 \left(\frac{-V_{GS} + \phi_j}{V_p} \right) + 2 \left(\frac{-V_{GS} + \phi_j}{V_p} \right)^{3/2} \right]. \tag{9.6}$$

The corresponding saturation voltage follows from

$$V_{DSS} = V_p + V_{GS} - \phi_j. \tag{9.7}$$

For low drain voltages V_p also determines the threshold voltage ($h(0) = a$)

$$V_T = \phi_j - V_p. \tag{9.8}$$

Substituting (9.8) in Eq. (9.6) and taking a Taylor series expansion around $V_{GS} \approx V_T$, it can be shown that

$$I_{DSS} = \frac{W\mu_n \varepsilon_s}{2aL}(V_{GS} - V_T)^2. \tag{9.9}$$

This relation is similar to the enhancement-mode MOSFET equation with the channel depth a replacing the insulator thickness. In fact, this simple result has been used to model the JFET in circuit simulation. However, for most types of JFET this approach is too inaccurate.

The channel conductance in saturation is zero for the idealized situation, since I_{DSS} is not a function of V_{DS}. For practical devices, in particular with a short channel length, the above results have to be corrected for the effects of series resistance, velocity saturation and reduction of effective channel length at increased values of V_{DS}. This will be discussed in the next section.

9.1.2 A Model for Short-Channel Transistors

Owing to the distance between gate junction and source and drain contact regions (whose minimum value is determined by lithographic and breakdown voltage conditions) and the relatively low channel dope, series resistances are present near the source and drain ends as illustrated in Fig. 9.1 a. To correct for the voltage drop between the source and drain contacts and the channel, a resistance R_s and R_d can be put in series with the source and drain terminal. Although for analog applications the increase of CPU time owing to the extra number of nodes may be acceptable, the effect of a source resistance can be easily taken into account in the saturated drain-current expression.

In this case the effect of drain resistance is not important and in Eq. (9.6) only the V_{GS} terms have to be replaced by

$$(V_{GS} - I_{DSS}R_s).$$

When R_s is relatively small, its effect can be considered to be of second order. Taking a series expansion of Eq. (9.6) and keeping only first-order terms in $I_{DSS}R_s$, the above relation can be rewritten as

$$I_{DSS} \approx I_p \frac{\left[1 - 3\left(\frac{-V_{GS} + \phi_j}{V_p}\right) + 2\left(\frac{-V_{GS} + \phi_j}{V_p}\right)^{3/2}\right]}{1 + 3\frac{R_s I_p}{V_p}\left[1 - \left(\frac{-V_{GS} + \phi_j}{V_p}\right)^{1/2}\right]}. \tag{9.10}$$

Next we define the latter result to be valid for the linear region as well. Although the neglect of R_d causes an error, this shortcoming is acceptable for most JFET applications. In addition, the above incorporation of R_s in the transistor equations facilitates the determination of this parameter from measurements.

Using a velocity saturation model similar to that discussed for MOSFETs (compare section 7.2.1.3) it can be shown [9.2, 9.3] that, similar to MOSFETs, the original drain-current expression has to be divided by a term

9.1 The Drain Current of the Junction-Gate FET

$$1 + \frac{\mu_n V_{DS}}{v_s L},$$

where v_s is the limiting velocity. Naturally this additional term causes the saturation voltage to become lower than that given by Eq. (9.7). Usually this voltage is obtained from the condition $\partial I_{DS}/\partial V_{DS} = 0$. In this case the saturation voltage is found from solving a cubic equation, which is acceptable for simulation of most analog designs.

Owing to the similar situation at pinch-off condition, for channel-length modulation a MOSFET expression like Eq. (7.89) can be applied.

Following the various approaches discussed above, the complete model equations can now be given.

For the linear region the drain current reads

$$I_{DS} = I_p \frac{3(u^2 - u_0^2) - 2(u^3 - u_0^3)}{[1 + \theta_A(1 - u_0)][1 + \theta_C V_p(u^2 - u_0^2)]}, \tag{9.11}$$

where

$$u^2 = \frac{V_{DS} - V_{GS} + \phi_j}{V_p}, \tag{9.12}$$

$$u_0^2 = \frac{-V_{GS} + \phi_j}{V_p}, \tag{9.13}$$

$$\theta_A = \frac{3R_s I_p}{V_p} = R_s \frac{q\mu_n N_D aW}{L} \tag{9.14}$$

and

$$\theta_C = \frac{\mu_n}{v_s L}. \tag{9.15}$$

Note the similarity of the form of the denominator with the corresponding MOSFET expression.

The drain saturation voltage follows from a solution of the cubic equation

$$u_{SAT}^3 + 3u_{SAT}\left(\frac{1}{\theta_C V_p} - u_0^2\right) + 2u_0^3 - \frac{3}{\theta_C V_p} = 0. \tag{9.16}$$

Unfortunately, since the coefficients of this equation change sign with the variation of gate bias, no direct analytic solution can be given. Therefore (9.16) has to be solved by iteration.

For drain voltages in excess of u_{SAT} the drain current is given by

$$I_{DSS} = I_p \frac{3(u_{SAT}^2 - u_0^2) - 2(u_{SAT}^3 - u_0^3)}{[1 + \theta_A(1 - u_0)][1 + \theta_C V_p(u_{SAT}^2 - u_0^2)]}$$

$$\times \left[1 + \alpha \ln\left(1 + \frac{u^2 - u_{SAT}^2}{\alpha V_0}\right)\right]. \tag{9.17}$$

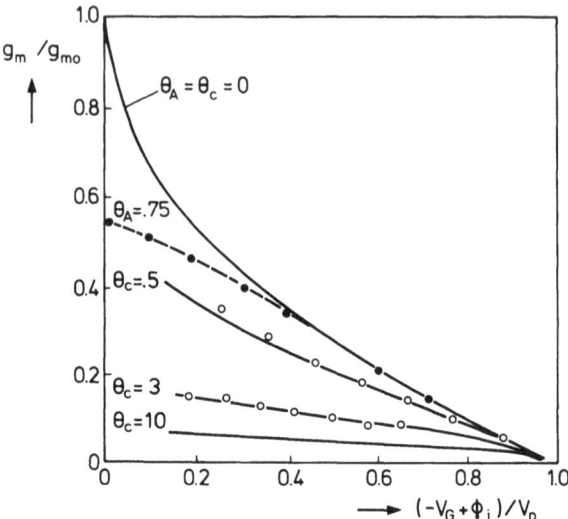

Fig. 9.3 Normalized JFET transconductance as a function of normalized gate bias for the saturation region. In addition the effect of series resistance and velocity saturation is shown. Dots are measured values

The parameters in the above model are the current constant I_p, the pinch-off voltage V_p, the series resistance parameter θ_A, the velocity saturation parameter θ_C and the channel-length modulation parameter α. Although formally the built-in voltage ϕ_j is a parameter too, in practice, since the model equations are rather insensitive to changes in its value, ϕ_j can be fixed at a value of 0.70 Volt. Of course, in order to maintain continuity of the drain conductance at u_{SAT}, a numerical manipulation is required for the last term of Eq. (9.17).

In order to show the effects of series resistance and velocity saturation on the JFET characteristics, in Fig. 9.3 the saturated transconductance according to (9.17) has been plotted as a function of normalized gate bias. In addition, as a normalization factor for the transconductance the value

$$g_{mo} = \frac{q\mu_n N_D aW}{L}$$

is recognized as the conductance of the undepleted channel. The reason for using the transconductance rather than the drain current lies in the fact that only this quantity has been compared with experimental results in a few cases [9.14, 9.15]. This is not impractical, since for analog applications the transconductance is more important than the absolute value of the current. While velocity saturation reduces the transconductance in the entire gate bias region, the series resistance mainly affects the value at low gate bias, where the channel resistance becomes low. As a result, in practice the transconductance never shows the steep rise at low values of V_{GS}, as pre-

9.2 The Drain Current of the MESFET

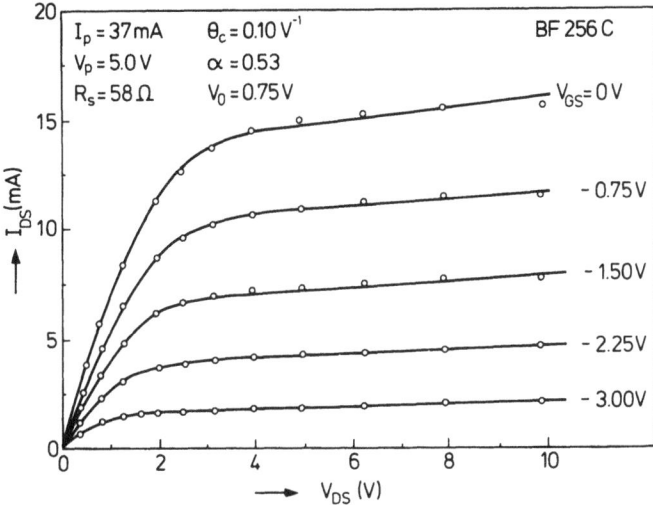

Fig. 9.4 Comparison of measured and calculated JFET characteristics

dicted from the ideal $R_s = 0$ plots. The above calculated results are confirmed by experimental results, which are represented by the dots in Fig. 9.3. Finally, Fig. 9.4 gives a comparison of the measured characteristics of a commercial Si JFET with calculated results according to the above equations. Except for the drain conductance, an excellent fit is obtained. However, the deviations at high values of I_{DS} and V_{DS} might be caused by heating. In addition, the low value of the velocity saturation parameter might be due to effects of drain series resistance (compare section 7.2.3).

9.2 The Drain Current of the MESFET

9.2.1 Review of Empirical Models

In contrast to the JFET, the modelling of the MESFET has been the subject of many papers. This certainly reflects its importance as a high-frequency IC component. In order to describe the earlier transition from the linear to the saturation region due to velocity overshoot in GaAs MESFETs, a number of authors [9.4, 9.9, 9.10] have introduced the hyperbolic tangent formula as a multiplication factor

$$\tanh(\lambda V_{DS}),$$

where λ is a parameter.

Note that for very low values of V_{DS} this function reduces to λV_{DS}, thus describing the linear region. However, the cited authors differ in the description of the dependence of the saturated drain current on gate bias. The results vary from a simple square-law I_{DS}-V_{GS} relationship [9.4], via an

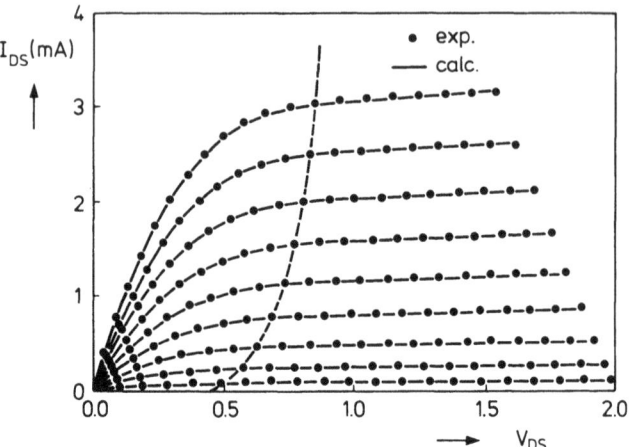

Fig. 9.5 Typical MESFET characteristics with dashed line as an indication of the saturation point

exponential relationship [9.9] to an expression quite similar to that of short-channel enhancement MOSFETs [9.10]. Furthermore, in order to model the non-zero drain conductance the entire current expression is usually multiplied by a factor

$(1 + \alpha V_{DS})$.

Fig. 9.5 illustrates the effect of the hyperbolic tangent function upon the I_{DS}-V_{DS} characteristics [9.10]. The dashed line, which represents the onset of saturation, is indeed quite different from that produced by Si devices. As a further refinement to the previous models a static feedback effect on threshold voltage has also been introduced [9.11]. Finally, the exponential V_{GS} dependence has been combined with a similar V_{DS} dependence, which has the features of the hyperbolic tangent function, but is computationally less expensive [9.12].

Since a physical basis for most terms of the model equations is lacking and a large variety of V_{DS} and V_{GS} terms has been combined in different order, all models cited above have to be considered as empirical.

As an example we present here the last cited model [9.12]:

— Saturation region ($V_{DS} > (V_{GS} - V_T)/k$),

$$I_{DSS} = \beta(1 + \alpha k V_{DS})(V_{GS} - V_T)^{1+p}. \tag{9.18}$$

— Linear region ($V_{DS} \leq (V_{GS} - V_T)/k$),

$$I_{DS} = \beta(1 + \alpha k V_{DS}) k V_{DS} p^{-p}[(1 + p)(V_{GS} - V_T) - k V_{DS}]^p. \tag{9.19}$$

The model is adjusted to ensure current and drain conductance continuity. Early saturation is achieved by choosing the parameter $k > 1$.

9.2 The Drain Current of the MESFET

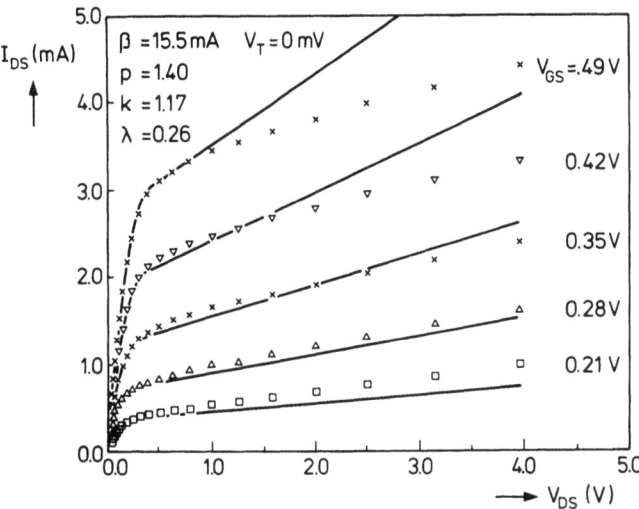

Fig. 9.6 Comparison of measured (dots) and calculated characteristics (fully drawn lines) of a 0.6 μm MESFET

Despite the fantastic accuracy claimed by several authors (compare Fig. 9.5), none of the above models is able to fit the measured characteristics of a submicron enhancement-mode MESFET with an average accuracy within 8–10%. Fig. 9.6 gives a comparison between Eqs. (9.18) and (9.19) and the measured current for a device with a channel length of 0.6 μm. Not only are the deviations shown typical for all models, but the results in fact are no better than those obtained with a short-channel MOSFET enhancement model [9.13]. Somewhat better results can be obtained by optimizing on absolute rather than relative error. However, since the latter approach leads to better results in a wide bias range, it has been used throughout this book. The main defects are the poor description of the saturation region, which is often enlarged by the occurrence of low-frequency current drift [9.16] and the transition between linear region and saturation. The latter is caused by the fact that for all models the derivative of the drain conductance with respect to the drain source voltage is not continuous. Although this is not strictly required for circuit simulation, this deficiency causes major problems for analog applications. In the next section it will be shown how the MESFET model (9.18) can be improved.

9.2.2 An Improved Model

Since in the previous models channel-length modulation is characterized by multiplying the drain current by the term $(1 + \alpha V_{DS})$, consequently the dependence of the drain conductance on gate bias becomes proportional to a term $(V_{GS} - V_T)^{1+p}$. However, owing to the value $p \approx 1.5$ found from optimizing the measured values of the drain current, the model then predicts the drain conductance as depending on gate bias with a power higher than square-law. In reality the measured drain conductance varies more or less linearly with gate bias. Therefore, the parameter α has to be divided at least by a factor $(V_{GS} - V_T)$. Unfortunately, no physical mechanism is known to explain this necessary reduction of α.

The discontinuity of the derivative of the drain conductance at V_{DSS}, mentioned previously, can be eliminated by applying a numerical manipulation to the definition of saturation voltage. Instead of clamping the drain-source voltage abruptly to the saturation voltage (compare the conditions of Eqs. (9.18) and (9.19)), a smoothing function can be applied.

Finally, although in GaAs MESFET technology the source resistance is reduced by applying a recessed gate technique, the resulting value is still not negligible. Its effect on the drain current can be taken into account similar to the modelling of the JFET (compare Eq. (9.11)).

Following the above procedures, the improved MESFET model takes the form [9.13]:

— Gate-driving voltage,

$$V_{GT} = V_{GS} - V_T. \qquad (9.20)$$

— Transition from the linear to the saturation region,

$$V_{DSS} = V_{GT}/k. \qquad (9.21)$$

— The drain current

$$I_{DS} = \frac{\beta k p^{-p} V'_{DS}[(1+p)V_{GT} - kV'_{DS}]^p}{1 + R_s\beta(p+1)V_{GT}^p} + \alpha V_{GT} V_{DS}, \qquad (9.22)$$

where

$$V'_{DS} = \frac{V_{GT}}{k} - \frac{1}{2}\{V_{SAT} + (V_{SAT}^2 + V_{sm}^2)^{1/2}\}, \qquad (9.23)$$

$$V_{SAT} = \frac{V_{GT}}{k} - V_{DS}, \qquad (9.24)$$

$$V_{sm}^2 = eV_{GT}V_{DS}. \qquad (9.25)$$

and e is a modelling constant.

Fig. 9.7 gives a comparison between the measured and calculated characteristics for the same device as in Fig. 9.6. Comparing both figures the improvement becomes clear.

The above model has been extended to the subthreshold region, by redefining the effective gate drive (Eq. (9.20)) similar to a procedure introduced for modelling the MOSFET subthreshold region (compare Eq. (7.88)). In addition, since normally-off MESFETs are being operated with the Schottky-gate diode in the forward direction, at high values of gate drive the gate

9.3 Charges and Capacitances

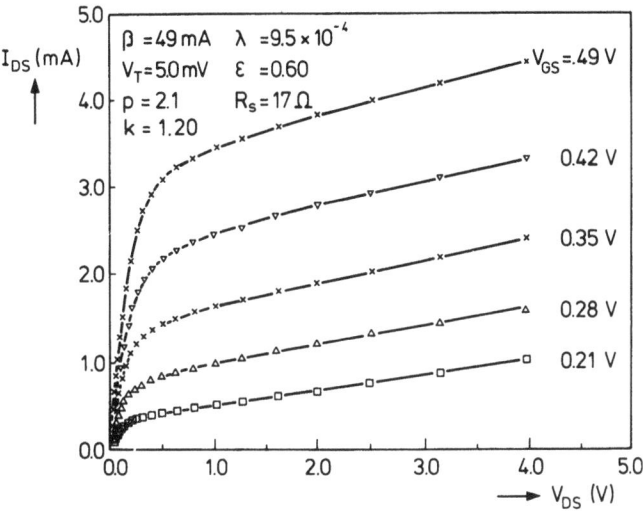

Fig. 9.7 Comparison of the measurements of Fig. 9.6 with characteristics calculated according to improved model

current can no longer be neglected. Therefore, a gate-diode model has to be added to an equivalent circuit of the MESFET.

9.3 Charges and Capacitances

As far as the authors know, no charge model based on a correct partitioning of the channel charge (compare section 6.6.2) has been derived for the devices considered. Owing to the inherent mathematical problems, this is not surprising. Therefore, mainly empirical models have been proposed [9.4, 9.10]. According to the latter cited approach the gate charge is given by

$$Q_G = 2C_{GS0}\phi_j\left[1 + \left(1 - \frac{V_1}{\phi_j}\right)^{1/2}\right] + C_{GD0}V_2, \qquad (9.26)$$

where

$$V_1 = \tfrac{1}{2}[V_{GS} + V_{GD} + \{(V_{GS} - V_{GD})^2 + K\}^{1/2}], \qquad (9.27)$$

$$V_2 = \tfrac{1}{2}[V_{GS} + V_{GD} - \{(V_{GS} - V_{GD})^2 + K\}^{1/2}] \qquad (9.28)$$

and K is a smoothing constant.

Here C_{GS0} is the gate-to-source capacitance for zero-gate/source bias and C_{GD0} is the gate-to-drain capacitance for zero-gate/drain bias. The voltage V_1 stands for the larger of the two values of V_{GD} and V_{GS}, and V_2 for the smaller of the two. These voltages have been introduced to obtain a smooth change of the gate charge when the device changes from normal bias ($V_{DS} > 0$) to reverse bias ($V_{DS} < 0$). In fact, the first term of (9.26) can be derived for an ideal long-channel MOSFET without bulk charge. Using the above gate charge it can be shown that the resulting small-signal capacitances C_{GS} and C_{GD} are not much different from measured values. Next, the increment of the source charge is calculated according to the somewhat artificial definition

$$\Delta Q_S = \tfrac{1}{2}[Q_G(V_{GS} + \Delta V_{GS}, V_{GD} + \Delta V_{GD}) - Q_G(V_{GS}, V_{GD} + \Delta V_{GD})$$
$$+ Q_G(V_{GS} + \Delta V_{GS}, V_{GD}) - Q_G(V_{GS}, V_{GD})], \qquad (9.29)$$

with a similar definition for ΔQ_D. Nevertheless it has been stated [9.10] that the displacement currents in depletion-type devices calculated from the above expression are fairly accurate. However, no real proof has been given. Since the above procedure is very empirical and in fact based on a long-channel MOSFET expression, it will lead to major errors in enhancement-

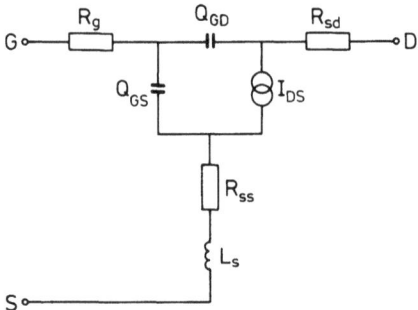

Fig. 9.8 Equivalent circuit of an enhancement-type MESFET

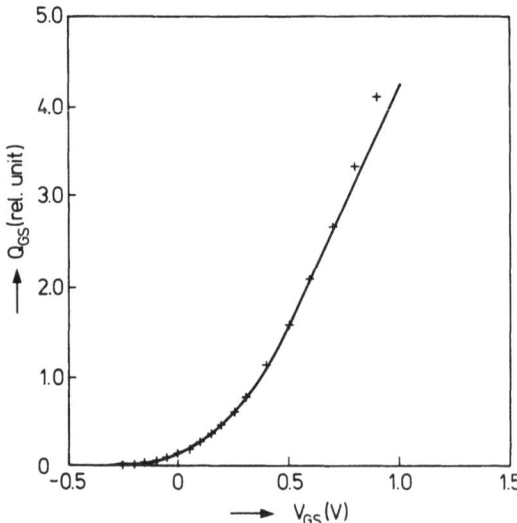

Fig. 9.9 The gate charge of an enhancement-type MESFET as a function of gate bias

type MESFETs, where the gate may operate in the forward direction. Hence for the analysis of h.f. circuits based on these devices, it is likely to be more useful to apply a modified MOSFET equivalent circuit like the one shown in Fig. 9.8 [9.17].

In this scheme, the charges Q_{GS} and Q_{GD} are given by an empirical (two-terminal) equation describing measured charge data as, for instance, given in Fig. 9.9. This figure has been obtained from measuring the gate capacitance characteristic. Note that the charge considered increases rapidly with gate bias. Due to this fact often the cut-off frequency of a MESFET decreases with an increase of drain current.

The other elements are parasitic resistances and inductances, which are not easy to determine. For instance, the gate resistance represents the effect of the 2-D transmission line of the gate configuration. On the other hand, the resistances R_{ss} and R_{sd} can be determined from a careful analysis of the $I_{DS} - V_{DS}$ characteristics (see the previous section).

References

[9.1] W. Shockley: A Unipolar Field-Effect Transistor. Proc. Inst. Radio Engrs. 40, 1365–1376 (1952).
[9.2] J. R. Hauser: Unipolar Transistors. In: Fundamentals of Silicon Device Technology. Prentice-Hall, New York (1968), pp. 269–349, chapt. 2.
[9.3] F. N. Trofimenkoff: Field-Dependent Mobility Analysis of the Field-Effect Transistor. Proc. IEEE 53, 1765–1766 (1965).
[9.4] W. R. Curtice: A MESFET Model for Use in the Design of GaAs Integrated Circuits. IEEE Transactions on Microwave Theory and Techniques *MTT-28*, 448–456 (1980).
[9.5] J. G. Ruch, G. S. Kino: Measurement of the Velocity-Field Characteristics of Gallium Arsenide. Applied Physics Letters 10, 40–42 (1967).
[9.6] P. Jeppesen, B. Jeppson: Computer Simulation of LSA Oscillators with High Doping to Frequency Ratio. Proc. IEEE 57, 795–796 (1969).
[9.7] C. S. Chang, H. R. Fetterman: Electron Drift Velocity vs. Electric Field in GaAs. Solid-State Electronics 29, 1295–1296 (1986).
[9.8] D. Boccon-Gibod: Modèle Analytique et Schéma Équivalent du Transistor à Effet de Champ en Arsénure de Gallium. Acta Electronica 23, 99–109 (1980).
[9.9] S. E. Sussmann-Fort, J. C. Hantgan, F. L. Huang: A Spice Model for Enhancement- and Depletion Mode GaAs FETs. IEEE Transactions on Microwave Theory and Techniques *MTT-34*, 1115–1119 (1986).
[9.10] H. Statz, P. Newman, I. W. Smith, R. A. Pucel, H. A. Haus: GaAs FET Device and Circuit Simulation in SPICE. IEEE Transactions on Electron Devices *ED-34*, 160–169 (1987).
[9.11] T. Kacprzak, A. Materka: Compact DC Model of GaAs FETs for Large-Signal Computer Calculation. IEEE Journal of Solid-State Circuits *SC-18*, 211–213 (1983).
[9.12] L. E. Larson: An Improved GaAs MESFET Equivalent Circuit Model for Analog Integrated Circuit Applications. IEEE Journal of Solid-State Circuits *SC-22*, 567–574 (1987).
[9.13] R. Dekker: A Pragmatic Circuit Model for GaAs MESFETs. M. Sc. Thesis, Fac. of Electr. Engin., Eindhoven University of Technology (1988).
[9.14] K. Lehovec, R. Zuleeg: Voltage-Current Characteristics of GaAs JFETs in Hot Electron Range. Solid-State Electronics 13, 1415–1426 (1970).

[9.15] A. S. Grove: Physics and Technology of Semiconductor Devices. John Wiley & Sons, New York (1967), p. 253.
[9.16] T. Ducourant, M. Rocchi: Modelling of the Drain Lag Effect in GaAs MESFETS. ESSDERC 88. Journal de Physique *C2*, 313–316 (1987).
[9.17] J. A. van Steenwijk: Internal Philips Technical Note (1988).

Parameter Determination 10

A good set of parameters is as important as a good model; the most accurate model will perform badly if the parameter values are not correct. Unfortunately, the set of parameter values is not unique, which means that there are other parameter sets possible that give more or less the same fit to the measured characteristics. This is mainly caused by the fact that the various device phenomena, as described by certain parameters, cannot always be distinguished clearly from each other in the measured characteristics. As examples we may mention quasi-saturation and high injection in the base, or Early effect and avalanche multiplication in bipolar transistors, and static feedback and channel length modulation in MOS transistors. This also means that usually not all compact model parameters are independent from each other: the value given to one parameter may influence the value of another. The final judgement whether a parameter set is good enough lies in the fit to the measurements and in their physical plausibility.

In this chapter we will first discuss the main optimization strategies for parameter extraction, then we will describe some specific measurements for bipolar (series resistances) models. The chapter will be concluded by worked-out examples of the extraction from measured characteristics of a complete parameter set of both a bipolar and a MOST model.

10.1 General Optimization Method

A simple, yet general model equation can be written in the form

$$I = f(V, P_1, \ldots, P_M). \tag{10.1}$$

This represents an I-V characteristic at a given temperature with M parameter values. We suppose that we have N measured points ($N > M$): $I_1 - V_1$, $I_2 - V_2$, ..., $I_N - V_N$. Substitution of these points in the model equation (10.1) then gives N calculated values f_l ($l = 1, \ldots, N$) of the model function f. We now consider the following function of the M parameters [10.1, 10.2, 10.3],

$$S(P_1, \ldots, P_M) = \sum_{l=1}^{N} \left(\frac{I_l - f_l}{I_l}\right)^2. \tag{10.2}$$

S is a measure for the relative error between model and experiment. Those terms with small values of I_l on the right-hand side of Eq. (10.2) may receive too much emphasis, so we can then replace I_l in the denominator by a lower limit $I_{\min}$: I_l becomes $\max(I_l, I_{\min})$ [10.2]. However, in the further procedure we will disregard this for the sake of simplicity.

We want now to minimize the quantity $S(P_1, \ldots, P_M)$; this is known as the least-squares method.

Thus we place

$$\frac{\partial S}{\partial P_i} = -2 \sum_{l=1}^{N} \frac{I_l - f_l}{I_l} \cdot \frac{1}{I_l} \cdot \frac{\partial f_l}{\partial P_i} = 0. \tag{10.3}$$

Introducing $r_l = 1 - f_l/I_l$ as the residue in the l-th measured point, we have

$$\frac{\partial r_l}{\partial P_i} = -\frac{1}{I_l} \frac{\partial f_l}{\partial P_i}.$$

Eq. (10.3) now becomes

$$\sum_{l=1}^{N} r_l \frac{\partial r_l}{\partial P_i} = 0, \quad i = 1, \ldots, M. \tag{10.4}$$

Eq. (10.4) represents M equations, in general non-linear, with M unknowns (the parameter values $P_1, \ldots, P_M$).

The residues r_l form the components of an N-dimensional residual vector $\mathbf{R}$, the parameters $P_1, \ldots, P_M$ are the components of an M-dimensional parameter vector $\mathbf{P}$. The $N \times M$ matrix J with elements $\partial r_l/\partial P_i$ is called the Jacobian matrix of the functions $\mathbf{R}(\mathbf{P})$. Its transposed matrix, of dimension $M \times N$, is denoted by J^T. Eq. (10.4) can be rewritten in vector form as

$$J^T \mathbf{R}(\mathbf{P}) = \mathbf{G}(\mathbf{P}) = 0. \tag{10.5}$$

The vector $G(P)$ has as components the M functions

$$g_i = \sum_{l=1}^{N} r_l \frac{\partial r_l}{\partial P_i}$$

with $i = 1, \ldots, M$.

Let the vector $\mathbf{A}$ be the solution of Eq. (10.5):

$$P_1 = a_1, \quad P_2 = a_2, \ldots, \quad P_M = a_M.$$

As Eq. (10.5) is in general a non-linear set of equations, we cannot solve it by simple matrix inversion methods, but have to apply an iteration method as approximation. We will write the result of the k-th iteration as $\mathbf{P}^{(k)} = \mathbf{A} + \mathbf{E}^{(k)}$, where the error vector $\mathbf{E}^{(k)}$ has the elements $e_{k1}, e_{k2}, \ldots, e_{kM}$. The M functions $g_i(\mathbf{P}^{(k)})$ of the k-th iteration are now approximated by Taylor's formula with the Lagrangian form of the remainder,

10.1 General Optimization Method

$$g_i(\mathbf{P}^{(k)}) \approx g_i(\mathbf{A}) + e_{k1}\frac{\partial g_i(\mathbf{A} + \Theta \mathbf{E}^{(k)})}{\partial P_1} + \cdots + e_{kM}\frac{\partial g_i(\mathbf{A} + \Theta \mathbf{E}^{(k)})}{\partial P_M},$$
(10.6)

with $0 < \Theta < 1$. The functions $g_i(\mathbf{A}) \equiv 0$.
As a further approximation we now put $e_{ki} \approx P_i^{(k)} - P_i^{(k+1)}$ and $\mathbf{A} + \Theta \mathbf{E}^{(k)} \approx \mathbf{P}^{(k)}$. Eq. (10.6) then becomes

$$g_i(\mathbf{P}^{(k)}) \approx (P_1^{(k)} - P_1^{(k+1)})\frac{\partial g_i(P^{(k)})}{\partial P_1} + \cdots + (P_M^{(k)} - P_M^{(k+1)})\frac{\partial g_i(P^{(k)})}{\partial P_M}.$$
(10.7)

By defining J_G as the Jacobian matrix of $G(P)$ with elements $\partial g_i/\partial P_j$ we can rewrite Eq. (10.7) in vector notation,

$$\mathbf{G}(\mathbf{P}^{(k)}) \approx J_G(\mathbf{P}^{(k)})(\mathbf{P}^{(k)} - \mathbf{P}^{(k+1)})$$

or

$$\mathbf{P}^{(k+1)} \approx \mathbf{P}^{(k)} - [J_G(\mathbf{P}^{(k)})]^{-1}\mathbf{G}(\mathbf{P}^{(k)}).$$
(10.8)

Eq. (10.8) is a recursion formula for the subsequent iterations of the parameter vector **P**. The elements $\partial g_i/\partial P_j$ of the $M \times M$ matrix J_G follow from the definition of g_i,

$$g_i = \frac{\partial r_1}{\partial P_i}r_1 + \cdots + \frac{\partial r_N}{\partial P_i}r_N,$$

$$\frac{\partial g_i}{\partial P_j} = \frac{\partial r_1}{\partial P_i}\cdot\frac{\partial r_1}{\partial P_j} + \cdots + \frac{\partial r_N}{\partial P_i}\frac{\partial r_N}{\partial P_j} + r_1\frac{\partial^2 r_1}{\partial P_i \partial P_j} + \cdots + r_N\frac{\partial^2 r_N}{\partial P_i \partial P_j}.$$

The first part of the latter equation is given by the Jacobian matrix of the functions $\mathbf{R}(\mathbf{P})$ and becomes in vector notation $J^T J$. The second part is

$$\sum_{l=1}^{N} r_l \frac{\partial^2 r_l}{\partial P_i \partial P_j}.$$

It becomes small when $\mathbf{R}(\mathbf{P})$ is almost linear or when R becomes small. In the Levenberg-Marquardt approximation [10.4]

$$\sum_{l=1}^{N} r_l \frac{\partial^2 r_l}{\partial P_i \partial P_j}$$

is replaced by $\lambda_k (J^T J)_{ii}$, where λ_k is a non-negative real constant, depending on the iteration number, and $(J^T J)_{ii}$ is the diagonal matrix of $J^T J$. The recursion formula then finally becomes

$$\mathbf{P}^{(k+1)} \approx \mathbf{P}^{(k)} - [J^T J + \lambda_k (J^T J)_{ii}]^{-1} J^T \mathbf{R}(\mathbf{P}^{(k)}).$$
(10.9)

If we choose $\lambda_k = 0$, Eq. (10.9) becomes the Gauss-Newton iteration formula. This is usually a rapid iteration method, but convergence is not secured. Large values of λ_k make the convergence more secure, but slower. Another

284   10 Parameter Determination

trick to enhance convergence is to take $\lambda_k = 0$ and check whether $S(\mathbf{P}^{(k+1)}) < S(\mathbf{P}^{(k)})$; if not, the iteration step is halved, and so on.

To start the iteration process we need an initial guess $\mathbf{P}^{(0)}$ for the parameter values; the more accurate this guess is, the more rapid the process will be. With a bad guess for $\mathbf{P}^{(0)}$ convergence may not even be obtained!

The evaluation of the Jacobian J can be very time consuming, because it must be carried out at each iteration step. The most accurate and fast way is to do it algebraically. However, this is model-dependent: For complex models with many parameters it requires a lot of code writing, because we have to differentiate the model equations with respect to the parameters (see Eq. (10.3)). This makes model changes very cumbersome.

Numerical evaluation of J is model-independent, but less accurate and will sometimes lead to convergence problems. In spite of this drawback, the numerical evaluation is commonly used. The elements $\partial r_l/\partial P_i$ of J are calculated by means of a finite difference scheme. To save time, we can evaluate J numerically only for each (let us say) 10th iteration step. For the steps in between we use the following approximation [10.5],

$$\left(\frac{\partial r_l}{\partial P_i}\right)^{(k+1)} = \left(\frac{\partial r_l}{\partial P_i}\right)^{(k)} + \frac{\left\{r_l^{(k)} - r_l^{(k-1)} - \sum_{i=1}^{M}\left(\frac{\partial r_l}{\partial P_i}\right)^{(k)}(P_i^{(k)} - P_i^{(k-1)})\right\}(P_i^{(k)} - P_i^{(k-1)})}{\|\mathbf{P}^{(k)} - \mathbf{P}^{(k-1)}\|^2}$$

or in vector notation

$$J^{(k+1)} = J^{(k)} + \frac{\{R^{(k)} - R^{(k-1)} - J^{(k)}(P^{(k)} - P^{(k-1)})\}(P^{(k)} - P^{(k-1)})}{\|\mathbf{P}^{(k)} - \mathbf{P}^{(k-1)}\|^2}$$

(10.10)

The denominator is the norm of the difference vector $\mathbf{P}^{(k)} - \mathbf{P}^{(k-1)}$.

10.1.1 The Linear Case

As stated previously Eq. (10.4) represents in general M non-linear equations with M unknowns. It is sometimes possible, however, to rearrange certain model equations so that they become linear with respect to some parameters. The parameter extraction is then possible by simple matrix inversion and no iterations are required.

To illustrate the method we will give two examples. The first example is for the bipolar E-M model. The collector current with Early effect is given by Eq. (4.6),

$$I_c = I_s \frac{\exp(V_{b_1 e_1}/U_T) - \exp(-V_{c_1 b_1}/U_T)}{1 + \frac{V_{b_1 e_1}}{V_{ear}} - \frac{V_{c_1 b_1}}{V_{eaf}}}.$$

In the forward mode at low currents we have $V_{b_1 e_1} \approx V_{be}$ and $V_{c_1 b_1} \approx V_{cb} \gg 0$, so

10.2 Specific Bipolar Measurements

$$I_c \approx I_s \frac{\exp(V_{be}/U_T)}{1 + \frac{V_{be}}{V_{ear}} - \frac{V_{cb}}{V_{eaf}}}$$

or

$$\frac{\exp(V_{be}/U_T)}{I_c} = \frac{1}{I_s}\left(1 - \frac{V_{cb}}{V_{eaf}}\right) + \frac{V_{be}}{I_s V_{ear}}.$$

This equation is linear in the parameters I_s^{-1}, $(I_s V_{eaf})^{-1}$ and $(I_s V_{ear})^{-1}$. These parameters can easily be obtained from measuring V_{be}, V_{cb} and $I_c^{-1}\exp(V_{be}/U_T)$. In theory only 3 measured points are needed to find I_s, V_{eaf} and V_{ear}, but in practice more points will be used to increase the accuracy. The second example concerns a MOST model in the linear current region, for low V_{DS} values [10.6]. The drain current in that region is given by

$$I_D \approx \frac{\beta(V_{GS} - V_T)V_{DS}}{1 + \theta_A(V_{GS} - V_T) + \theta_B V_{SB}}.$$

The threshold voltage

$$V_T = V_{T0} + \gamma\sqrt{V_{SB} + 2\varphi_F} - \gamma\sqrt{2\varphi_F}.$$

$\varphi_F = U_T \ln(N/n_i)$ with $U_T = kT/q$. N is the dope concentration of the substrate. φ_F is considered as a constant and usually chosen as $\varphi_F = 350$ mV. The model is not very sensitive for the φ_F value.
The parameters of this model equation are V_{T0}, γ, β, θ_A and θ_B. We can rewrite the model equation for I_D as

$$V_{GS} \approx V_{T0} + \gamma\{\sqrt{V_{SB} + 2\varphi_F} - \sqrt{2\varphi_F}\} + \frac{1}{\beta}\frac{I_D}{V_{DS}} + \frac{\theta_B}{\beta}V_{SB}\frac{I_D}{V_{DS}}$$

$$+ \frac{\theta_A}{\beta^2}\left(\frac{I_D}{V_{DS}}\right)^2 + \text{rest term}.$$

The rest term is small in practice and can be omitted. This makes then the equation for V_{GS} linear in the parameters V_{T0}, γ, β^{-1}, θ_B/β and θ_A/β^2; it can easily be solved from measured values for $I_D(V_{GS}, V_{SB})$ with e.g. $V_{DS} = 100$ mV. A similar approach is also possible for the determination of some parameters in the saturation region [10.7].
For complex models with many parameters it is usually impossible to find all the parameters with this linearization method; the iterative optimization method therefore remains the most widely used parameter extraction method.

10.2 Specific Bipolar Measurements

Among the various methods of parameter extraction for the bipolar transistor models, the determination of the emitter, base and collector series resistances have always got special attention. This is because accurate values for these resistances will facilitate the determination of all those other parameters that depend on internal junction voltages.

Another special point of attention is the measurement of the cut-off frequency f_T, because this too is often problematic, especially if the current gain is low (e.g. with lateral *pnp* transistors).

10.2.1 Measurements of Series Resistances

We will first discuss the d.c. measurements, then the a.c. ones. There are 4 d.c. measurements in common use.

1. From the Gummel plots (see e.g. Fig. 4.4 a) at high currents we measure the voltage deviation ΔV from the straight line extrapolation. This deviation in the $I_c(V_{be})$ curve may include high-injection effects and/or quasi-saturation, so it is better to take the measured $I_b(V_{be})$ curve, see Fig. 10.1. This ΔV is given by

$$\Delta V = I_e R_e + I_b R_b = I_c R_e + I_b (R_e + R_b),$$

where $R_b = R_{bc} + R_{bv}$. R_{bv} is usually bias-dependent (cf. section 3.9.2.). Rewriting gives

$$\frac{\Delta V}{I_c} = R_e + \frac{R_{bv}}{h_{FE}} + \frac{R_e + R_{bc}}{h_{FE}}. \tag{10.11}$$

If the bias dependence of R_{bv} is caused by conductivity modulation due to high injection in the base, the ratio R_{bv}/h_{FE} may be considered as constant [10.8]. If we plot $\Delta V/I_c$ as a function of h_{FE}^{-1}, we get a straight line, whose intersection point gives $R_e + R_{bv}/h_{FE}$ and whose slope gives $R_e + R_{bc}$. The calculation of R_{bv} from the emitter geometry and the sheet resistance of the

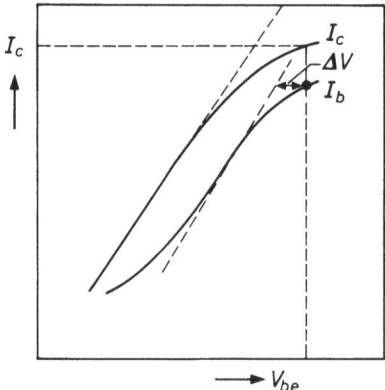

Fig. 10.1 Voltage deviation ΔV of the $I_b(V_{be})$ curve, due to the voltage drop across the emitter and base series resistances

10.2 Specific Bipolar Measurements

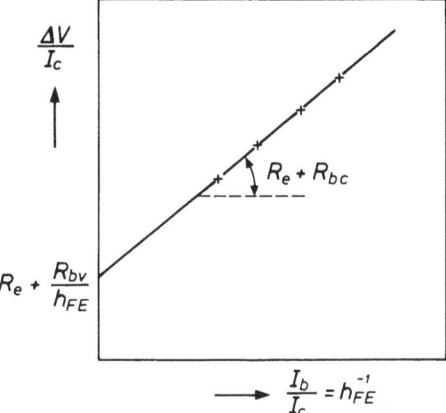

Fig. 10.2 The quotient of the voltage deviation ΔV and the collector current I_c, as a function of the inverse of the current gain h_{FE}^{-1} at high currents. The slope gives $R_e + R_{bc}$, the intersection $R_e + R_{bv}/h_{FE}$

pinched base (section 3.9.2) at low currents then leads to values for R_e and R_{bc}. See also Fig. 10.2.

The method only works if we can calculate R_{bv} accurately enough and if R_{bv}/h_{FE} remains constant in the current range where h_{FE} falls off. Situations where quasi-saturation or current crowding dominate must therefore be avoided (see sections 3.7 and 3.9.2).

2. The open collector method. The transistor under test is driven by a base current source, while the open collector-emitter voltage (with $I_c = 0$) is measured [10.9]. Fig. 10.3 gives the measurement set-up and the result.

The collector is used as a voltage probe for the internal voltage: $V_{ce} \approx$ constant $+ I_b R_e$. The measurement is in fact performed under hard saturation conditions (see section 3.7), at low V_{ce} values, and the "constant" in the equation for V_{ce} is not constant [10.10], but bias-dependent, so that we can make errors in the order of 100%!

Moreover, in integrated circuits the transistors have a substrate and the base current will be split into a part that flows directly to the emitter and a part that flows in the direction of the substrate and that returns via the buried layer, the collector epilayer and the collector-base junction [10.11]; see Fig. 4.9 for the current paths. In these situations the V_{ce} value also depends on R_{bc} and R_{epi}! From the foregoing it will be clear that this method is not very well suited for the determination of absolute values; for a quick comparison between samples of the same type, however, it can be used.

It is not necessary to measure $V_{ce}(I_b)$ with $I_c = 0$; if we select two values I_{c_1} and I_{c_2} such that $I_{c_1}/I_{b_1} = I_{c_2}/I_{b_2}$, the voltage difference ΔV_{ce} in Fig. 10.3 should give the value of the fixed collector resistance R_{cc} if R_e is already known,

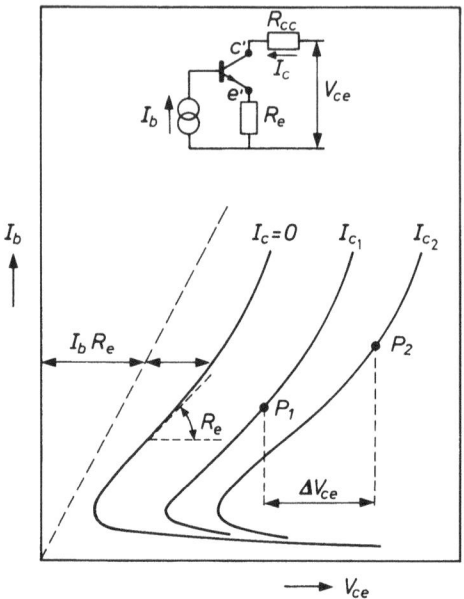

Fig. 10.3 Measurement setup and results of the open collector method for the determination of R_e and R_{cc}. The current gain is the same in P_1 and P_2

$$\Delta V_{ce} = (I_{c_2} - I_{c_1}) \left\{ R_{cc} + R_e \left(1 + \frac{I_{b_1}}{I_{c_1}}\right) \right\}. \tag{10.12}$$

From the remarks made above, it will be clear that this R_{cc} value too is subject to large errors.

3. A more accurate method for the determination of R_{cc} consists in operating the transistor in (hard) saturation and keeping the substrate current constant [10.12]. According to Eq. (4.48) this implies that $V_{c_1 b_2}$ is kept constant; see also Fig. 4.9. We then have

$$V_{cb} = I_c \left(R_{cc} - \frac{R_{bv}}{h_{FE}} \right) + V_{c_1 b_2} \approx I_c R_{cc} + V_{c_1 b_2}.$$

Thus selection of different bias points (I_c, V_{cb}) while keeping I_{sub} constant then gives the value of R_{cc}.

4. The on-resistance in hard saturation. This method is used under operational conditions similar to those in method 2. By disregarding the voltage drop across the epilayer, which is heavily flooded by injected carriers, we obtain [10.13] from Fig. 10.3

$$V_{ce} \approx V_{c'e'} + I_c R_{cc} + I_e R_e. \tag{10.13}$$

The order of magnitude of $V_{c'e'}$ in Eq. (10.13) is 10–20 mV. A drawback is that we rather measure $R_{cc} + (1 + h_{FE}^{-1}) R_e$ than R_{cc} and R_e separately.

10.2 Specific Bipolar Measurements

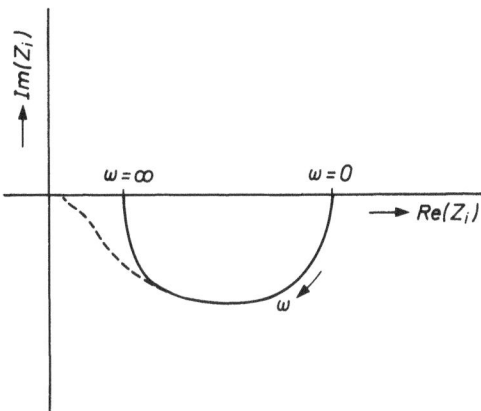

Fig. 10.4 The behaviour of the input impedance Z_i in the complex frequency plane. The intersection points with the real axis are used for the determination of r_e and $r_{bc} + r_{bv}$

The major a.c. measurements for the determination of series resistances are the input impedance method for r_e and r_b, the third harmonic method for r_c and the noise figure method for r_e and r_b.

1. Input impedance in common emitter ($\tilde{v}_{ce} = 0$) [10.14, 10.15]. The input impedance Z_i as a function of frequency follows a semi-circle in the complex frequency plane, see Fig. 10.4,

$$Z_i = Y_{ie}^{-1} = \frac{1}{g_{ie} + j\omega C_{ie}}.$$

Fig. 3.22 shows the relevant small-signal equivalent circuit of a vertical npn transistor. From this circuit it follows that the intersection with the real axis in the polar diagram of Fig. 10.4 is given by

$$Z_i = r_{bc} + r_{bv} + (\beta_f + 1)r_e + \frac{\beta_f}{g_m} \tag{10.14}$$

for $\omega = 0$.

Repeating the measurement of Z_i for various bias points and plotting $(Z_i - \beta_f/g_m)$ versus $(\beta_f + 1)$ then gives r_e and $r_{bc} + r_{bv}$, provided that r_{bv} does not vary with β_f. The latter means that high injection in the base should be avoided.

At high frequencies ($\omega \to \infty$) the intersection point with the real axis becomes

$$Z_i = r_{bc} + r_{bv} + r_e; \tag{10.15}$$

the influence of $C_{T_{ex}}$ and C_{T_c} is neglected here. Eqs. (10.14) and (10.15) together can be used to find r_e and $r_{bc} + r_{bv}$ at a fixed bias point, without using β_f variations. However, the problem is that we usually cannot neglect

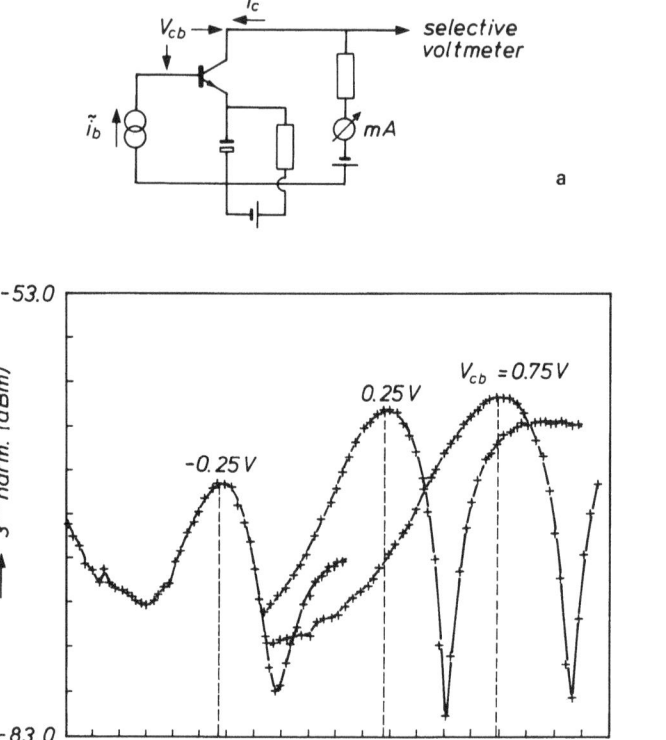

Fig. 10.5 a: Setup for the measurement of the third harmonic distortion; b: The amplitude of the third harmonic as a function of the bias conditions

the capacitances $C_{T_{ex}}$ and C_{T_c}: They cause a distortion of the semicircle and a shift of the intersection with the real axis, see Fig. 10.4.

2. Third harmonic distortion measurements are used for the determination of the collector series resistance $(R_{cc} + R_{cv})$ [10.16]. Fig. 10.5 a shows the principle of the measuring circuit: The grounded emitter device under test is driven by a low frequency (e.g. 10 kHz) small-signal base current source with very low non-linear distortion. The third harmonic distortion in the collector current is measured as a function of the d.c. collector current at various V_{cb} values. At the onset of quasi-saturation this distortion shows a maximum (see Fig. 10.5 b). At these maxima we have $V_{cb} + V_{dc} = I_c(R_{cv} + R_{cc})$, so $(R_{cv} + R_{cc})$ can be determined. The device under test is current driven and the distortion of the collector current is caused by the non-linear behaviour of the a.c. current gain h_{fe}. The h_{fe} fall-off at high collector currents introduces then the maximum of the third harmonic. However, we must be sure that this fall-off is caused by quasi-saturation (this has a strong V_{cb} depen-

10.2 Specific Bipolar Measurements

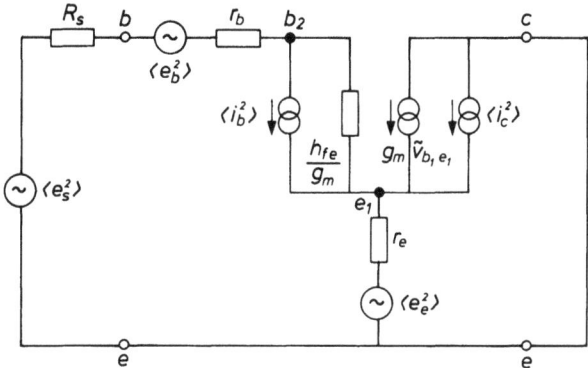

Fig. 10.6 Simplified equivalent noise circuit of a bipolar transistor

dence, see section 3.7). If it is caused by high injection in the base, the collector resistance has nothing to do with it and the method fails. Hot carrier flow in the collector epilayer makes R_{cv} non-ohmic (see section 3.7.2), but this does not invalidate the method: it can even be used to determine the parameter I_{hc} also (see section 4.3.2). Instead of a current drive a voltage drive for the base can also be used, but then the distortion produced by the emitter-base junction may interfere with the collector distortion, resulting in erroneous values for $(R_{cv} + R_{cc})$.

3. From the measurement of the noise figure in the "white" noise range, the base plus emitter series resistances $(r_{bc} + r_{bv} + r_e)$ can be obtained [10.17]. The measurement is carried out at rather low frequencies (e.g. 5 MHz), so we can disregard the influence of the capacitances. Fig. 10.6 shows a simplified version of the equivalent noise circuit of Fig. 3.24.
The external noise source is given with $e_s = \sqrt{4kTR_s \Delta f}$, the source resistance is R_s. The internal noise sources are: the thermal noise of the base resistance $e_b = \sqrt{4kTr_b \Delta f}$, the thermal noise of the emitter series resistance $e_e = \sqrt{4kTr_e \Delta f}$ and the shot noise sources $i_b = \sqrt{2qI_b \Delta f}$ and $i_c = \sqrt{2qI_c \Delta f}$.
From the circuit in Fig. 10.6 it can be derived that the noise figure F follows from

$$F - 1 = \frac{1}{R_s}\{(1 + h_{fe}^{-1})(r_b + r_e + \tfrac{1}{2}g_m^{-1}) + \tfrac{1}{2}g_m(r_b + r_e)^2\}$$

$$+ \frac{g_m}{h_{fe}}(r_b + r_e) + \frac{g_m}{2h_{fe}}R_s, \qquad (10.16)$$

where $g_m = qI_c/kT$.
For low values of R_s (e.g. 10Ω) the first term dominates in Eq. (10.16). At a given bias condition g_m and h_{fe} are known, so we can calculate $(r_b + r_e)$ from the measured F value, if R_s is also known.

The interpretation of F becomes more difficult than Eq. (10.16) suggests, when current crowding is present (see section 3.12). Applying Eq. (10.16) under these circumstances gives then too low a value for $r_b + r_e$. If the sidewall injection of the emitter-base junction dominates, which may occur for small emitter sizes, the transistor noise will increase [10.18] and the use of Eq. (10.16) leads to too high a value for $(r_b + r_e)$.

Although many methods exist for the measurement of the series resistances, as we have seen in this section, a reliable, accurate and foolproof procedure is still lacking. The accuracy of some methods is not satisfactory, others can only be applied if certain conditions are fulfilled. This makes it understandable, that sometimes it is preferred to calculate a series resistance rather than to measure it.

10.2.2 Measuring the Cut-Off Frequency f_T

The cut-off frequency f_T is defined by means of the frequency dependence of the a.c. current gain in common emitter, see section 3.11. It is usually determined by the extrapolation of the 6 dB per octave slope of the h_{fe} fall-off. However, for low values of the zero-frequency value β_f of the current gain ($\beta_f < 10$), the 6 dB slope may be hardly noticeable, and this method fails.

An alternative definition [10.19] can help in this situation. At low frequencies it follows from Eq. (3.114) that

$$h_{fe}^{-1} \approx \frac{1}{\beta_f} + j\omega\tau_f\left(1 + \frac{1}{3\beta_f}\right).$$

f_T is now defined by

$$f_T^{-1} = 2\pi \lim_{\omega \to 0} \frac{d}{d\omega} \text{Im}(h_{fe}^{-1}). \tag{10.17}$$

The definition of Eq. (10.17) is applicable at lower values of β_f than the 6 dB per octave definition, but at very low β_f values it also fails.

These difficulties are usually not encountered with vertical *npn* transistors, unless they are in heavy saturation. However, with lateral *pnp* transistors with their inherently low h_{fe} values (cf. chapter 5), these problems are rather common.

10.3 Example of Parameter Extraction for a Bipolar Transistor Model

It will now be shown how the methods discussed in the previous sections of this chapter can be applied to the determination of a complete transistor parameter set. The transistor in this example has $A_{em} = 12 \ \mu m^2$, $W_{epi} =$

10.3 Example of Parameter Extraction for a Bipolar Transistor Model

Table 10.1.

I	II	III	IV	V	VI
C_{0e}, p_e, V_{de} C_{0c}, p_c, V_{dc} C_{0s}, p_s, V_{ds}	Q_{b0} x_{cjc} (x_{cje})	I_s, β_f I_{bf}, V_{lf} I_k R_e	I_{ss}, β_{rx} I_{br}, V_{lr} I_{ks}	R_{epi} I_{hc} (V_{dc})	η τ_{ne}, m_τ

$1\,\mu\mathrm{m}$ and $N_{\mathrm{epi}} = 3 \times 10^{16}\,\mathrm{cm}^{-3}$. We will use the Mextram model of section 4.3 (vertical npn transistor). Each model has its own strategy for the parameter extraction; for Mextram we must first determine the depletion capacitance parameters (column I in Table 10.1). Once these parameters are known we can obtain Q_{b0} and x_{cjc} from the Early effects (column II). In columns III and IV are the parameters, extracted from the forward and reverse Gummel plots, respectively. The quasi-saturation region of the (I_c, V_{ce}) characteristics give the collector epilayer parameters (column V), the f_T curves the transit time parameters (column VI).
The series resistances of the base and collector (R_{bc}, R_{bv}, R_{cc}) are directly measured (see section 10.2) or calculated.

10.3.1 The Depletion Capacitances

The parameters of the depletion capacitances are determined first. These capacitances are measured at low frequencies (100 kHz to 1 MHz) as functions of the respective junction voltages. Fig. 10.7 shows $C_{T_e}(V_{be})$. The

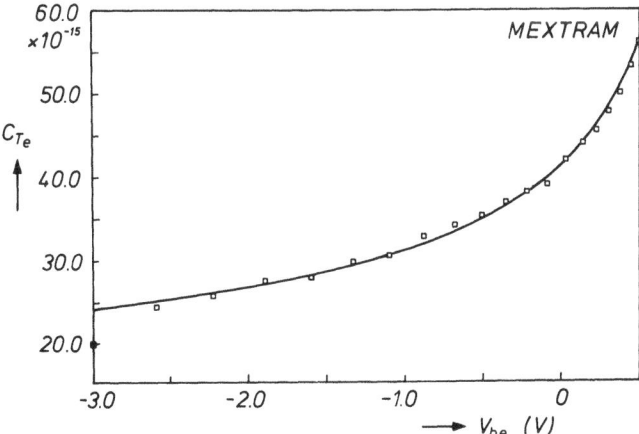

Fig. 10.7 The emitter depletion capacitance C_{T_e} as a function of V_{be}. Fully drawn line: model calculations, squares: experimental

parameters V_d and p are found to be correlated, so V_d is usually set to a fixed value; C_0 and p are obtained by a non-linear optimization method, as described in section 10.1. The formula $V_d = 2U_T \ln(N_{\text{dope}}/n_i)$ can be used as a guideline for setting the fixed values.

The result for C_{T_e} is: $C_{0e} = 41.3$ fF, $p_e = 0.352$ and $V_{de} = 0.850$ volts (set). C_{T_c} and C_{T_s} are treated in a similar way, which results in $C_{0c} = 73.8$ fF, $p_c = 0.357$, $V_{dc} = 0.730$ volts, and $C_{0s} = 1.51$ pF, $p_s = 0.373$, $V_{ds} = 0.600$ volts. The voltage range must be chosen carefully: not too much forward bias (< 500 mV) in order to prevent large current flow, the reverse bias in accordance with the application range. The reverse bias for the C_{T_e} requires special attention: It may damage the low current behaviour of the transistor permanently (h_{FE}!).

This way of parameter extraction for the depletion capacitances is, of course, not restricted to the Mextram model, but applies to all models.

10.3.2 Early Effects

The next step is the measurement of the forward and reverse Early effects. For the reverse Early effect measurement the emitter acts as collector and the collector as emitter. From Eq. (4.30) it follows that

$$I_e = I_n = \frac{I_r}{1 + \frac{Q_{T_e} + Q_{T_c}}{Q_{b0}}} = \frac{Q_{b0} I_r}{Q_{b0} + Q_{T_c}} \cdot \frac{1}{1 + \frac{Q_{T_e}}{Q_{b0} + Q_{T_c}}}$$

$$= I_e(0) \frac{1}{1 + \frac{Q_{T_e}}{Q_{b0} + Q_{T_c}}} . \qquad (10.18)$$

During the measurement V_{cb} is a fixed forward bias ($V_{cb} = -0.680$ volts in this case), whereas the reverse bias V_{be} is varied; see Fig. 10.8. $I_e(0)$ is the emitter current at $V_{be} = 0$ volts. Because the depletion capacitances are already known, Q_{T_e} and Q_{T_c} can be calculated. Q_{T_c} is of minor importance here because $Q_{T_c} \ll Q_{b0}$. Non-linear optimization gives the value of Q_{b0}. Strictly speaking we should not have used the total Q_{T_e}, but only the bottom part $x_{cje} Q_{T_e}$ of it (see also section 4.3.3). However, for the separation of bottom and sidewall components we need several emitter geometries. If these are not available we can estimate x_{cje} from the given geometry, as has been done here ($x_{cje} = 0.7$). If nothing is known, the best thing is to take $x_{cje} = 1$ as default value.

In a similar way the forward Early effect is measured. We now have

$$I_c = \frac{Q_{b0} I_f}{Q_{b0} + x_{cje} Q_{T_e}} \cdot \frac{1}{1 + \frac{x_{cjc} Q_{T_c}}{Q_{b0} + x_{cje} Q_{T_e}}} . \qquad (10.19)$$

10.3 Example of Parameter Extraction for a Bipolar Transistor Model

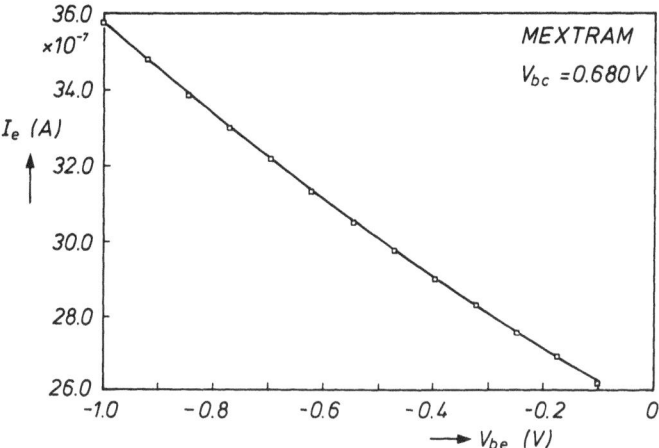

Fig. 10.8 Measurement (squares) of the reverse Early effect: $I_e(V_{be})$ for a given V_{bc}, to determine the parameter Q_{b0}

As the parameters Q_{b0} and x_{cje} are already known from the reverse Early measurement, the optimization program brings us the value of the parameter x_{cjc}, which denotes the active part of the collector depletion charge (see section 4.3.3). The results are here $Q_{b0} = 8.11 \cdot 10^{-14}$ Coul., $x_{cje} = 0.7$ and $x_{cjc} = 6.1 * 10^{-2}$.

For models, where the Early effects are described by means of Early voltages as in Eq. (4.6), these voltages can be obtained by linear extrapolation of the measured curves $I_e(V_{be})$ and $I_c(V_{cb})$. V_{ear} is then obtained from the curve in Fig. 10.8, for the reverse biased e-b junction. It should be noted that the value of V_{ear} thus found is usually much too high for the normal forward behaviour, when the e-b junction is forward biased.

10.3.3 The Gummel Plots

The measurement of $I_c(V_{be})$ and $I_b(V_{be})$ at a fixed value of V_{cb} (here $V_{cb} = 1$ volt), in the forward mode of operation, gives us many parameters. From $I_c(V_{be})$ at low current values we obtain I_s, provided that the Early effect is correctly incorporated. See Fig. 10.9. From $I_b(V_{be})$ we obtain β_f, I_{bf} and V_{lf} in the low current range. The results for our example are: $I_s = 6.11 \times 10^{-18}$ A, $\beta_f = 109$, $I_{bf} = 6.03 \times 10^{-15}$ A and $V_{lf} = 0.444$ volts. The base resistance parameters R_{bc} and R_{bv} can be measured according to one of the methods outlined in section 10.2.1, or they can be calculated if the geometry and sheet resistances are known. With R_{bc} and R_{bv} known, R_e is determined from $I_b(V_{be})$ at high current. The high injection parameter I_k may be obtained from $I_c(V_{be})$ at high currents; its value is correlated with the values for R_e, R_{bc} and R_{bv}. Another possibility for I_k is its determina-

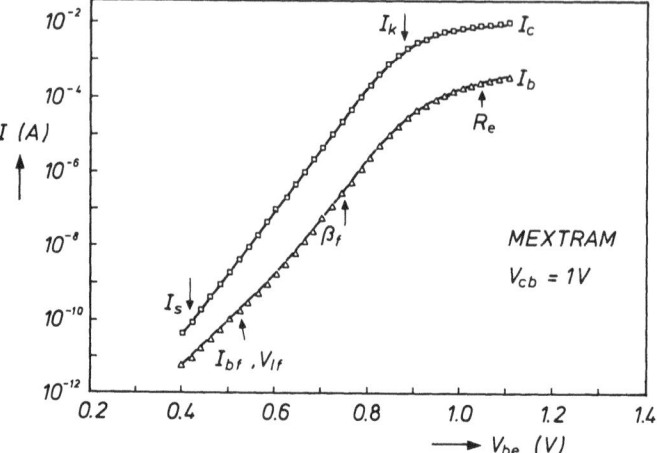

Fig. 10.9 The Gummel plots $I_c(V_{be})$ and $I_b(V_{be})$ in the forward mode of operation, at constant V_{cb}. The arrows indicate the various parts of the curves that are used for the determination of the related parameters

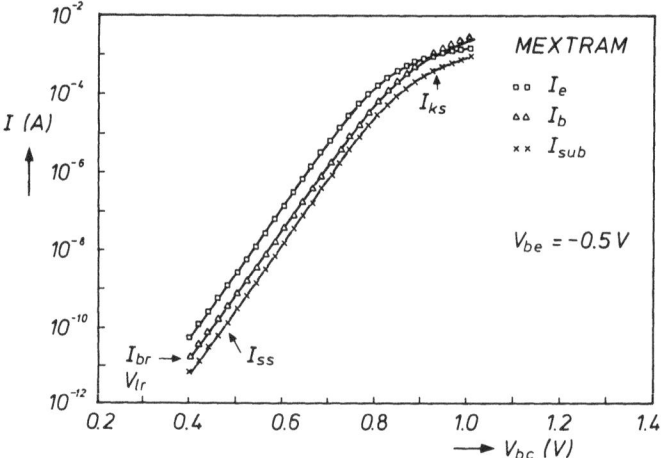

Fig. 10.10 The Gummel plots $I_e(V_{bc})$, $I_b(V_{bc})$ and $I_{sub}(V_{bc})$ in the reverse mode of operation, at constant V_{be}. The significance of the arrows is as in Fig. 10.9

tion from the h_{FE} fall-off at high currents, but here the presence of quasi-saturation effects can influence the I_k value.

The results are $R_{bc} = 326\,\Omega$, $R_{bv} = 500\,\Omega$ (calculated), $R_e = 1.7\,\Omega$ and $I_k = 10.5\,\text{mA}$.

In the reverse mode we measure I_e, I_b and I_{sub} as a function of the forward bias V_{cb}, with $V_{be} = -0.5$ volts (see Fig. 10.10).

At low current levels we determine the parameters $I_{ss} = 8.02 \times 10^{-19}\,\text{A}$, $\beta_{rx} = 6.64$, $I_{br} = 1.06 \times 10^{-13}\,\text{A}$, $V_{lr} = 0.634$ volt. The parameter I_{ks} can be

10.3 Example of Parameter Extraction for a Bipolar Transistor Model

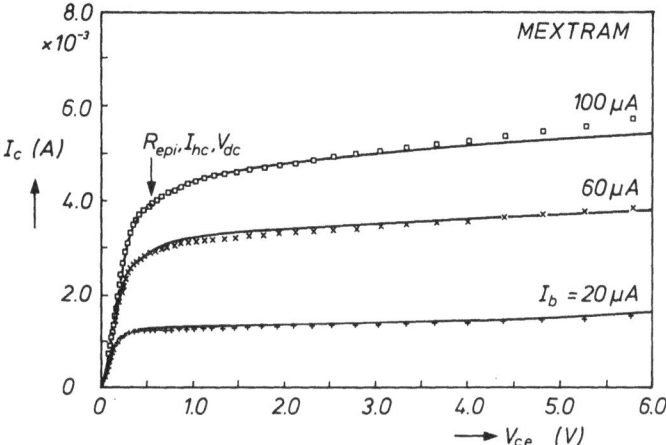

Fig. 10.11 Comparison between experiment (□ × +) and model calculation (fully drawn lines) of the $I_c(V_{ce})$ characteristics

determined from $I_{sub}(V_{cb})$ at high current levels when we know R_{bc}, R_{bv} and R_{cc}, but R_{cc} is not yet determined. It is obtained here from the on-resistance in hard saturation (Eq. (10.13)). We get $R_{cc} = 34.2\,\Omega$ and then find $I_{ks} = 4.03$ mA.

10.3.4 The Quasi-Saturation

The parameters for the description of the quasi-saturation are R_{epi}, V_{dc} and I_{hc}, see section 4.3.2. They can be obtained by optimization of the $I_c(V_{ce})$ characteristics below the knee voltage or from third-harmonics measurements (see section 10.2.1). Here the first method was applied, which resulted in $R_{epi} = 169\,\Omega$, $I_{hc} = 8$ mA and $V_{dc} = 0.730$ volts. Hot carrier effects in the epilayer are hardly present in this transistor, so the value of I_{hc} is not very accurate. This is not so serious because the parameter I_{hc} has a small influence on the fit of these characteristics.

V_{dc} is also a parameter of the collector depletion capacitance, but it was not really extracted from the $C_{T_c}(V_{cb})$ plot: Its value was set to $V_{dc} = 0.730$ volts without any harm to the fit. The overall result of the $I_c(V_{ce})$ characteristics is shown in Fig. 10.11.

10.3.5 The Cut-Off Frequency f_T

Fig. 10.12 shows $f_T(I_c)$ at various values of V_{cb}. The rising part of these curves is mainly determined by the depletion charges Q_{T_e} and Q_{T_c} for which the parameters are already determined. The maximum values of f_T depend

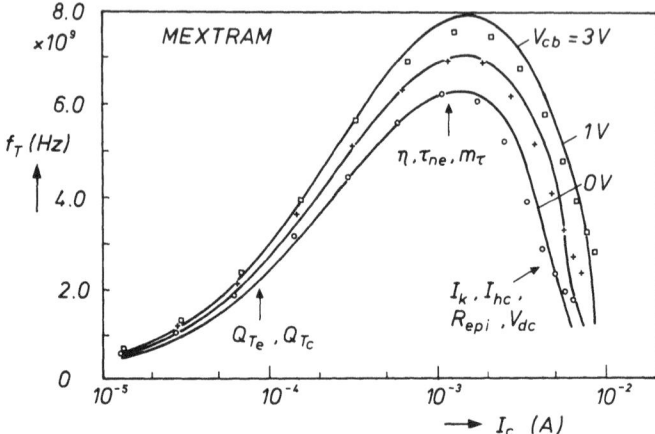

Fig. 10.12 The measured (o + ◇) and calculated (fully drawn lines) cut-off frequency f_T as a function of I_c and V_{cb}. Arrows indicate those parts where the respective parameters have large influence

mainly on the charges Q_{be} and Q_e (Eqs. (4.33) and (4.34)). The parameters involved here are η, τ_{ne} and m_τ. The parameter η depends on the doping profile in the active base region, as outlined in section 3.3.3. Its value is difficult to find from f_T measurements because it strongly correlates with the other parameters τ_{ne} and m_τ. For thick base regions η is rather low (2–3), but it increases when the base width becomes narrow. For our high-frequency transistor we have chosen $\eta = 6$. The other parameter values τ_{ne} and m_τ are then found from optimization: $\tau_{ne} = 2.85$ ps and $m_\tau = 1$ (this is the theoretical limit for m_τ).

The fall-off in this case is dominated by high injection in the base rather than by quasi-saturation. Thus the parameter I_k is more important here than the parameters R_{epi}, V_{dc} and I_{hc}.

10.3.6 Concluding Remarks

The parameter list of the Mextram model contains two additional parameters: x_{ijb} and N_{epi}. The parameter x_{ijb} denotes the sidewall component of the ideal base current. Its determination requires the measurement of transistors with various emitter dimensions, which are not given in this example. Thus we have put $x_{ijb} = 0$. The parameter N_{epi} is not really necessary because it is related to the parameter V_{dc}: $V_{dc} = 2U_T \ln(N_{\text{epi}}/n_i)$ (Eq. (3.69)). If we use N_{epi} instead of V_{dc} as a parameter for the quasi-saturation, V_{dc} is then available for the depletion capacitance C_{T_c}, but this is not really necessary. In general, the temperature dependence of the parameters is obtained by repeating the measurements for the parameter extraction at different temperatures, e.g. in the range $-50°C$ to $+150°C$. Thus we can obtain the value of the exponent a_b of the sheet resistance of the pinched base (Eq. (3.129)). With this value we calculate the exponent m_b, whereafter the bandgap voltage in the base follows from Eq. (3.130),

$$\ln(I_s T^{-m_b}) = -qV_{gb}/k_B T. \tag{10.20}$$

V_{gb} is the slope of the semilogarithmic plot of $I_s T^{-m_b}$ against $q/k_B T$. The other bandgap voltages are extracted in the same way. Parameters such as η, x_{cjc} and x_{cje} are temperature-independent.

The parameter extraction can be judged as satisfactory if the overall error of the model predictions with respect to the measured characteristics is less than 10%. The parameter values must be physically plausible.

10.4 Parameter Determination for MOSFETs

10.4.1 Enhancement Devices

In order to be able to explain the procedures for obtaining MOSFET parameters, first we summarize all quantities required for a correct simulation of implanted, short-channel devices. Successively we have

V_{T0} —zero-bias threshold voltage (V)
γ_i, γ —body effect coefficients ($V^{1/2}$)
V_{sx} —characteristic voltage for threshold voltage implant (V)
β_0 —unit area gain factor (AV^{-2})
θ_A —mobility reduction factor owing to gate-induced field (V^{-1})
θ_B —mobility reduction factor owing to back-bias (V^{-1})
θ_C —mobility reduction factor owing to lateral field (V^{-1})
η —static feedback factor
α —channel length modulation factor
V_p —characteristic voltage for channel length modulation (V)
s, s_i —drain-induced barrier lowering factor ($V^{-1/2}$, V^{-1})
I_0 —unit area subthreshold current constant (A), (optional)
m —subthreshold slope factor ($V^{1/2}$)
t —fitting parameter for subthreshold
a —substrate current proportionality constant
V_a —ionization rate parameter (V)
C_{0x} —unit area gate insulator capacitance (F)
V_{FB} —flat-band voltage (V)

These are the electrical parameters by which the characteristics of a device with given channel length and width can be described. However, since in the case of small devices, several parameters depend on the chosen geometry, an additional set of geometry relations for the above set of parameters is required. Either by physical argument (compare sections 6.3.5, 6.3.6, 6.4.1, 6.4.3, 6.5.1) or empirical investigation, in practice it is found that the following simple relation applies to most parameters

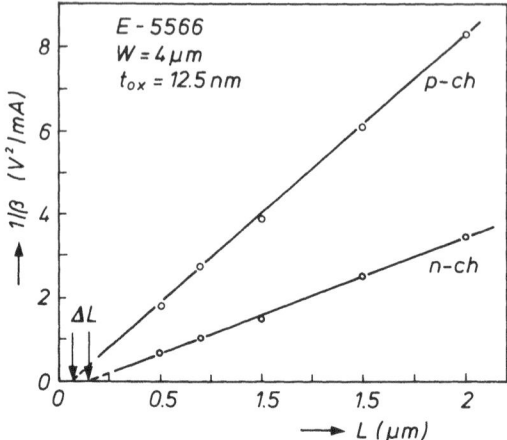

Fig. 10.13 Determination of channel length correction by plotting the (inverse) gain factor as a function of drawn gate length

$$P = P_0 + P_L(L - \Delta L)^{-1} + P_W(W - \Delta W)^{-1}. \qquad (10.21)$$

In the above relation L, W are the drawn length and width (mask dimensions) of a device, and ΔL, ΔW corrections to the above quantities owing to process steps such as etching, underdiffusion, oxide encroachment, etc.
Usually for a given process ΔL and ΔW are determined by measuring the gain constant β as a function of various drawn values of channel length and width. Fig. 10.13 gives an example of measuring ΔL for a submicron process with LDD-type junctions.
In both cases the result inspires confidence in the approach followed. However, it has been reported [10.20] that in case of low-dose LDD-type junctions the above procedure is unreliable (see next section). In addition, it should be remarked that the gain factor can only be correctly determined in conjunction with other parameters (e.g. θ_A, θ_B, etc.). Since the geometry dependence of parameters is subject of the next chapter, the discussion in this section is limited to the determination of the electrical parameters.
Although for a device with given geometry in principle the above electrical parameters can be obtained simultaneously from global optimization, it is found that it is more practical to split the parameters into five groups. Otherwise competition between parameters may cause the drain current to be optimized at the cost of some unphysical parameter value. Table 10.2 gives a classification, which is quite natural.
The first group is determined from the strong inversion regime at low drain bias ($V_{DS} \leq 100$ mVolt), the second group from the drain current and drain conductance characteristics at higher drain bias, the third group from the drain and back bias-dependent subthreshold characteristics, the fourth group from the substrate current characteristics and the last group from

10.4 Parameter Determination for MOSFETs

Table 10.2

I	II	III	IV	V
V_{TO}	θ_C	m	a	C_{Ox}
γ_i	η	(I_0)	V_a	V_{FB}
γ	α	s		
V_{SX}	V_p	s_i		
β		(t)		
θ_A				
θ_B				

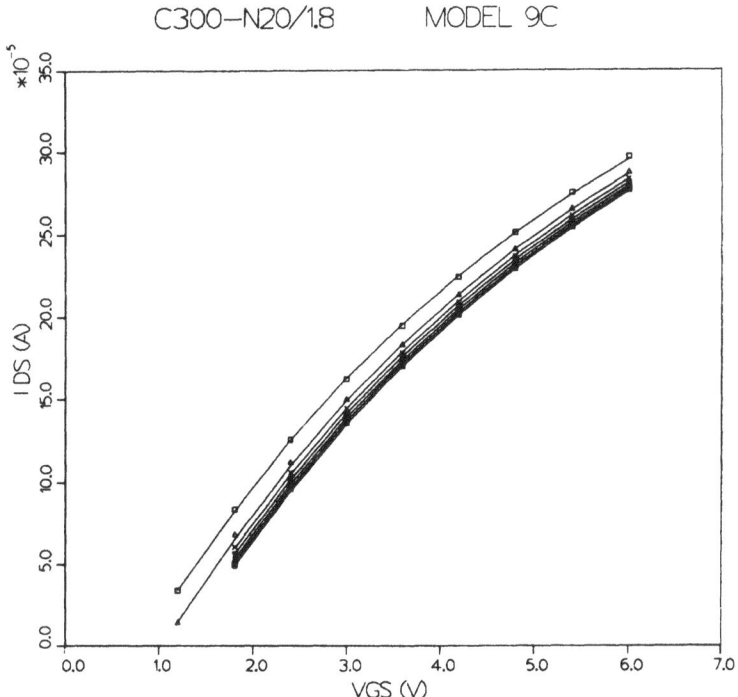

Fig. 10.14 Measured (dots) and calculated characteristics (fully drawn lines) of an n-channel, LDD-type MOSFET at a drain-source bias voltage of 0.10 Volts and a back-bias varying between 0 Volt and 4.5 Volts in steps of 0.75 Volts. The corresponding parameters are: $L_{eff} = 1.6\,\mu m$, $V_{TO} = 0.77\,Volt$, $\beta = 1.04\,mA/V^2$, $\gamma_i = 0.62\,V^{1/2}$, $\gamma = 0.12\,V^{1/2}$, $V_{SX} = 0.66\,Volt$, $\theta_A = 0.14\,V^{-1}$ and $\theta_B = 0.024\,V^{-1/2}$

the gate-bulk capacitance in the subthreshold and accumulation mode of operation.

Fig. 10.14 gives the measured drain current characteristics (dots) at low drain bias for an implanted n-channel MOSFET ($W = 20\,\mu m$, $L - \Delta L \approx 1.6\,\mu m$). For comparison the simulated characteristics after optimization (fully drawn lines) are also given. The simulated values have been calculated according

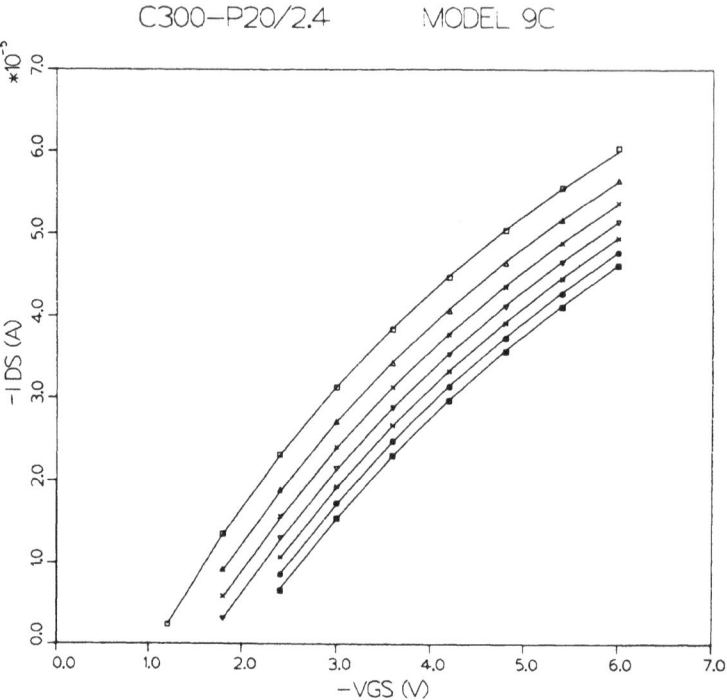

Fig. 10.15 Measured (dots) and calculated (fully drawn lines) characteristics for a p-channel MOSFET under the same bias conditions mentioned in Fig. 10.14. The corresponding parameters are: $V_{T0} = 1.02\text{V}$, $\beta = 0.22$ mA/V^2, $\gamma = 0.61$ V$^{1/2}$, $\theta_A = 0.16$ V^{-1} and $\theta_B = 0.18$ V$^{-1/2}$

to the model presented in section 7.3.2. From the above comparison the following parameter values have been obtained:

$$V_{T0} = 0.77 \text{ V}, \quad \beta = 1.04 \text{ mA/V}^2, \quad \gamma_i = 0.62 \text{ V}^{1/2}, \quad \gamma = 0.12 \text{ V}^{1/2},$$
$$V_{SX} = 0.66 \text{ V}, \quad \theta_A = 0.14 \text{ V}^{-1}, \quad \theta_B = 0.024 \text{ V}^{-1/2}.$$

Similar results are presented in Fig. 10.15 for a (buried) p-channel device ($W = 20$ μm, $L - \Delta L = 2.3$ μm). In this case the corresponding parameter values are $V_{T0} = 1.02$ V, $\beta = 0.22$ mA/V^2, $\gamma = 0.61$ V$^{1/2}$, $\theta_A = 0.16$ V^{-1} and $\theta_B = 0.18$ V$^{-1/2}$. Generally the accuracy is quite satisfactory (average deviation is less than 2%). Typical of the n-channel device are the two γ parameters and the low value of θ_B. However, owing to their high value of θ_B, the p-channel characteristics are less parallel at higher values of backbias [10.21]. In addition, the no longer small values of θ_A for both types of devices already indicate the presence of series resistance. When the difference between the body coefficients is small or can be disregarded for practical reasons, the parameters of the first group can also be determined directly from a few selected measured values and by rewriting the usual current expression in a suitable form (see section 10.1.1). This is a useful procedure for process control measurements or wafer statistics investigations.

10.4 Parameter Determination for MOSFETs

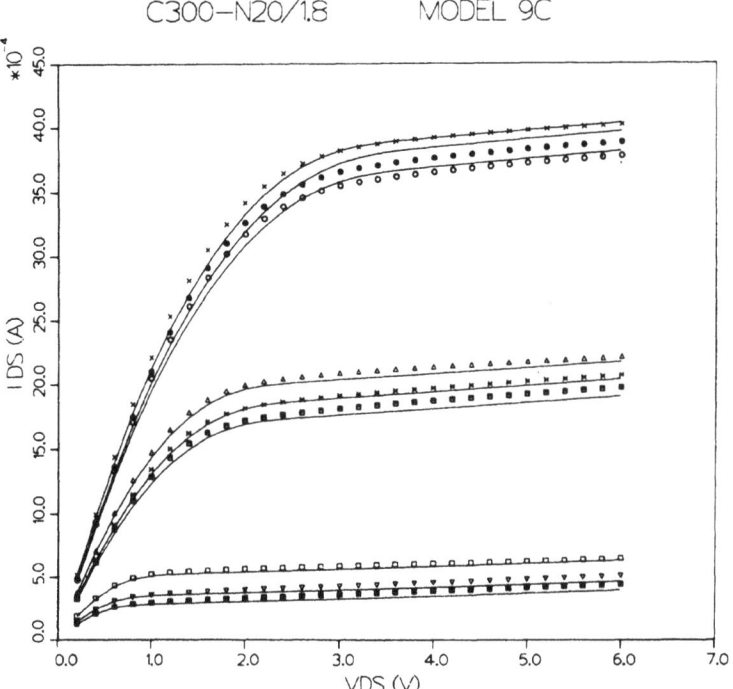

Fig. 10.16 Measured (dots) and calculated characteristics (fully drawn lines) for the same device used in Fig. 10.14. For each gate bias value the characteristics are shown for back-bias values of 0, -2 and -5 Volts. Corresponding parameters additional to those of Fig. 10.14 are: $\theta_C = 0.098$ V^{-1}, $\eta = 0.036$, $\alpha = 1$ and $V_p = 1.1 * 10^5$ Volts

Using the ΔL and ΔW-values from the gain factor plot, the geometry dependence of the parameters considered can be established. Figs. 11.11 and 11.12 give the result for n-channel devices corresponding to Fig. 10.14. Except for θ_B an almost linear relation down to submicron dimensions is found.

For the same device shown in Fig. 10.14, Fig. 10.16 gives the measured high current characteristics at several values of back-bias (dots) together with the calculated (modelled) values (fully drawn lines). In addition Fig. 10.17 gives the measured and modelled drain conductance characteristics. The latter has been obtained by numerical differentiation of Fig. 10.16. The associated parameter values additional to those of Fig. 10.14 are $\theta_C = 0.098$ V, $\eta = 0.036$, $\alpha = 1$ and $V_p = 1.1 * 10^5$ Volts.

Note that for this LDD-type device, the effect of α on the characteristics is zero. In addition it should be mentioned that the value of the parameter η has been determined, giving the measured data of the drain conductance a higher priority than the absolute values of the drain current. Otherwise the latter quantity may be optimized at the cost of large errors in the drain conductance. This fact and the complicated effect of the depletion charge on

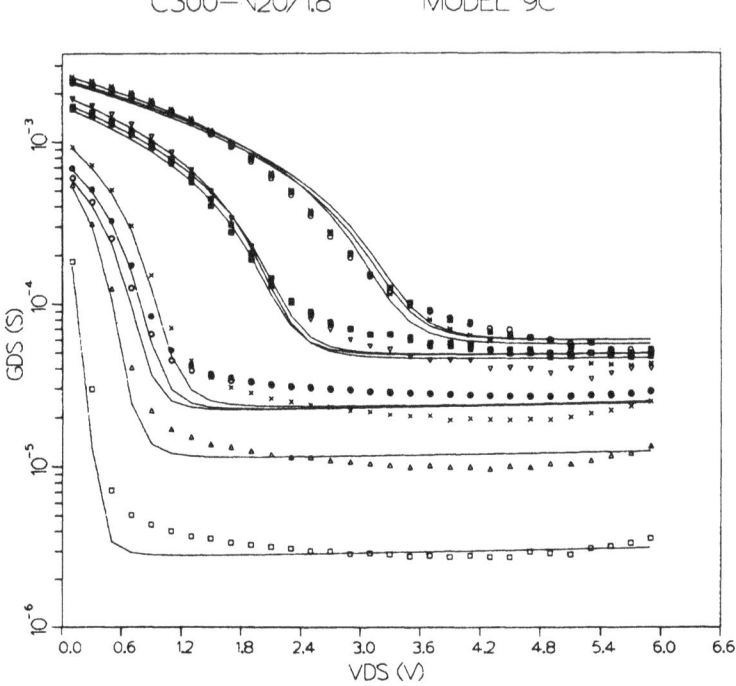

Fig. 10.17 Measured (dots) and calculated drain conductance characteristics (fully drawn lines) corresponding to Fig. 10.16. Gate bias values applied have been indicated

the saturated current in the presence of two very different values of body coefficient (compare section 7.2.1.2) mean that the average deviation between measured and modelled current characteristics is 4%. If the above requirements are dropped, as for instance is the case in digital applications, a higher accuracy can be achieved, even for micron-sized devices.

Next for the same p-channel device shown in Fig. 10.15, Figs. 10.18 and 10.19 give the drain current and drain conductance characteristics. From the comparison between measured (dots) and simulated values, in this case the following additional parameter values have been obtained: $\theta_c = 0.025$ V^{-1}, $\eta = 0.0096$, $\alpha = 0.034$ and $V_p = 1.20$ V.

Note that the conductance characteristics of Fig. 10.19 are quite different from those presented in Fig. 10.17. This is caused by the fact that, for a device with conventional junctions, the channel length modulation effect is more dominant in the saturation mode. Furthermore, owing to the smaller hole mobility, the velocity saturation parameter has a smaller value than for n-channel devices.

Next in Fig. 10.20 the subthreshold characteristics have been plotted at several values of drain and back-bias. From the comparison between measured and modelled values (fully drawn lines) the following parameter values

10.4 Parameter Determination for MOSFETs

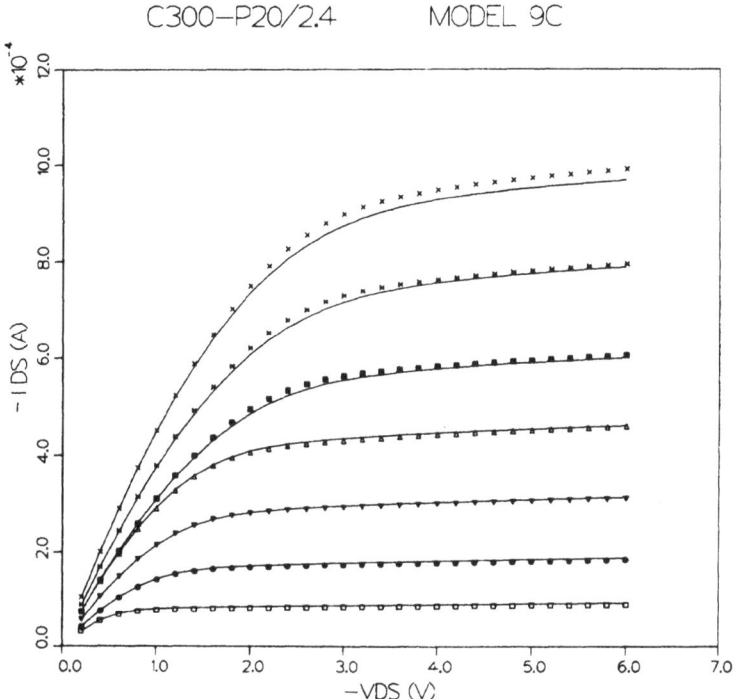

Fig. 10.18 Measured (dots) and calculated characteristics (fully drawn lines) for the same device used in Fig. 10.15. For each gate bias value the characteristics are shown for back-bias values of 0, 2 and 5 Volts. Corresponding parameters additional to those of Fig. 10.15 are: $\theta_C = 0.025 \text{ V}^{-1}$, $\eta = 0.0096$, $\alpha = 0.034$ and $V_p = 1.20$ Volts

additional to those of Figs. 10.14 and 10.16 have been obtained: $s_i = 0.0072$ V^{-1}, $m = 0.45$ V and $t = 0.53$. For the p-channel device considered previously, a similar comparison from Fig. 10.21 yields $s = 0.0012 \text{ V}^{-1/2}$, $m = 0.28$ $\text{V}^{1/2}$ and $t = 1.55$. Usually the average deviation between measured and calculated results is larger than in the former cases (15%). However, the largest deviations occur at high values of back-bias and for most applications this deficiency is not important. For instance, in analog applications description of the transconductance as a function of gate drive (slope description in Fig. 10.20) is much more important than the absolute value of the (subthreshold) current. Note the complicated effect of drain-induced barrier lowering. For the n-type device at low back-bias the subthreshold characteristics are hardly affected by an increase in drain bias, but at $V_{SB} = 5$ Volts, the considered effect becomes very large. This result is due to the fact that the depletion region has moved into the low-doped substrate beyond the threshold implant. Since the p-channel device has been realized in a much higher doped n-well, the above effect is absent.

Using the substrate current measured as a function of gate and drain bias, the characteristic parameters can be obtained. The result is shown in Fig.

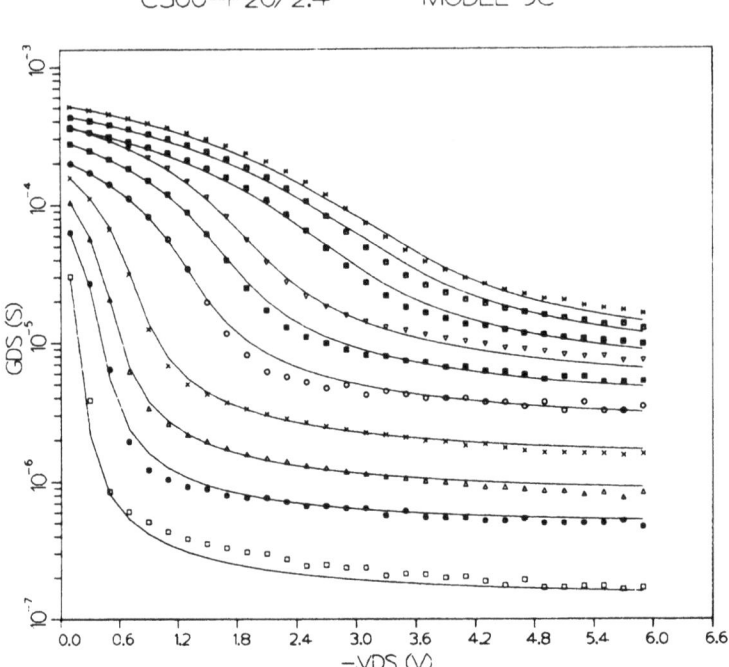

Fig. 10.19 Measured (dots) and calculated drain conductance characteristics (fully drawn lines) corresponding to Fig. 10.18. Gate bias values applied have been indicated

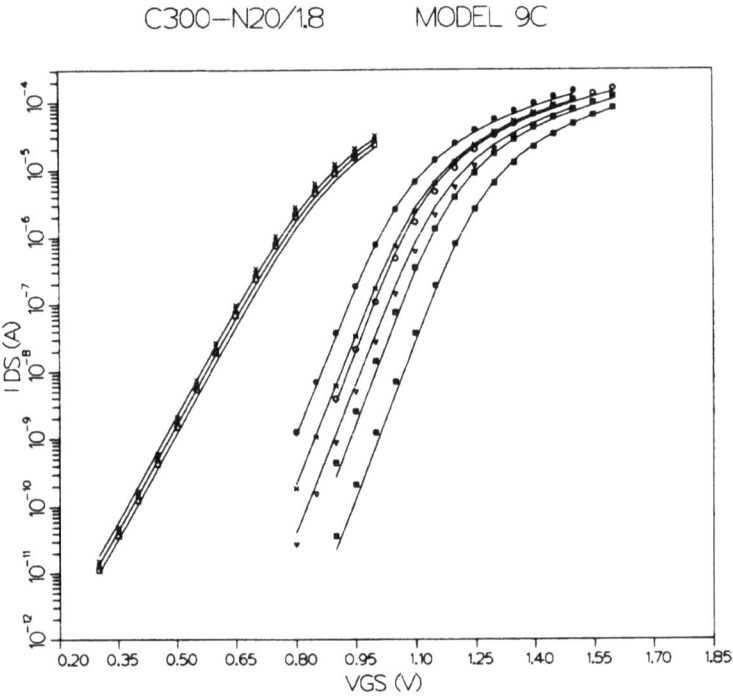

10.4 Parameter Determination for MOSFETs

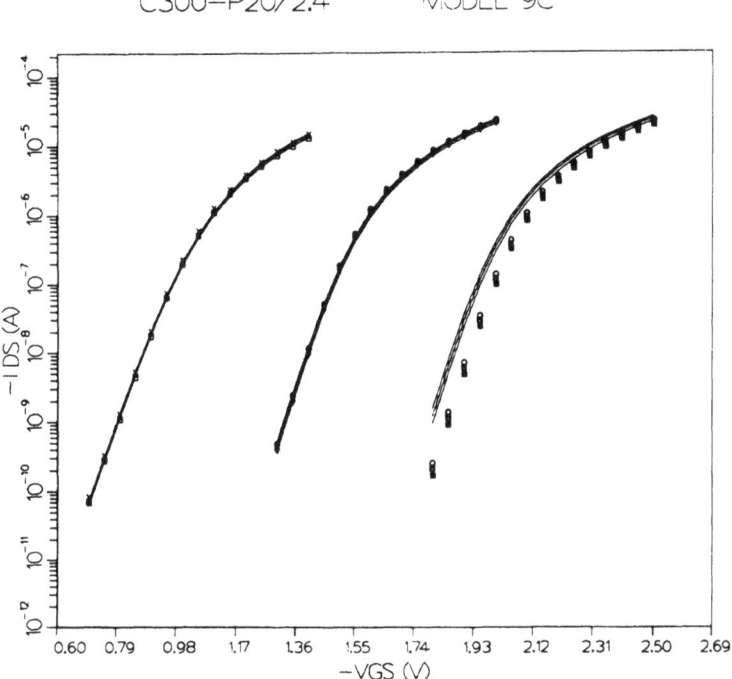

Fig. 10.21 Measured (dots) and calculated subthreshold characteristics (fully drawn lines) for the same device used in Figs. 10.15 and 10.18. For each value of back-bias (0, 2 and 5 Volts), the characteristics were taken at a drain-source voltage of 1, 3.5 and 6 Volts. Corresponding parameters additional to those of Fig. 10.15 are: $s = 0.0012$ $V^{-1/2}$, $m = 0.28$ $V^{1/2}$ and $t = 0.55$

10.22. The corresponding parameters have a value $a = 2.80$ and $V_a = 30.8$ V. Despite the semi-empirical model of section 7.2.4 the values calculated according to Eq. (7.87) fit the measured ones quite well (average error is within 10% over the whole range of interest). Naturally, owing to the sensitivity of the substrate current to the field near the drain end of the channel, the usefulness of the model is restricted to the practical bias range shown in the figure. At higher values of drain bias, often applied during stress experiments, the model remains valid, but the values of a and b change slightly.

Usually the intrinsic charge parameters C_{ox} and V_{FB} are obtained by measuring the gate-bulk capacitance of a large MOSFET as a function of gate bias. The behaviour in the deep subthreshold region is particularly important. For details we refer to textbooks [10.22]. Other parameters closely

◁ Fig. 10.20 Measured (dots) and calculated subthreshold characteristics (fully drawn lines) for the same device used in Figs. 10.14 and 10.16. For each value of back-bias (0, −2 and −5 Volts), the characteristics were taken at a drain-source voltage of 1, 3.5 and 6 Volts. Corresponding parameters additional to those of Fig. 10.14 are: $s_i = 0.0072$ V^{-1}, $m = 0.45$ V and $t = 0.53$

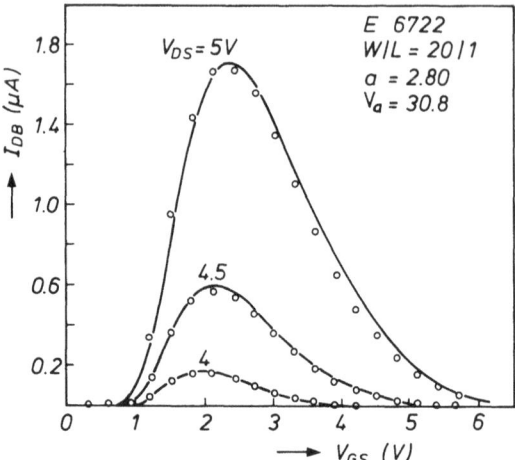

Fig. 10.22 Measured (dots) and calculated substrate current characteristics (fully drawn lines) for an *n*-channel, LDD-type MOSFET ($L_{\text{eff}} = 0.80$ µm). Corresponding parameters have been indicated

related to the previous ones are a number of parasitic capacitances, such as the gate-source/drain junction spreading capacitances (see section 6.7.2) and the source/drain-bulk junction capacitance. Usually the first-mentioned capacitances are determined by measuring the total gate-source or drain capacitance in the off-state mode, and taking devices with a large width (e.g. $W = 100$ µm). However, owing to 2-D effects in small devices, these capacitances are not constant (compare section 6.7.2). Finally, the peripheral and areal parts of the junction capacitances can be obtained from special structures with junctions of varying length and width.

10.4.2 Depletion Devices

For the compact model for depletion devices described in section 8.2.3 two additional parameters are needed:

V_{FB} —flat band voltage (V)
r_0 —gain ratio.

Note that in this case V_{FB} is not a parameter for a charge model; on the contrary this parameter is the boundary between the depletion and the accumulation mode of operation.

Similar to the procedure for enhancement devices, the parameters can be classified in several groups. The parameters V_{T0}, γ, β, V_{FB}, r_0 and θ_A can be determined from conductance measurements at low values of drain bias. On the other hand the parameters θ_C and η are determined from the general characteristics. Although the subthreshold and substrate current characteristics do exist, owing to lack of interest no practical model is currently

10.5 Specific MOSFET Measurements

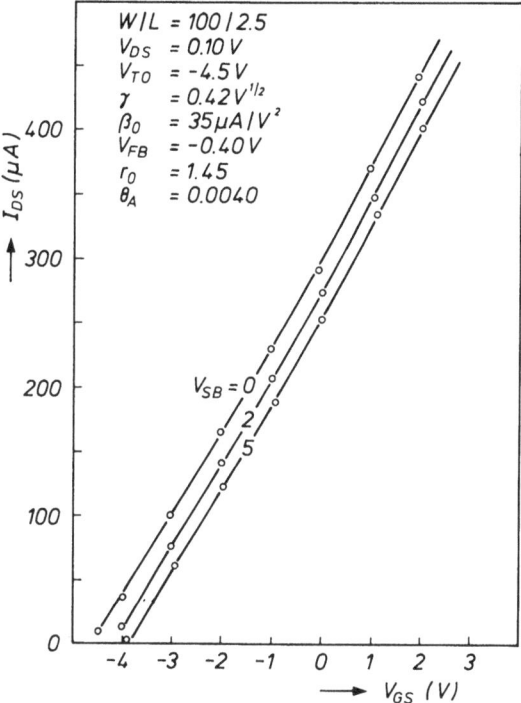

Fig. 10.23 Measured (dots) and calculated characteristics (fully drawn lines) of an n-channel, depletion type MOSFET. Corresponding parameters have been indicated

available. Fig. 10.23 gives a measured conductance characteristic at different values of back-bias.

Values calculated according to the model presented in section 8.2.3 together with the corresponding parameters have been added. Obviously, for this group an optimization routine, which has been applied here, gives the most accurate results. A faster method is to determine the unknown parameter values from six well-chosen measured values of the conductance and to rewrite the conductance equations in a suitable form. However, in this case a first guess for V_{FB} is required to guarantee that one half of the measured values is taken at $V_{GS} < V_{FB}$ and the other half at $V_{GS} > V_{FB}$. Fortunately for most processes $-0.8\text{ V} < V_{FB} < -0.3\text{ V}$.

For a plot of the general characteristics and the complete list of parameters we refer to Fig. 8.8.

10.5 Specific MOSFET Measurements

In contrast to bipolar devices, until recently series resistances did not impose a serious limit to the performance of MOSFETS. However, in the presence of lightly doped junctions, the source and drain series resistance can no longer be neglected in micron-sized devices. In addition, it has been reported that in this case the usual method of obtaining the length correction ΔL from a plot of β^{-1} versus drawn length L (see Fig. 10.13) becomes erroneous [10.20]. In particular, if R_s is rather large, its value may become dependent on the bias condition. This may cause errors in determining the value of the gain factor β. As a result, nonlinearities in the plot of β^{-1} versus L are observed.

To circumvent the above problem, an alternative measuring method has been proposed [10.24]. It is based on rewriting the drain current expression, which takes into account the effect of the source and drain resistance. When $V_{SB} = 0$ and V_{DS} has a small value, it follows from (7.82) that the total resistance $R = V_{DS}/I_{DS}$ is given by

$$R - R_s(V_{GT}) = \frac{1 + \theta_A V_{GT}}{\mu_s C_{Ox} W V_{GT}} [L - \Delta L(V_{GT})]. \tag{10.22}$$

In this equation μ_s is the zero-bias channel mobility and $V_{GT} = V_{GS} - V_T$. In addition, $R_s = R_{ss} + R_{sd}$ and ΔL may depend on the gate bias. Using a constant gate drive V_{GT}, R will vary linearly with L. This is illustrated in Fig. 10.24. A slight increase of V_{GT} will change the dependence of R on L according to the same figure and from the intercept the value of ΔL and R_s at the gate bias used can be obtained. Repeating this approach for other V_{GT} values, ΔL and R_s have been measured as a function of gate bias. The result is given in Fig. 10.25.

Although both quantities vary considerably with gate bias, it has to be noted that the above results apply to devices with relatively deep, low-dose $(1 * 10^{13}/\text{cm}^2)$ junctions. Therefore it is questionable whether the above effect is serious in devices with a higher implant dose ($\geqslant 4 \times 10^{13}/\text{cm}^2$) as required for higher reliability.

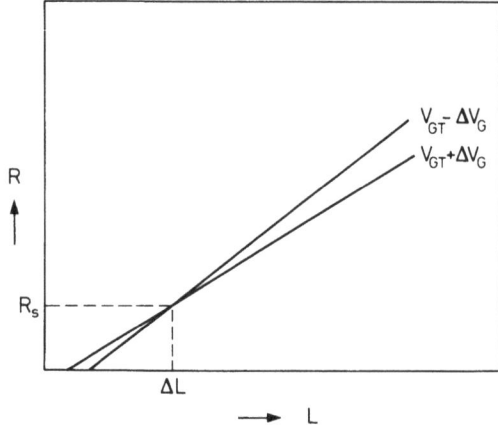

Fig. 10.24 Total resistance of LDD-type MOSFET as a function of drawn channel length for two closely separated gate biases

10.5 Specific MOSFET Measurements

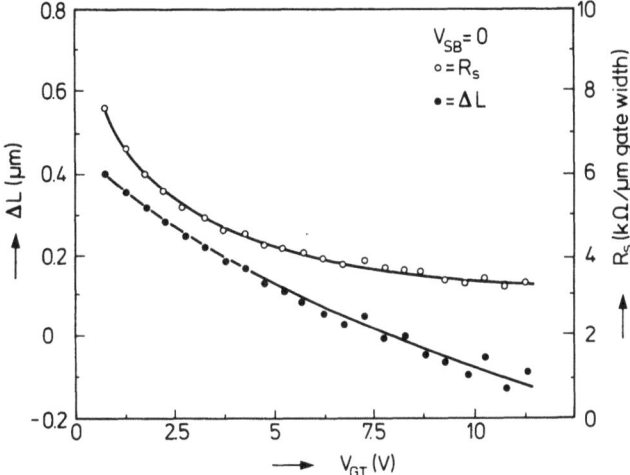

Fig. 10.25 Length correction factor and series resistance of LDD-type MOSFET (with low n^- implantation dose). The two parameters show a strong dependence on the gate overdrive $(V_{GT} = V_{GS} - V_T)$ [10.24]

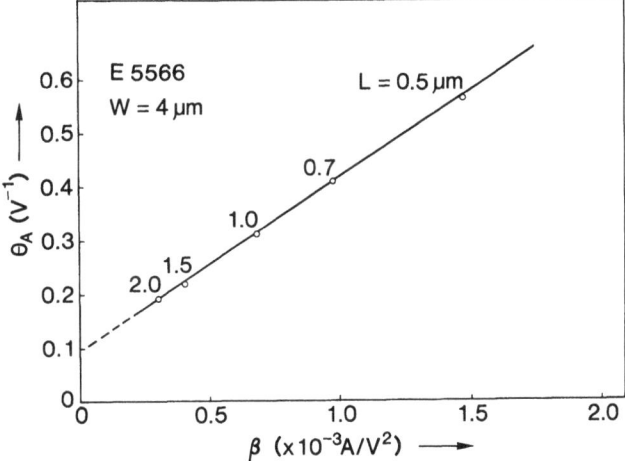

Fig. 10.26 Model parameter θ_A versus gain factor β. The slope yields the average value of $(R_{ss} + R_{sd})$ [10.25]

In fact the results of Fig. 10.13 apply to LDD-type devices too. Since in this case the gain factor has been determined from measured data using the full model expression (7.82), a perfect straight line in the β^{-1} versus L plot is observed. In addition, since the other parameters θ'_A and θ'_C (compare section 7.2.3) are obtained simultaneously, their value can be used to determine the value of R_{ss} and R_{sd}. This is shown in Figs. 10.26 and 10.27, where θ'_A and θ'_C have been plotted as functions of β. From the slopes, the resistance values

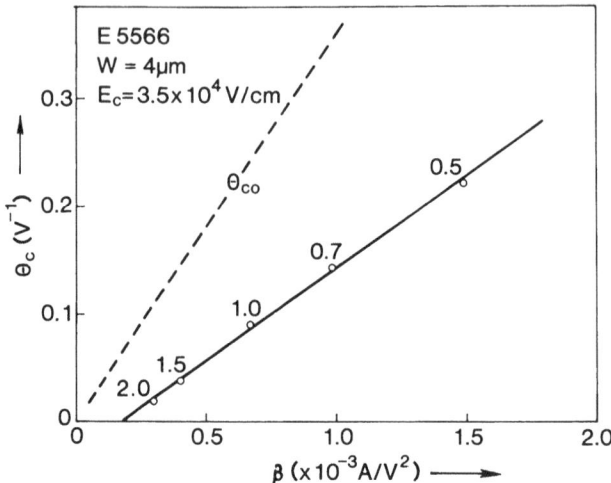

Fig. 10.27 Model parameter θ_C versus gain factor. From the slope and the velocity saturation field (E_c) the average value of R_{sd} is obtained [10.25]

can be calculated [10.25]. For the devices considered, $R_s = 1.25$ kΩ/μm gate width and $R_{sd} = 0.55$ kΩ/μm gate width. Naturally, since the measurements at higher gate drive have a larger weight in the determination of θ'_A and θ'_C, the resulting resistance values may represent more the average value at a gate drive > 2 Volts. On the other hand, if R_s had followed the dependence shown in Fig. 10.25, strong deviations from the straight lines of Figs. 10.26 and 10.27 would more likely have been observed.

References

[10.1] S. J. Wang, J. Y. Lee, C. Y. Chang: An Efficient and Reliable Approach for Semiconductor Device Parameter Extraction. IEEE Trans. Comp.-Aided Des. *CAD-5*, 170 (1986).

[10.2] D. E. Ward, K. Doganis: Optimized Extraction of MOS Model Parameters. IEEE Trans. CAD Int. Circ. Syst. *CAD-1*, 163 (1982).

[10.3] K. Doganis, D. L. Scharfetter: General Optimization and Extraction of IC Device Model Parameters. IEEE Trans. Electr. Dev. *ED-30*, 1219 (1983).

[10.4] D. W. Marquardt: An Algorithm for Least-Squares Estimation of Non-Linear Parameters. J. Soc. Indust. Appl. Math. *11*, 431 (1963).

[10.5] C. G. Broyden: In: Numerical Methods for Unconstrained Optimization (W. Murray, ed.). Academic Press, New York (1972).

[10.6] F. M. Klaassen, W. de Groot, F. L. van de Markt: Computer Algorithm to Determine MOS Process Parameters. Philips Res. Repts. *31*, 84 (1976).

[10.7] H. P. Tuinhout, S. Swaving, J. J. M. Joosten: A Fully Analytical MOSFET Model Parameter Extraction Approach. Proc. IEEE Int. Conf. Microelectr. Test Structures, Long Beach (1988), p. 79.

[10.8] T. H. Ning, D. D. Tang: Method for Determining the Emitter and Base Series Resistances of Bipolar Transistors. IEEE Trans. Electr. Dev. *ED-31*, 409 (1984).

[10.9] L. J. Giacoletto: Measurements of Emitter and Collector Series Resistances. IEEE Trans. Electr. Dev. *ED-19*, 692 (1972).

[10.10] J. Choma, Jr.: Error Minimization in the Measurement of Bipolar Collector and Emitter Resistances. IEEE J. Solid-St. Circ. *SC-11*, 318 (1976).

[10.11] H. G. Rudenberg: On the Effect of Base Resistance and Collector-to-Base Overlap on the Saturation Voltage of Power Transistors. Proc. IRE *46*, 1304 (1958).

[10.12] W. D. Mack, M. Horowitz: Measurement of Series Collector Resistance in Bipolar Transistors. IEEE J. Solid-St. Circ. *SC-17*, 767 (1982).

[10.13] W. Filensky, H. Beneking: New Technique for Determination of Static Emitter and Collector Series Resistances of Bipolar Transistors. Electr. Ltrs. *17*, 50 (1981).

[10.14] W. M. C. Sansen, R. G. Meyer: Characterization and Measurement of the Base and Emitter Resistances of Bipolar Transistors. IEEE J. Solid-St. Circ. *SC-7*, 492 (1972).

[10.15] A. Neugroschel: Measurement of the Low-Current Base and Emitter Resistances of Bipolar Transistors. IEEE Trans. Electr. Dev. *ED-34*, 817 (1987).

[10.16] H. C. de Graaff, R. J. van der Wal: Measurement of the Onset of Quasi-Saturation in Bipolar Transistors. Solid-St. Electr. *17*, 1187 (1974).

[10.17] Sh. T. Hsu: Noise in High-Gain Transistors and Its Application to the Measurement of Certain Transistor Parameters. IEEE Trans. Electr. Dev. *ED-18*, 425 (1971).

[10.18] H. R. Claessen, J. A. M. Geelen, H. C. de Graaff: The Influence of the Emitter Sidewall Injection on Transistor Noise Figure. In: Solid-State Devices (G. Soncini, P. U. Calzolari, eds.). Elsevier, Amsterdam (1988).

[10.19] H. K. Gummel: On the Definition of the Cut-Off Frequency f_T. Proc. IEEE *57*, 2159 (1969).

[10.20] J. Y. Sun, M. R. Wordeman, S. E. Laux: On the Accuracy of Channel Length Characterization of LDD MOSFETs. IEEE Trans. Electr. Dev. *ED-33*, 1556 (1986).

[10.21] F. M. Klaassen, W. Hes: Compensated MOSFET Devices. Solid-State Electronics *28*, 359 (1985).

[10.22] S. M. Sze: Physics of Semiconductor Devices. John Wiley & Sons, New York (1982), chapt. 7.

[10.23] J. H. H. M. Quint, F. M. Klaassen, R. Petterson: 2-D and 3-D Capacitance Effects in MOS VLSI. Proceedings ESSDERC 87, Bologna. North-Holland, Amsterdam (1987), p. 417.

[10.24] G. J. Hu, C. Chang, Y. Chia: Gate-Voltage-Dependent Effective Channel Length and Series Resistance of LDD MOSFETS. IEEE Transactions on Electron Devices *ED-34*, 2469 (1987).

[10.25] F. M. Klaassen, P. T. J. Biermans, R. M. D. Velghe: The Series Resistance of Submicron MOSFETS and Its Effect on Their Characteristics. Journal de Physique *49*, C4-257 (1988).

11 Process and Geometry Dependence, Optimization and Statistics of Parameters

The parameters of a compact model are in general dependent on the geometry of the device and on the technological process steps. The geometry of the device is mainly characterized by the lateral dimensions of the junctions in bipolar transistors and by the channel length and width in MOS transistors. The dimensions must be known in silicon because only the real dimensions are electrically significant. They are determined by the dimensions of the mask, corrected for such effects as outdiffusion, underetching, misalignment of masks, encroachment of isolation oxides, etc.

In this chapter much attention will be paid to the "process block" approach. A process block is a computer program that enables the circuit designer to calculate the compact model parameters for a given device geometry and a given process. It has already proven its usefulness in bipolar circuit design; in MOS design it is not yet fully developed, partly because in MOST devices the geometrical dimensions are more dominant than process quantities such as sheet resistances.

Fig. 11.1 shows an example of an *e-b* junction which is partly walled by an isolating oxide. The electrically significant dimensions in silicon are L and Z, where

$$L = L_{mask} + \Delta L$$

and

$$Z = Z_{mask} + \Delta Z. \qquad (11.1)$$

The corrections (ΔL and ΔZ) may be either positive or negative; they have, for a given technological process, values which are specific for that process. The process dependence of the model parameters can be treated at three different levels:

1. The elementary level of the technological process flowchart data such as times and temperatures of diffusion steps, doses and energies of ion implantations, diffusivities of dopants, silicon dioxide growth rates, etc. [11.1, 11.2, 11.3].

2. The level of electrical process parameters, such as the flat-band voltage and oxide capacitance in MOS processes [11.4, 11.5] and the sheet resistances and breakdown voltages in bipolar processes [11.6].

11 Process and Geometry Dependence

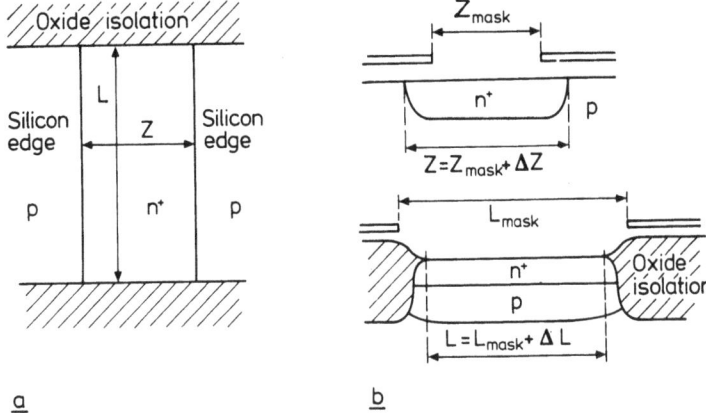

Fig. 11.1. Top view (*a*) and cross sections (*b*) of an emitter base junction, partly walled by an oxide isolation. The mask dimensions and the electrical dimensions in silicon are indicated

3. The level of the compact model parameters themselves, e.g. the threshold voltages V_T of MOS transistors or the current gain h_{FE} of bipolar transistors [11.7, 11.8, 11.9].

The process data, mentioned in 1, are statistical variables with mean values and spread. They determine the electrical process parameters and, ultimately, also the model parameters. The geometrical dimensions too are of a statistical nature because the corrections ΔL and ΔZ are statistical quantities. All these statistical variations can be considered in one chip, on one wafer, in one batch (lot of wafers) or in a whole process during a period of time.

From the foregoing it will be clear that many parameters of a compact model are correlated with each other and that for good statistical modelling these correlations must be taken into account. The "worst case" when all model parameters have their most unfavorable values, is in reality most unlikely.

Treating the model parameter dependences and correlations at the elementary level of the technological process data has the advantage of discussing primary causes, which are statistically independent. The disadvantage is that measured statistical data is scarce and even non-existent at the chip or wafer level. Such data can be readily obtained for the compact model parameters themselves, but here the separation between geometry and process influences is lost. A good compromise seems the use of electrical process parameters. Such parameters can be measured in special test structures (Process Control Modules) [11.6] and also be used for monitoring the process. Each wafer must then contain several (5 to 10) PCMs.

Owing to differences in approach this chapter will be divided into two parts, one for bipolar devices and one for MOS devices. However, generally we will discuss the geometrical scaling [11.10, 11.11], the performance optimization [11.12] and the statistical modelling largely in terms of electrical process parameters (PCM data).

11.1 Unity Parameters and Geometrical Scaling in Bipolar Modelling

Most parameters of the bipolar models mentioned in chapter 4 are geometry- and process-dependent. There are a few exceptions, like the field factor of the Mextram model and the transit times τ_f, τ_r, etc. which depend only on the vertical doping profile [11.11, 11.13]. With regard to the geometry dependence we will limit ourselves to rectangularly shaped junctions.

11.1.1 Geometry Dependence

For *pn* junctions as sketched in Fig. 11.1, the parameters related to currents, charges and capacitances can in general be described by the following expression,

$$P = u_b ZL + 2u_{si}L + 2u_{ox}Z + 4P_c. \tag{11.2}$$

Here P is an arbitrary model parameter, pertaining to a given *p-n* junction. Z and L are the electrical width and length of this junction. The quantity $u_b ZL$ is the bulk (or bottom) contribution, $2u_{si}L$ is the contribution of the silicon sidewall, $2u_{ox}Z$ that of the oxide sidewall. P_c is the contribution of a corner. The quantities u_b, u_{si} and u_{ox} are called density or unity parameters; they are independent of Z and L and only functions of certain process parameters, like sheet resistances and breakdown voltages. These process parameters are obtained from measurements on process control modules (PCM data) or, for not yet existing processes, from computer process and device simulations [11.2, 11.14].

If there is only one type of sidewall present we have in Eq. (11.2) $u_{si} \equiv u_{ox}$. For three oxide sides and one silicon sidewall Eq. (11.2) must be changed accordingly. The electrical dimensions Z and L are determined by the mask dimensions Z_{mask} and L_{mask} and the corrections ΔZ and ΔL. The mask dimensions are usually known, but ΔZ and ΔL are not. So the parameter P can be considered as a function of 6 unknowns: u_b, u_{si}, u_{ox}, ΔL, ΔZ and P_c. Rewriting Eq. (11.2) in terms of Z_{mask} and L_{mask} gives with the help of Eq. (11.1)

$$P = 2u_{si}\Delta L + 2u_{ox}\Delta Z + u_b \Delta L \Delta Z + 4P_c$$
$$+ (u_b \Delta L \cdot 2u_{ox})Z_{mask} + (u_b \Delta Z + 2u_{si} + u_b Z_{mask})L_{mask}. \tag{11.3}$$

We now measure several geometries with different values (at least 4) of Z_{mask} and L_{mask}. Plotting P as a linear function of e.g. L_{mask} with Z_{mask} as a parameter then gives n slopes and n intersection points, see Fig. 11.2. These slopes and intersection points are linear functions of Z_{mask},

$$\text{slope} = u_b \Delta Z + 2u_{si} + u_b Z_{mask}$$
$$\text{intersection} = 2u_{si}\Delta L + 2u_{ox}\Delta Z + u_b \Delta L \Delta Z + 4P_c$$
$$+ (u_b \Delta L + 2u_{ox})Z_{mask}.$$

11.1 Unity Parameters and Geometrical Scaling in Bipolar Modelling

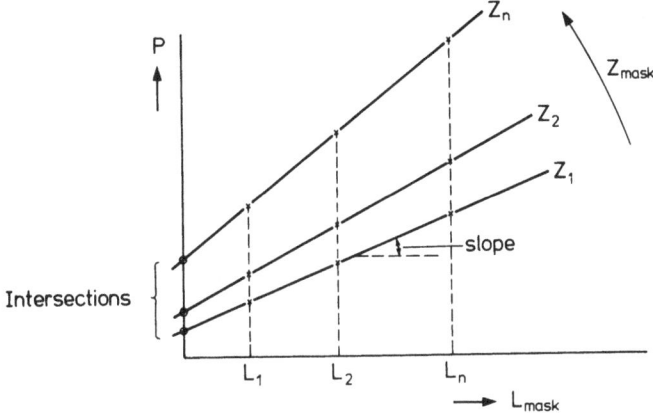

Fig. 11.2 The parameter P as linear function of the mask dimensions

Another linear regression on the slopes and intersection points results in 4 measured values:

$$\left.\begin{aligned} m_1 &= u_b, \\ m_2 &= 2u_{si} + u_b \Delta Z, \\ m_3 &= 2u_{ox} + u \Delta L, \\ m_4 &= 2u_{si} \Delta L \Delta Z + 2u_{ox} \Delta Z + u_b \Delta L \Delta Z + 4P_c. \end{aligned}\right\} \quad (11.4)$$

So we end with 6 unknowns and only 4 equations. Measuring more, different geometries does not help; the only way out is to choose the corrections ΔL and ΔZ.

The choice of ΔL and ΔZ is based on the following considerations:

— we can estimate the electrical dimensions in silicon by means of microcopic inspection (e.g. SEM);
— once chosen, ΔZ and ΔL determine the unity parameters u_{si}, u_{ox} and P_c (see Eq. (11.4)), so we can try to avoid negative sidewall components in order to maintain physical significance;
— all parameters related to the same junction should preferably have the same ΔL and ΔZ.

From the set of Eqs. (11.4) it becomes clear that u_b follows directly from the measurements ($u_b = m_1$), but u_{si}, u_{ox} and P_c can only be determined if ΔL and ΔZ are chosen. In spite of the above-mentioned considerations the determination of the sidewall and corner contributions remains a bit arbitrary. This can be illustrated once more by the example of a 2-dimensional computer simulation of a junction depletion capacitance. At zero bias we have

$$C_0 = c_b Z + 2c_{si} = c_b(Z_{mask} + \Delta Z) + 2c_{si}.$$

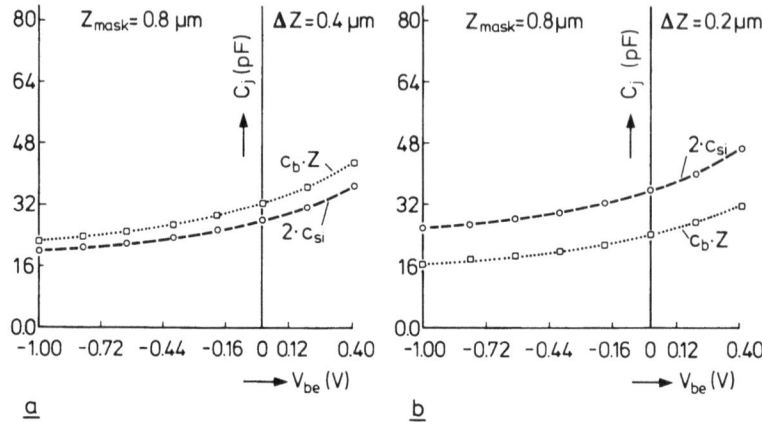

Fig. 11.3 The bottom ($c_b Z$) and sidewall ($2 c_{si}$) components of a junction depletion capacitance as a function of junction voltage, obtained from a 2D-computer simulation; a: for $\Delta Z = 0.4$ μm, b: for $\Delta Z = 0.2$ μm

See Fig. 11.3. The computer simulation shows that

$$C_0 = 59.5 * 10^{-12} \text{ F/cm} \quad \text{and} \quad c_b = 2.67 * 10^{-7} \text{ F/cm}.$$

If we take $Z_{mask} = 0.8$ μm and $\Delta Z = 0.4$ μm, we get $c_{si} = 1/2\{C_0 - (Z_{mask} + \Delta Z)c_b\} = 13.74 * 10^{-12}$ F/cm and, if $\Delta Z = 0.2$ μm, c_{si} becomes $16.4 * 10^{-12}$ F/cm.

In Fig. 11.3 we can also see that the choice of ΔZ has a slight influence on the voltage dependence of the bulk and sidewall components. The change in ΔZ, from 0.4 to 0.2 μm, is greatly exaggerated in the given example, in practice ΔZ can be estimated more accurately and the inherent arbitrariness does not seriously affect the applicability of the unity parameter approach. For not too small geometries we can moreover neglect the corner components ($P_c = 0$).

11.1.2 Process Dependence of Unity Parameters

The unity parameters u_b, u_{si} and u_{ox} depend on the process steps and are strongly related to certain process quantities, characterized by measured (or simulated) PCM data. To establish the relations between u_b, u_{si} and u_{ox} and these PCM data, we must rely on device physics (see chapter 3). The aim is to select a minimum number of process quantities ($p_1, p_2, \ldots$) that are preferably statistically uncorrelated. This minimum number, however, must be large enough to determine all the involved unity parameter.
We then vary these quantities over a certain range that covers all the expected variations. In this way we create a matrix batch, consisting of a set

11.1 Unity Parameters and Geometrical Scaling in Bipolar Modelling

of process variations where each set contains a number of geometries. Such a matrix batch can be made in real silicon or it can be composed by computer simulations. The relations between the unity parameters and the process quantities can usually be written of the form

$$\left. \begin{array}{l} u_b = A_b p_1^{b_{1b}} p_2^{b_{2b}}, \\ u_{si} = A_{si} p_1^{b_{1si}} p_2^{b_{2si}}, \\ u_{ox} = A_{ox} p_1^{b_{1ox}} p_2^{b_{2ox}}. \end{array} \right\} \quad (11.5)$$

Dependence on more than two process quantities is very rare in practice. As examples we will treat a few important model parameters: the collector saturation current I_s, the fixed base charge Q_{b0} and the zero bias values of the depletion capacitances C_{0e} and C_{0c}. The electrical process quantities (pinched base resistance ρ_{shpi}, the sheet resistance of the inactive base region (ρ_{shsp}) and the breakdown voltage (BV_{cb0}) of the base-collector junction. I_s, C_{0e} and Q_{b0} are closely related to the doping profile in the active base and therefore to the pinched base resistance ρ_{shpi} (see section 3.3). To include sidewall contributions we must also consider the doping profile of the inactive base, so ρ_{shsp} is important too. C_{0c} depends on the doping profile around the c-b junction and in the collector epilayer. BV_{cb0} is the related quantity here, and it is easy to measure.

The process variations of the process under consideration were

ρ_{shpi}:   3.5 kΩ – 16 kΩ,

ρ_{shsp}:   440 Ω – 550 Ω,

BV_{cb0}:   11 V – 25 V.

In total this matrix batch contained 12 wafers; each wafer had 10 PCMs and several transistors, with 9 different geometries. The transistor structure is a partly oxide-walled structure as given in Fig. 11.1.

The bird's beak of the thick oxide layer shortens the length of the junctions in the silicon (encroachment). For the emitter junction this encroachment (ΔL_e) varies between -3.2 and -4.2 µm; $\Delta Z_e \approx -0.2$ µm. For the base-collector junction $\Delta L_b = -2.8$ to 3.8 µm and $\Delta Z_b \approx +0.6$ µm. Underetching and outdiffusion have opposite effects on the ΔZ values: The smaller the value of Z itself, the larger the infuence of the etching.

From experiments the following relations between unity parameters and PCM data were found:

— The saturation current I_s is strongly correlated with ρ_{shpi},

$$J_b = 1.35 * 10^{-22} * \rho_{shpi}^{0.98} \text{ A/µm}^2,$$

$$J_{si} = 0, \quad \text{which is often found [11.15, 11.16]},$$

$$J_{ox} = 3.56 * 10^{-23} * \rho_{shpi}^{1.10} \text{ A/µm},$$

where ρ_{shpi} must be expressed in ohms. See Fig. 11.4. The formula for J_b

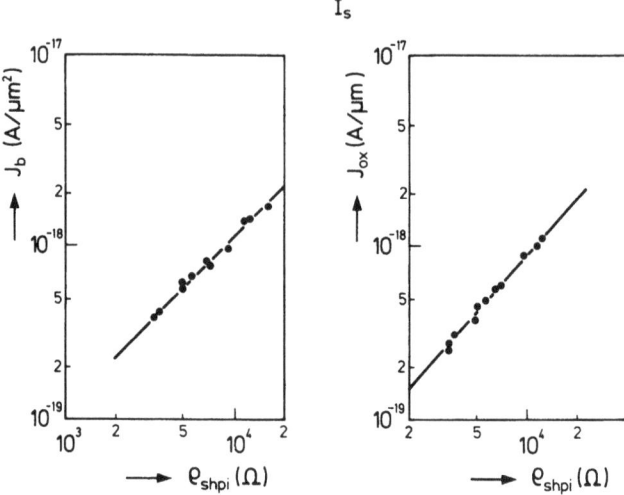

Fig. 11.4 The bottom (J_b) and oxide wall (J_{ox}) unity parameters of the model parameter I_s as a function of the sheet resistance of the pinched base. $J_{si} = 0$

agrees rather well with the theory of chapter 3, cf. Eqs. (3.23) and (3.24) and Fig. 3.4. I_s is given by $J_b ZL + 2ZJ_{ox} + 2LJ_{si}$.
— The fixed base charge Q_{b0} also depends on ρ_{shpi},

$$q_b = 4.78 * 10^{-11} * \rho_{shpi}^{-0.96} \text{ Coul}/\mu m^2,$$

$$q_{si} = 2.41 * 10^{-17} * \rho_{shpi}^{-0.78} \text{ Coul}/\mu m,$$

$$q_{ox} = 5.74 * 10^{-17} * \rho_{shpi}^{-1.02} \text{ Coul}/\mu m.$$

See Fig. 11.5.
— The zero bias value of the emitter depletion capacitance C_{0e} is correlated with ρ_{shpi} and ρ_{shsp},

$$c_b = 5.62 * 10^{-14} * \rho_{shpi}^{-0.35} \text{ F}/\mu m^2,$$

$$c_{si} = 1.09 * 10^{-13} * \rho_{shpi}^{+0.19} * \rho_{shsp}^{-0.89} \text{ F}/\mu m,$$

$$c_{ox} = 1.45 * 10^{-14} * \rho_{shpi}^{-0.32} \text{ F}/\mu m.$$

See Fig. 11.6.
— The zero bias value of the collector depletion capacitance C_{0c} is related to the collector breakdown voltage BV_{cb0},

$$c_b = 3.43 * 10^{-14} * BV_{cb0}^{-1.50} \text{ F}/\mu m^2 \quad \text{for} \quad BV_{cb0} < 19.5 \text{ Volts},$$

$$= 4.00 * 10^{-16} \text{ F}/\mu m^2 \quad \text{for} \quad BV_{cb0} > 19.5 \text{ Volts},$$

$$c_{si} = 0,$$

$$c_{ox} = 7.52 * 10^{-19} * BV_{cb0}^{1.87} \text{ F}/\mu m.$$

See Fig. 11.7.

11.1 Unity Parameters and Geometrical Scaling in Bipolar Modelling

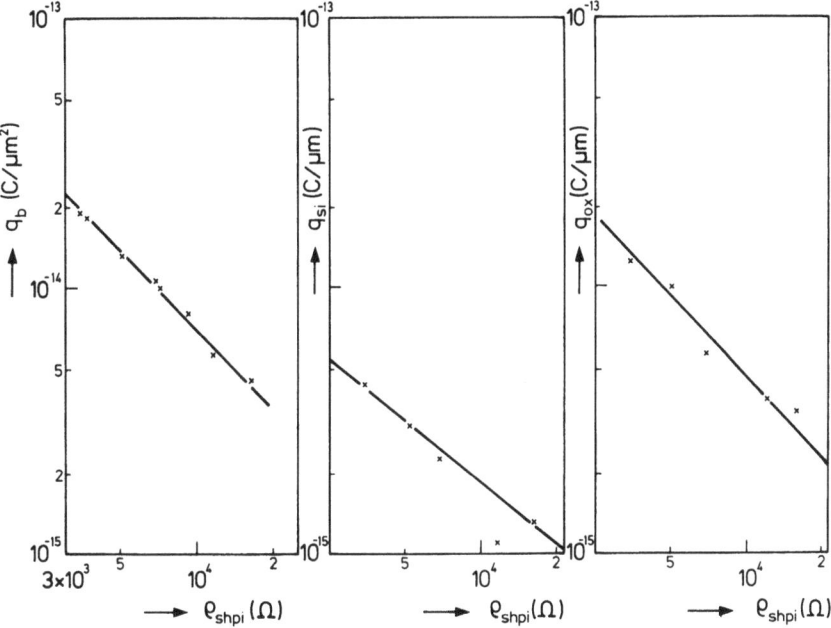

Fig. 11.5 The bottom (q_b), silicon edge (q_{si}) and oxide wall (q_{ox}) unity parameters of the model parameter Q_{b0}, as a function of the sheet resistance of the pinched base

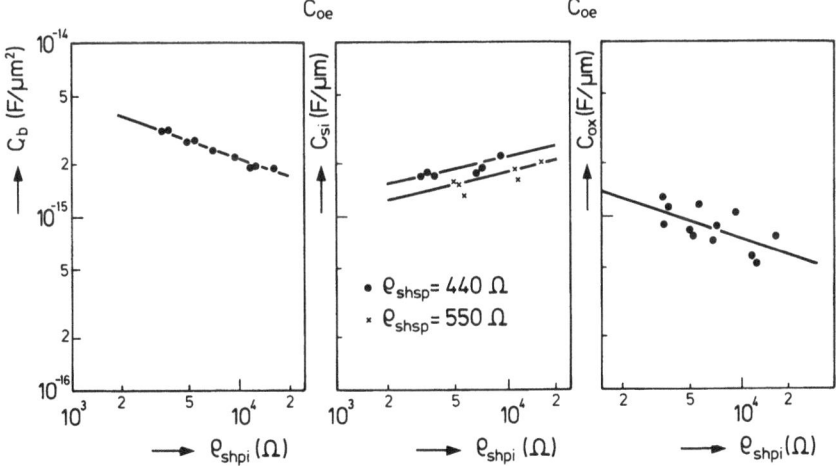

Fig. 11.6 The bottom (c_b), silicon edge (c_{si}) and oxide wall (c_{ox}) unity parameters of the model parameter C_{0e}, as a function of the sheet resistance of the pinched base. c_{si} also depends on the sheet resistance of the inactive base

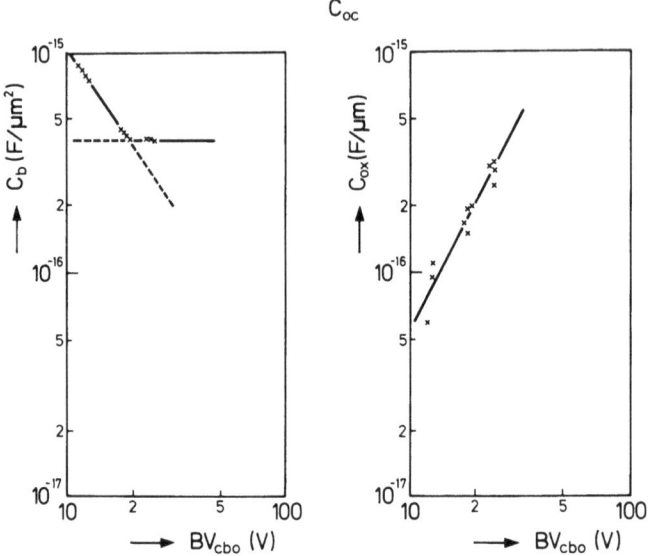

Fig. 11.7 The bottom (c_b) and oxide wall (c_{ox}) unity parameters of the model parameter C_{0c}, as a function of the collector-base breakdown voltage. $c_{si} = 0$

Parameters of a compound nature such as the Early voltages ($V_{ea} = Q_{b0}/C_T$, see section 3.6) or the current gain β_f have a more complicated geometry dependence than the parameters treated so far. E.g. for β_f, which is the ratio of the ideal components of I_c and I_b, we have

$$\beta_f = \frac{J_{cb}ZL + 2J_{csi}L + 2J_{cox}Z}{J_{bb}ZL + 2J_{bsi}L + 2J_{box}Z}.$$

By putting $J_{cb}/J_{bb} = \beta_\infty$ we get [11.16]

$$\beta_f = \beta_\infty \frac{1 + 2\dfrac{J_{csi}}{J_{cb}}\dfrac{1}{Z_e} + 2\dfrac{J_{cox}}{J_{cb}}\dfrac{1}{L_e}}{1 + 2\dfrac{J_{bsi}}{J_{bb}}\dfrac{1}{Z_e} + 2\dfrac{J_{box}}{J_{bb}}\dfrac{1}{L_e}}. \tag{11.6}$$

For smaller geometries β_f usually decreases but Eq. (11.6) also permits an increase for certain values of the unity parameters. The process dependence follows also from the current unity parameters, but a clear separation between geometry and process dependence is not possible here.

11.2 Bipolar Process Blocks and Circuit Optimization

We can combine the formulas of Eq. (11.2) and Eqs. (11.5) in a computer program that generates the required compact model parameters. The input data for such a program are then the geometrical dimensions of the tran-

11.2 Bipolar Process Blocks and Circuit Optimization

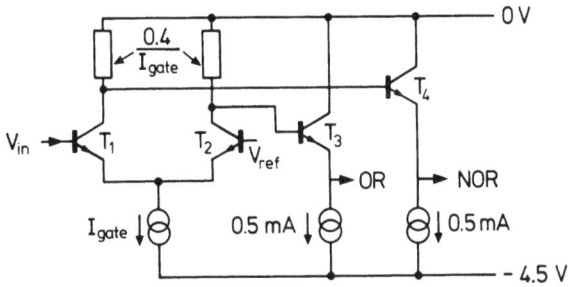

Fig. 11.8 Simplified circuit diagram of a basic ECL gate. The supply voltage is 4.5 V, the logic swing 0.4 V

sistors (Z and L) and the PCM data (p_1, p_2, etc.). These programs will be called process blocks and function as a preprocessor for the circuit simulation package. Each process has its own process block because the coefficients A_b, A_{si}, A_{ox}, b_{1b}, b_{2b}, etc., which are characteristic for a given process, must be known.

We can use process blocks for circuit optimization and statistical design in existing processes, but also for designing new processes by generating matrix batches with the help of process and device simulation packages.

In this section we will illustrate the use of a process block by means of optimizing the delay of an ECL gate [11.12]. We will limit ourselves to a nominal design and disregard the statistical fluctuations of the design parameters [11.17] here. The technological process involved is the same as in section 11.1.1. Fig. 11.8 gives a simplified circuit diagram of a basic ECL gate; T_1 and T_2 are the switching devices, T_3 and T_4 the output emitter followers.

We want to optimize the minimum gate delay with the following constraints: The supply voltage is 4.5 Volts, the logic swing is 400 mV and the emitter follower currents are 0.5 mA. The gate delay is calculated from the simulated results of a 12-stage ring-oscillator. The model used for the simulations is the Mextram model described in section 4.3.

As a starting point we take a device layout of one collector contact, one base contact and one emitter contact (*cbe*). The emitter mask dimensions are $Z_{emask} = 2.5$ µm and $L_{emask} = 8$ µm. The relevant PCM data is $\rho_{shpi} = 7$ kΩ, $\rho_{shsp} = 500$ Ω and $BV_{cbo} = 17$ Volts. In Fig. 11.9 we see the gate delay as a function of gate current. The minimum delay for *cbe* layout is 290 ps at a gate current of 400 µA. We first investigate the influence of a double base contact (*cbeb* layout). This decreases the base resistance, but it increases the base-collector junction area and thus C_{co}. The result, however, is extremely beneficial, as is shown in Fig. 11.9, "*cbeb*" curve. The second step is the use of a p^+ boost diffusion in the inactive base region. This lowers the resistance (R_{bc}), but it increases the depletion capacitance (C_{oc}). Here too, the overall result is positive, see Fig. 11.9 "boost" curve.

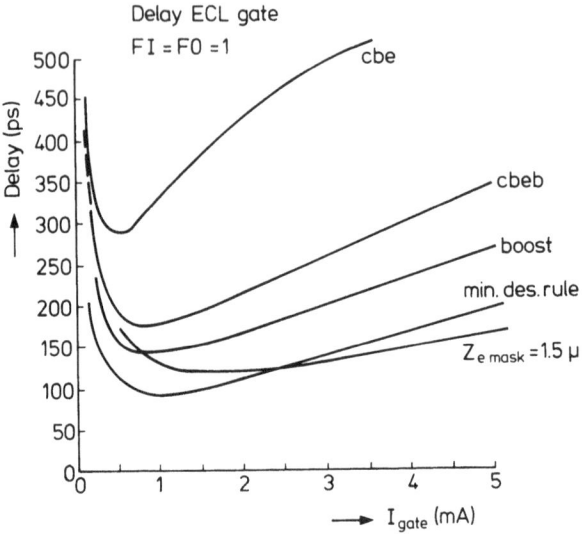

Fig. 11.9 Delay of an ECL gate as a function of the gate current I_{gate}. The various curves represent the subsequent steps of the optimization procedure

The third step consists of changing the length of the emitter stripe. This affects many parameters (I_s, Q_{b0}, C_{0e}, R_{bc}, R_{bv}, etc.). From simulations for several L_e values it became clear that $L_{emask} = 12$ μm gives the lowest minimum delay (165 ps at 400 μA), but the difference with $L_{emask} = 8$ μm is rather small (175 ps at 800 μA); because of lower power consumption and the smaller area the 8 μm emitter is preferred.

The fourth step is the optimization of the PCM data. The BV_{cb0} is increased from 17 to 19.5 Volts: this decreases the C_{0c}, but increases the epilayer resistance R_{cv}. The overall result gives a slight improvement. A further increase of BV_{cb0} is pointless, for C_{c0} does not change anymore (see section 11.1.1), while R_{cv} continues to rise.

The ρ_{shpi} also affects a great many parameters; its influence on the minimum delay is, however, rather limited. Variations in the range from 4 to 12 kΩ show a broad minimum around 7 kΩ. Similar exercises with ρ_{shsp} in the range 100–500 Ω show an optimal value for $\rho_{shsp} = 200$–300 Ω. So the optimal set of PCM data is $BV_{ce0} = 19.5$ Volts, $\rho_{shpi} = 7$ kΩ and $\rho_{shsp} = 250$ Ω.

The last step is the influence of the lateral dimensions. If we only change the width of the emitter stripe Z_{emask} to 1.5 μm, base resistance and capacitances will be reduced and the delay is decreased (see Fig. 11.9, "1.5 μm" curve). A further reduction of Z_{emask} increases the delay again because the sidewall capacitances begin to dominate and constitute a lower limit, whereas high injection effects in the base deteriorate the delay. Shrinking of all the lateral dimensions and not only of the emitter width, that is changing the minimum design rules also 1.5 μm as well, gives then another improvement and finally brings the minimum delay to 100 ps at 900 μA.

11.3 Geometry- and Process Dependence of MOSFET Parameters

Similar to bipolar devices we will limit ourselves to rectangularly shaped devices.

11.3.1 Geometry Dependence

From the discussions in the previous chapters 6 and 7 it can be concluded that most electrical parameters obey relatively simple scale relations. For instance, for the gain constant we have (compare formula (7.3c)

$$\beta = \mu C_{ox} \frac{Z}{L} = \mu C_{ox} \frac{Z_m - \Delta Z}{L_m - \Delta L}, \tag{11.7}$$

where Z_m, L_m are the drawn (mask) values of gate width and length, respectively, and ΔZ, ΔL are the process-induced width and length reduction, respectively. Note that for reasons of uniformity in this chapter the gate width has been denoted with the symbol Z (replacing W).

With some exceptions, for most other active MOST parameters, their geometry dependence can be modelled according to the general relation

$$P = P_0 + \frac{P_L}{L} + \frac{P_Z}{Z}. \tag{11.8}$$

For the threshold voltage V_T and the substrate effect coefficients γ, the above result has been derived in section 7.2.1.1, for the mobility parameters θ_A and θ_c in section 7.2.3, and for η in section 6.5.1. However, if current crowding at a source contact contributes to the MOSFET series resistance, for θ_A a more complicated relation is needed [11.4]. Another exception is formed by the subthreshold shift parameters s (compare section 7.2.1.1). Owing to the inherent strong dependence of the drain-induced barrier lowering effect on channel length, these parameters usually depend on L according to

$$s = s_0 + \frac{s_L}{L^2}. \tag{11.9}$$

Finally for the source/drain junction capacitances the same relations as discussed in the previous section are used (formula (11.2) with three oxide and one silicon sidewall).

Usually the key parameters ΔZ and ΔL are determined by measuring the gain factor β as a function of the mask values Z_m and L_m. Preferably β is measured together with several other parameters at low drain bias condition (see the parameter table 1 in chapter 10). This method allows to correct for several nonlinearities in the drain conductance slope versus gate voltage, which otherwise may introduce errors in the value of β. Fig. 11.10 illustrates how ΔL and ΔZ can be obtained from a linear plot of β^{-1} and β as a function

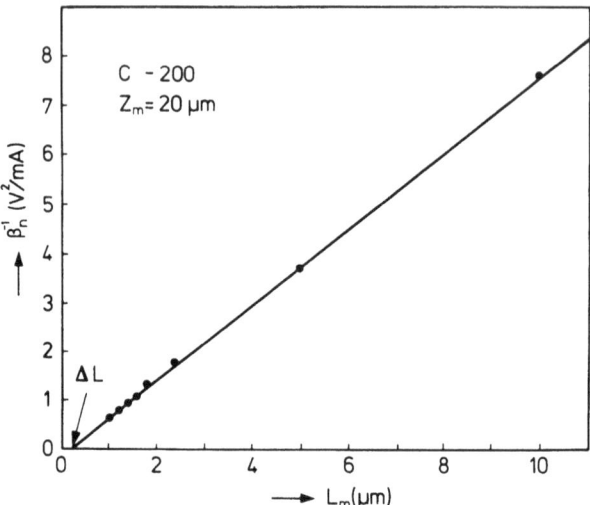

Fig. 11.10 a Inverse of gain constant as a function of drawn (mask) gate length. From the off-set at the X-axis the process-induced length reduction is determined

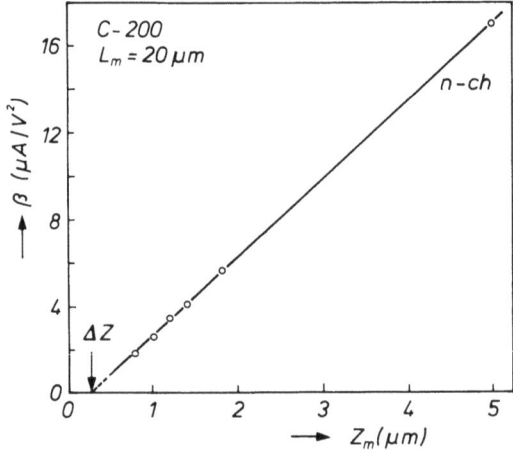

Fig. 11.10 b Gain constant as a function of drawn gate width. The off-set yields the width reduction ΔZ

of L_m and Z_m, respectively. The above results apply to n-channel, LDD-type devices from a 1.2 μm CMOS process. Although for this case nonlinear plots have been reported in the literature, no problem is observed here. On the contrary, the observed value of

$$\Delta L = 0.20 \ \mu m$$

can be almost completely ascribed to underdiffusion of the lightly doped

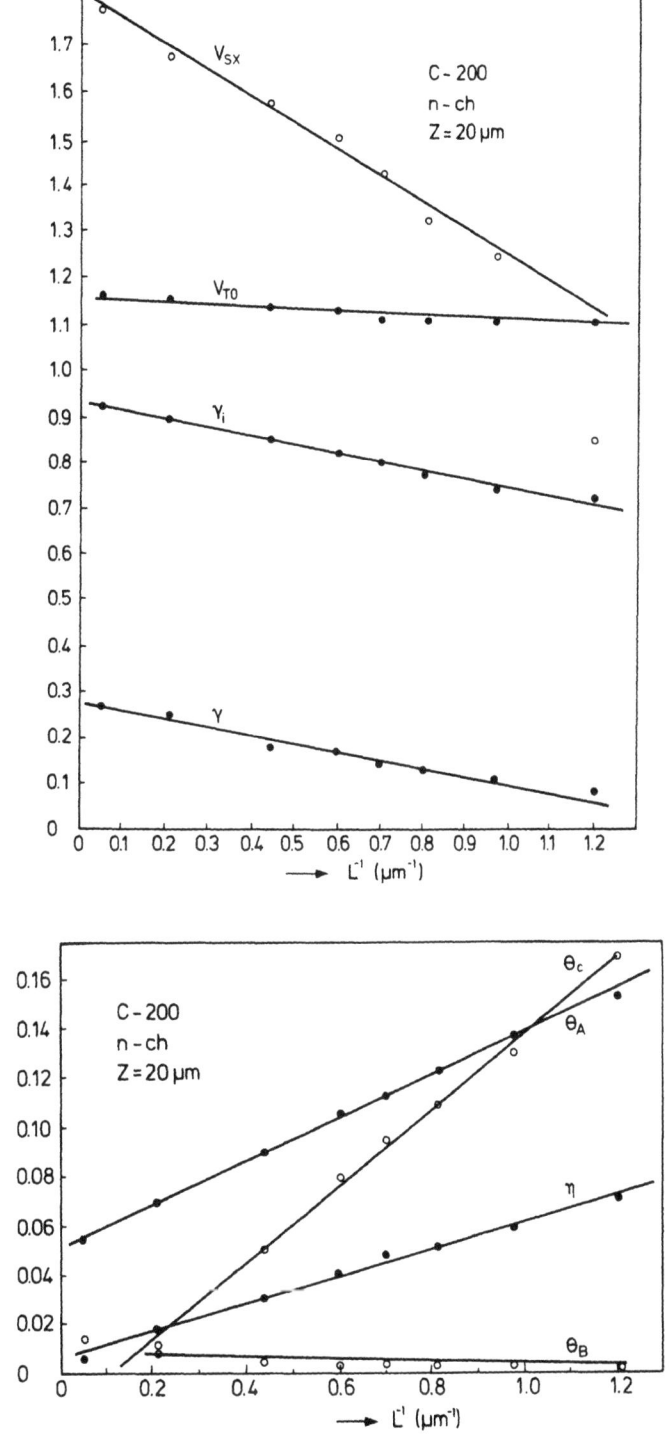

Fig. 11.11 Major MOSFET device parameters as a function of channel length

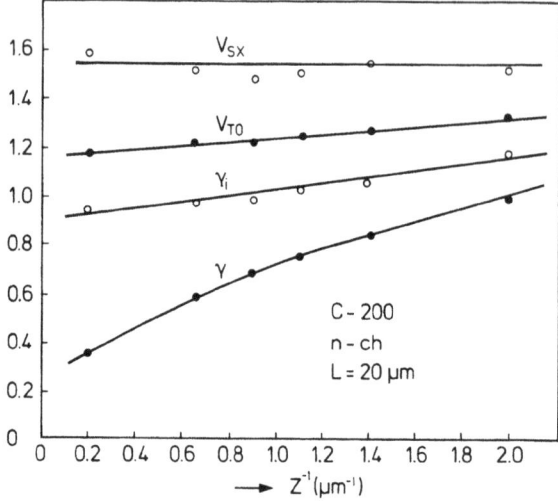

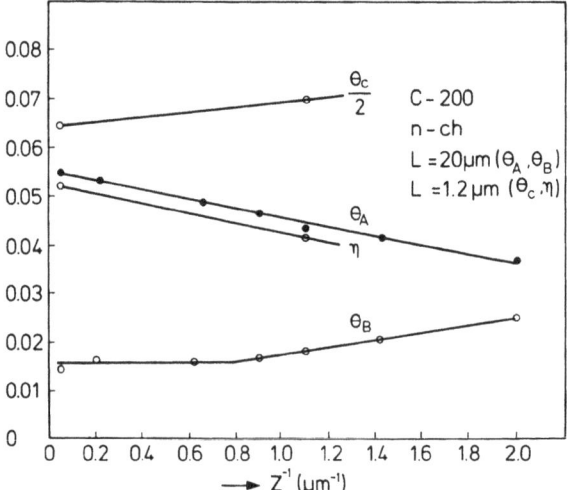

Fig. 11.12 Major MOSFET device parameters as a function of channel width

juntion implant. Owing to the fact that in this process the LOCOS bird's beak has been reduced by special process steps, the obtained value of

$$\Delta Z = 0.30 \ \mu m$$

is relatively small.

Next Fig. 11.11 gives the dependence of the major parameters on the effective channel length. In almost all cases the expected L^{-1} dependence is confirmed. In addition a negative sign of P_L for γ_i, γ and V_{SX} and a positive value for θ_A, θ_C and η can be easily explained. Perhaps the slight decrease of θ_B with reduced channel length is caused by a decrease of normal field effect on mobility owing to a reduction of effective depletion charge. The relatively large errors in the V_{SX} plot are mainly due to the fact that the MOSFET characteristics are less sensitive for variations in the value of V_{SX}. The relatively large increase of θ_A with reduced channel length is caused by the series resistances inherent to the LDD configuration. Note further that the θ_C curve does not pass through the origin. This is due to the fact that the original velocity saturation parameter has to be corrected for effects of the above series resistances and normal field, which becomes manifest in modern scaled processes (compare section 7.2.3).

For the same parameters Fig. 11.12 gives the dependence on the channel width. Since θ_C and η have only a significant value at shorter channels, for these devices a low value of L_m was taken as a reference (with consequently only limited data). Although the majority of the parameters follow a Z^{-1} dependence quite well, in particular at small value of channel width several deviations are found. These are mainly due to the outdiffusion of the channel stop implant into the active device region. This causes that in narrow devices the average value of the substrate doping is increased. Consequently the dependence of the lower substrate factor γ on channel width becomes rather complex. In addition the outdiffusion causes the effective channel width to decrease at higher back-bias (see section 6.4.3). In practice this effect is observed as an increase of the parameter θ_B for narrow channels.

11.3.2 Process Dependence

Generally the compact model parameters depend (with varying magnitude) on process variables like the gate oxide thickness, the oxide built-in charge and the threshold implant, but little on substrate doping and junction depth. Although in particular the threshold voltage and the substrate coefficients are directly affected by a change in the implant dose D_i (according to section 7.1.2 we have $V_T \sim D_i$, $\gamma_i \sim D_i^{1/2}$, $V_{SX} \sim D_i$), it has been found in practice that the process spreads in D_i are much smaller than those in the gate oxide thickness. Therefore the effect of D_i on the other parameters is not discussed here.

According to the analysis given in section 7.1.2, the dependence of threshold voltage and substrate coefficients on oxide thickness is straightforward. These parameters vary linearly with t_{ox}. Furthermore the boundary voltage V_{SX} is independent of t_{ox}. Nevertheless for V_{T0} a large difference exists between n-channel and p-channel devices. Whereas the variation of V_{T0} with t_{ox} is relatively large for n-type MOSFETs, owing to compensation effects in the depletion charge (see section 6.3.3), this dependence is almost non-existent in p-type MOSFETs. This is illustrated in Fig. 11.13a for the type of devices used in the previous figures.

Since a change in the value of the oxide thickness causes the normal field at the interface to change, the mobility parameter θ_A and the static feedback

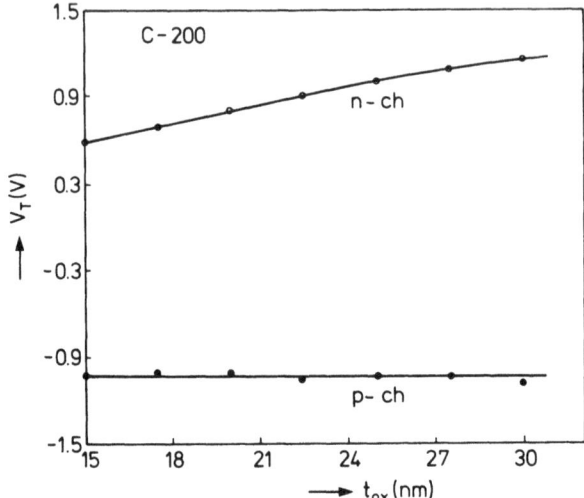

Fig. 11.13 a The threshold voltage of an *n*-type and *p*-type MOSFET as a function of gate oxide thickness

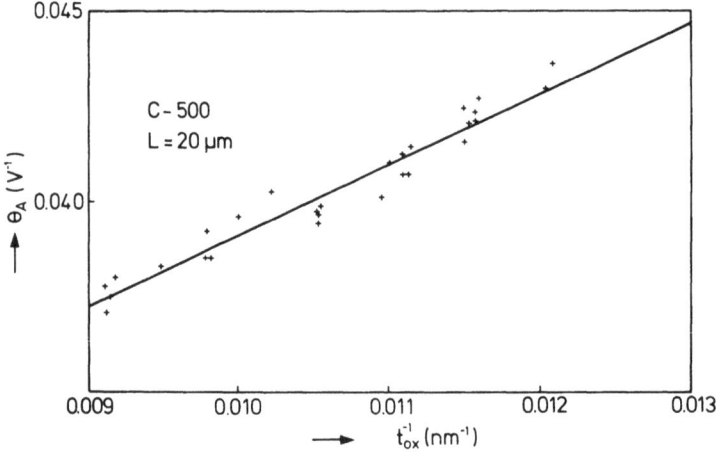

Fig. 11.13 b Mobility reduction parameter as a function of gate oxide thickness

parameter η are affected too, but the velocity saturation parameter is hardly affected. Figs. 11.13 b and 11.13 c give the dependence of θ_A and η on t_{ox}. Owing to the spread of the oxide charge in VLSI processing, the flatband voltage V_{FB} is an important process parameter. Fortunately its effect is mainly limited to the zero-bias threshold voltage V_{T0}. According to the basic theory V_{T0} varies linearly with V_{FB}. Generally the spread in V_{FB} causes severe limitations in the application of MOSFETs (see last section of this chapter).

11.4 Statistics: Definitions and Formulas

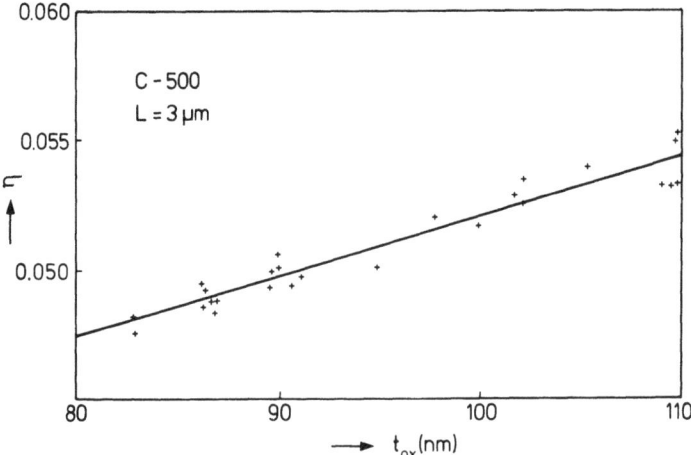

Fig. 11.13 c Static feedback factor as a function of gate oxide thickness

Unfortunatey for MOSFET devices a complete processblock has not yet been published.

11.4 Statistics: Definitions and Formulas

Everything in this section can also be found in textbooks on statistics [11.18, 11.19], but it seems useful to summarize here the basic definitions of the various statistical quantities and their mutual relations. All the values a device parameter may have on a wafer or in a batch of wafers form a population. Usually not all the devices of a given population are measured, but we take a sample out of that population. If the sample is not too small, its statistical quantities may serve as good estimators for the same quantities of the whole population. In the following we will consider only populations and we will disregard the (small) differences with samples.

The mean value or expectation of a stochastic variable x is defined as

$$E(x) = \bar{x} = \sum_{i=1}^{N} p(x_i) x_i$$

or

$$E(x) = \bar{x} = \int_{-\infty}^{+\infty} xp(x)\,dx$$

(11.10)

for discrete and continuous variables, respectively. Here $p(x)$ is the probability density function. An often encountered density function is what is known as the normal distribution

$$p(x) = \frac{1}{\sigma(2\pi)^{1/2}} \exp\left\{-\frac{(x-\bar{x})^2}{2\sigma^2}\right\}.$$

(11.11)

The parameter σ is called the standard deviation; the quantity σ^2 is the variance. A more general definition of variance is

$$\sigma^2 = E(x - \bar{x})^2 = E(x^2) - \{E(x)\}^2 = \overline{x^2} - \bar{x}^2. \qquad (11.12)$$

The bar denotes mean values or expectations: the often used sympol μ is avoided here because throughout this book μ denotes carrier mobilities. For normal distributions the range $x = x \pm \sigma$ comprises about 68% of the total population, 95% lies between $x = x \pm 2\sigma$ and 99.7% lies between $x = x \pm 3\sigma$. So for most practical purposes $x \pm 3\sigma$ can be regarded as the extremes.

If $g(x, y)$ is a function of two stochastic variables a Taylor expansion around the values $\bar{x}$ and $\bar{y}$ gives

$$g(x, y) = g(\bar{x}, \bar{y}) + (x - \bar{x})\frac{\partial g}{\partial x}\bigg|_{\bar{x},\bar{y}} + (y - \bar{y})\frac{\partial g}{\partial y}\bigg|_{\bar{y},\bar{x}}$$

$$+ \frac{1}{2}(x - \bar{x})^2 \frac{\partial^2 g}{\partial x^2}\bigg|_{\bar{x},\bar{y}} + \frac{1}{2}(y - \bar{y})^2 \frac{\partial^2 g}{\partial y^2}\bigg|_{\bar{y},\bar{x}}$$

$$+ (x - \bar{x})(y - \bar{y})\frac{\partial^2 g}{\partial x\, \partial y}\bigg|_{\bar{x},\bar{y}} + \cdots.$$

The mean value g then becomes approximately

$$E\{g(x, y)\} \approx g(\bar{x}, \bar{y}) + \frac{1}{2}\sigma_x^2 \frac{\partial^2 g}{\partial x^2}\bigg|_{\bar{x},\bar{y}} + \frac{1}{2}\sigma_y^2 \frac{\partial^2 g}{\partial y^2}\bigg|_{\bar{x},\bar{y}} + \sigma_{xy}^2 \frac{\partial^2 g}{\partial x\, \partial y}\bigg|_{\bar{x},\bar{y}}. \qquad (11.13)$$

σ_x^2 and σ_y^2 are the variances of x and y, respectively; σ_{xy}^2 is called the covariance of x and y. It is defined as

$$\sigma_{xy}^2 = E\{(x - \bar{x})(y - \bar{y})\} = \overline{xy} - \bar{x}\bar{y}. \qquad (11.14)$$

Note that σ_{xy}^2 may have negative values.
The correlation coefficient r_{xy} is defined as

$$r_{xy} = \frac{\sigma_{xy}^2}{\sigma_x \sigma_y}. \qquad (11.15)$$

If a linear functional relationship exists between x and y the correlation is complete and $r_{xy} = \pm 1$. If x and y are statistically uncorrelated $r_{xy} = \sigma_{xy} = 0$. For situations in between we often use a linear regression equation that couples the value y_j of the stochastic variable y with a certain probability to the value x_j of the variable x,

$$y_j = \bar{y} + r_{xy}\frac{\sigma_y}{\sigma_x}(x_j - \bar{x}) + e_j, \qquad (11.16)$$

where e_j is the error, for which we have $\bar{e} = 0$ and $\sigma_e^2 = \sigma_y^2(1 - r_{xy}^2)$, provided that x and y have a two-dimensional normal distribution. The variance of

11.5 Bipolar Statistical Modelling

the function $g(x, y)$ is

$$\sigma_g^2 = \overline{g^2} - \bar{g}^2 \approx \sigma_x^2 \left(\frac{\partial g}{\partial x}\bigg|_{\bar{x},\bar{y}}\right)^2 + \sigma_y^2 \left(\frac{\partial g}{\partial y}\bigg|_{\bar{x},\bar{y}}\right)^2$$

$$+ 2\sigma_{xy}^2 \left(\frac{\partial g}{\partial x}\bigg|_{\bar{x},\bar{y}}\right)\left(\frac{\partial g}{\partial y}\bigg|_{\bar{x},\bar{y}}\right). \tag{11.17}$$

Eq. (11.17) follows from the Taylor expansion of $g(x, y)$.
Application of Eqs. (11.13) and (11.17) to the function $g = x + y + z$ gives

$$\left.\begin{array}{l} E(g) = \bar{g} = \bar{x} + \bar{y} + \bar{z} \\ \text{and} \\ \sigma_g^2 = \sigma_x^2 + \sigma_y^2 + \sigma_z^2 + 2\sigma_{xy}^2 + 2\sigma_{xz}^2 + 2\sigma_{yz}^2. \end{array}\right\} \tag{11.18}$$

If x, y and z are uncorrelated, $\sigma_{xy} = \sigma_{xz} = \sigma_{yz} = 0$ and σ_g^2 simplifies into

$$\sigma_g^2 = \sigma_x^2 + \sigma_y^2 + \sigma_z^2.$$

For the function $g = ax^m y^n$ we get

$$\left.\begin{array}{l} \bar{g} \approx a\bar{x}^m \bar{y}^n \left\{1 + \frac{1}{2}m(m-1)\frac{\sigma_x^2}{\bar{x}^2} + \frac{1}{2}n(n-1)\frac{\sigma_y^2}{\bar{y}^2} + mn\frac{\sigma_{xy}^2}{\bar{x}\bar{y}}\right\} \\ \text{and} \\ \sigma_g^2 \approx a^2 \bar{x}^{2m} \bar{y}^{2n}\left(m^2 \frac{\sigma_x^2}{\bar{x}^2} + n^2 \frac{\sigma_y^2}{\bar{y}^2} + 2mn \frac{\sigma_{xy}^2}{\bar{x}\bar{y}}\right). \end{array}\right\} \tag{11.19}$$

With x and y uncorrelated and $\sigma_x/\bar{x}, \sigma_y/\bar{y} \ll 1$ (e.g. $\sigma_x/\bar{x} < 0.2$ and $\sigma_y/\bar{y} < 0.2$) we may write

$$\left(\frac{\sigma_g}{\bar{g}}\right)^2 \approx m^2 \left(\frac{\sigma_x}{\bar{x}}\right)^2 + n^2 \left(\frac{\sigma_y}{\bar{y}}\right)^2.$$

11.5 Bipolar Statistical Modelling

Compact model parameters can be regarded as statistical variables with mean values and variances as a measure of their spread. The model parameters are statistically not independent; their correlation can, similar to their process dependence, be described at three different levels:

— The compact model parameters are correlated with the elementary technological process data (temperature, diffusion times, etc.).
— Correlation with electrical process parameters (PCM data such as sheet resistances).
— Correlation with one or two compact model parameters (I_s, β_f).

The main emphasis here will be on the second possibility because it fits very well with the use of the process blocks (see section 11.2).

11.5.1 Process Blocks and Statistical Models

From Eq. (11.2) it follows that with $P_c = 0$ a compact model parameter P can be written as the sum of three contributions,

$$P = P_b + P_{si} + P_{ox},$$

where

$$P_b = u_b Z L,$$
$$P_{si} = 2 u_{si} L,$$
$$P_{ox} = 2 u_{ox} Z.$$

If we further assume that the unity parameters u_b, u_{si} and u_{ox} are functions of e.g. two electrical process parameters p_1 and p_2 (see Eq. (11.5)), then the model parameter P can be regarded as a function of the stochastic variables p_1, p_2, Z and L. The dimensions Z and L are stochastic, because

$$\bar{Z} = Z_{mask} + \overline{\Delta Z},$$
$$\bar{L} = L_{mask} + \overline{\Delta L},$$
$$\sigma_Z = \sigma_{\Delta Z},$$
$$\sigma_L = \sigma_{\Delta L}.$$

The quantities Z_{mask} and L_{mask} are not considered here as statistical variables. The mean value of P follows from Eqs. (11.18) and (11.19),

$$\bar{P} = \overline{P_b} + \overline{P_{si}} + \overline{P_{ox}}$$
$$\approx \overline{u_b} \bar{Z} \bar{L} + 2\overline{u_{si}} \bar{L} + 2\overline{u_{ox}} \bar{Z}. \tag{11.20}$$

With Eq. (11.5) we get for

$$\overline{u_b} \approx A_b \bar{p}_1^{b_{1b}} \bar{p}_2^{b_{2b}},$$
$$\overline{u_{si}} \approx A_{si} \bar{p}_1^{b_{1si}} \bar{p}_2^{b_{2si}}, \tag{11.21}$$
$$\overline{u_{ox}} \approx A_{ox} \bar{p}_1^{b_{1ox}} \bar{p}_2^{b_{2ox}}.$$

We have made use here of the fact that the unity parameters are independent of Z and L and that Z and L are themselves uncorrelated [11.20, 11.21]. It is furthermore assumed that the variances are small: $\sigma_p/\bar{p} < 0.2$. According to Eq. (11.17) the variance of P can be written as

$$\sigma_P^2 \approx \sigma_{p_1}^2 \left(\frac{\partial P}{\partial p_1}\bigg|_{\bar{p}_1}\right)^2 + \sigma_{p_2}^2 \left(\frac{\partial P}{\partial p_2}\bigg|_{\bar{p}_2}\right)^2 + \sigma_Z^2 \left(\frac{\partial P}{\partial Z}\bigg|_{\bar{Z}}\right)^2$$
$$+ \sigma_L^2 \left(\frac{\partial P}{\partial L}\bigg|_{\bar{L}}\right)^2 + 2\sigma_{p_1 p_2}^2 \frac{\partial P}{\partial p_1}\bigg|_{\bar{p}_1} \frac{\partial P}{\partial p_2}\bigg|_{\bar{p}_2}. \tag{11.22}$$

The process parameters may be correlated, so $\sigma_{p_1 p_2}^2$ is not necessarily zero.

11.5 Bipolar Statistical Modelling

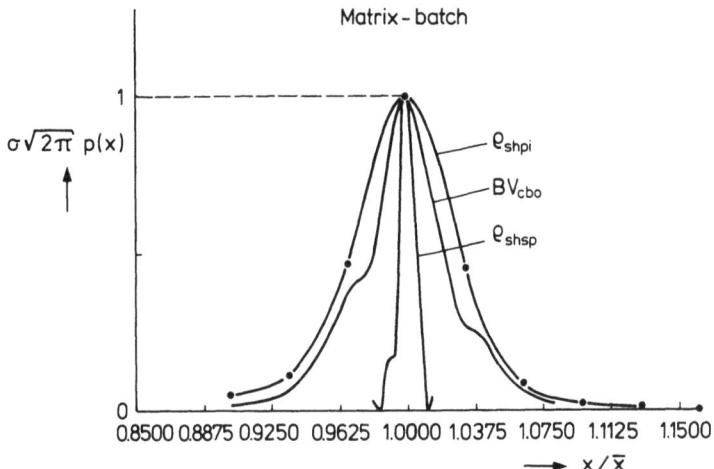

Fig. 11.14 Normalized distribution functions of the spread over a typical wafer for three different electrical process quantities (PCM data)

Eq. (11.21) shows that the spread of the parameter value P is determined by the spread of the process quantities and the spread of the geometrical dimensions, together with the sensitivities of P with respect to these variables ($\partial P/\partial p_1$, $\partial P/\partial p_2$, etc.). Spread can be distinguished in the spread between matched pairs, spread in one wafer, in one batch or over a complete process.

We will give an example, using the process described in section 11.1.2. Fig. 11.14 shows the normalized distribution functions of the PCM data ρ_{shpi}, ρ_{shsp} and BV_{cbo} for a typical wafer. These curves resemble more or less the normal distribution function for different variance values,

$$\sigma_{\rho_{shpi}}/\bar{\rho}_{shpi} \approx 2.6\%, \qquad \sigma_{\rho_{shsp}}/\bar{\rho}_{shsp} \approx 0.41\%$$

and

$$\sigma_{BV_{cbo}}/\overline{BV}_{cbo} \approx 1.7\%.$$

ρ_{shpi} and ρ_{shsp} are nearly uncorrelated: $r = +0.042$.
The process block approach can be used for the prediction of mean values. We will show this for a wafer with the following mean values,

$$\bar{\rho}_{shpi} = 7020\ \Omega, \qquad \bar{\rho}_{shsp} = 441\ \Omega, \qquad \overline{BV}_{cbo} = 16.8\ \text{Volts}.$$

The mean values of the emitter size are $\bar{Z} = 2.3\ \mu\text{m}$ and $\bar{L} = 4.89\ \mu\text{m}$. The parameters of the depletion capacitance of the emitter-base junction (C_{0e}, V_{de} and p_e) are now calculated from the given mean values of the PCM data for that wafer. We use for C_{0e} the relations already given in section 11.1.2. For the prediction of the complete $C(V)$ curve we have used additional information about V_{de} and p_e,

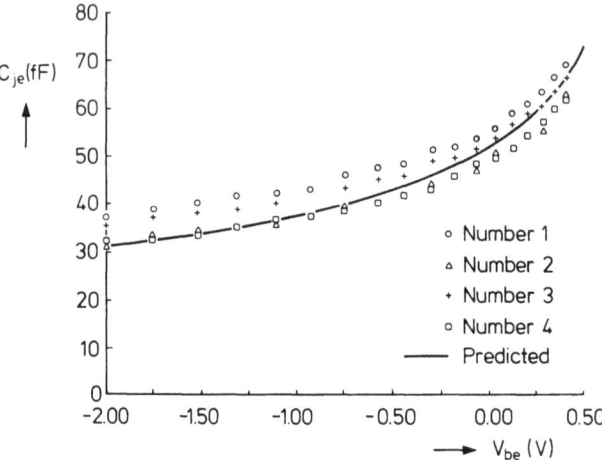

Fig. 11.15 Measurements (o △ + □) and prediction (——) of the emitter depletion capacitance versus junction voltage. The prediction uses the PCM data

$$V_{de} = 0.983 - 5.9 * 10^{-6} * \rho_{shpi},$$

$$p_e = \frac{0.247}{\ln\left(\dfrac{V_{de}}{V_{de} - 0.4}\right)}.$$

The predicted $C(V)$ curve is confronted with measurements and the result is shown in Fig. 11.15.

For another wafer the spread of the parameter C_{0e} is analyzed. From PCM measurements we know that

$$\overline{\rho}_{shpi} = 8315\,\Omega, \qquad \sigma_{shpi} = 333\,\Omega,$$
$$\overline{\rho}_{shsp} = 442\,\Omega, \qquad \sigma_{shsp} = 1.77\,\Omega,$$
$$\overline{Z} = 2.3\,\mu\text{m}, \qquad \sigma_Z = 0.07\,\mu\text{m},$$
$$\overline{L} = 21\,\mu\text{m}, \qquad \sigma_L = 0.11\,\mu\text{m}.$$

Using Eq. (11.19) and neglecting any covariances we get

$$\sigma_{\rho_{shpi}} \frac{\partial C_{0e}}{\partial \rho_{shpi}} = 0.825\,\text{fF},$$

$$\sigma_{\rho_{shsp}} \frac{\partial C_{0e}}{\partial \rho_{shsp}} = 0.385\,\text{fF},$$

$$\sigma_L \frac{\partial C_{0e}}{\partial L} = 1.10\,\text{fF},$$

$$\sigma_Z \frac{\partial C_{0e}}{\partial Z} = 3.45\,\text{fF}$$

11.5 Bipolar Statistical Modelling

Table 11.1

	Calculated $\sigma_x/\bar{x}$	Measured $\sigma_x/\bar{x}$
I_s	4.9%	5.1%
C_{0e}	1.7%	1.6%
C_{0c}	8.5%	8.8%
Q_{b0}	5.1%	4.7%
τ_{Ne}	5.0%	4.5%

and
$$\sigma_{C_{0e}} = \sqrt{(0.825)^2 + (0.385)^2 + (1.10)^2 + (3.45)^2} = 3.75\,\text{fF}.$$

So the spread within one wafer is for this emitter size (2.3 × 21 μm) mainly due to the spread of the smallest dimension. The same holds for the spread between matched pairs.

Measured over several batches the spreads in ρ_{shpi} and ρ_{shsp} are much larger,
$$\sigma_{\rho_{shpi}} = 1500\,\Omega \quad\text{and}\quad \sigma_{\rho_{shsp}} = 62.5\,\Omega.$$

Each of these spreads separately make $\sigma_{C_{0e}} = 6.0$ fF and 2.75 fF, respectively. Together they give $\sigma_{C_{0e}} = \sqrt{(6.0)^2 + (2.75)^2} = 7.54$ fF. The spread in geometry has hardly any influence now! The spread for other parameters can be calculated in a similar way. Table 11.1 gives for the spread in one wafer (the one with $\bar{\rho}_{shpi} = 8315\,\Omega$, $\bar{\rho}_{shsp} = 442\,\Omega$ and $\overline{BV}_{cb0} = 12.4$ Volt) the calculated and measured values of $\sigma_x/\bar{x}$ of the following Mextram parameters I_s, C_{0e}, C_{0c}, Q_{b0} and τ_{Ne}. The measured values were obtained from a sample of 50 devices.

These results show that it is quite possible to predict the mean value and variance of compact model parameters from statistical data about the device geometry and the PCM values.

In the following subsections a short discussion will be given of the other methods of statistical modelling.

11.5.2 Correlation Between Compact Model Parameters

Instead of using the geometry and PCM data as a starting point for statistical modelling, we can also use the compact model parameters themselves. We must then find out how these model parameters are correlated, e.g. by measuring the device characteristics, followed by parameter extraction. The correlation is expressed by means of linear regression equations as given by Eq. (11.16). In [11.2] the example is given of a set of Gummel-Poon parameters, which are all correlated with the saturation value I_s of the collector current. This is possible because I_s has a functional dependence on ρ_{shpi} and we have seen already (section 11.1.2) that ρ_{shpi} dominates many other

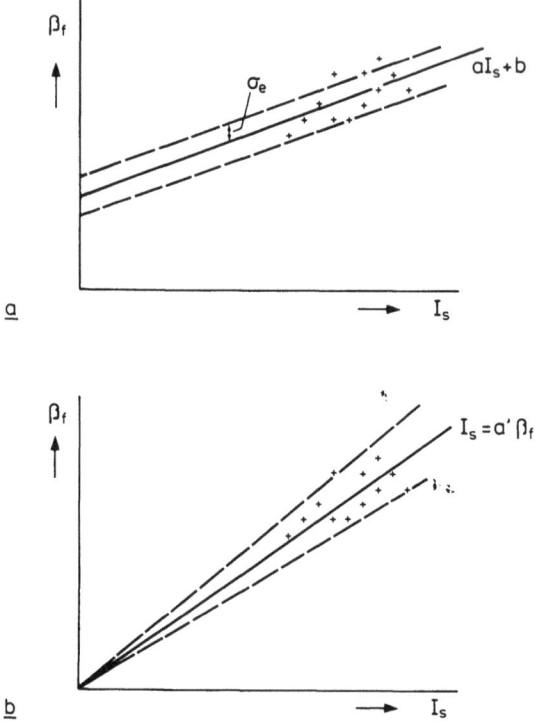

Fig. 11.16 Scatter plot of the model parameters β_f and I_s with two slightly different linear regression equations. The dashed lines indicate the error margins

parameters. So we get from Eq. (11.16)

$$P = \bar{P} + rPI_s \frac{\sigma_P}{\sigma_{I_s}}(I_s - \bar{I}_s) + e$$

or

$$P = aI_s + b. \tag{11.23}$$

The values for a and b are found from a scatter plot (see Fig. 11.16a for the correlation between β_f and I_s). The mean value and the variance (spread) of the parameter P are now

$$\left.\begin{array}{l} \bar{P} = a\bar{I}_s + b \\[4pt] \text{and} \\[4pt] \sigma_P^2 = a^2 \sigma_{I_s}^2 + \sigma_e^2 + 2a\sigma_{eI_s}^2. \end{array}\right\} \tag{11.24}$$

σ_e is the variance of the error term in Eq. (11.16), σ_{eI_s} the covariance of the error with I_s, because these two may be correlated.

A similar result was obtained with the parameters of an extended Ebers-Moll model [11.8], where many model parameters were correlated with β_f,

11.5 Bipolar Statistical Modelling

$$P = a'\beta_f(1 + e), \tag{11.25}$$

where e is the error term. Fig. 11.16 b illustrates Eq. (11.25) for I_s and β_f. From Eq. (11.25) the mean value and variance follow,

$$\bar{P} = a'\bar{\beta}_f,$$
$$\sigma_p^2 = a'^2(\sigma_{\beta_f}^2 + \beta_f^2 \sigma_e^2). \tag{11.26}$$

Here e and β_f are considered as independent.
In Monte-Carlo simulations we can use Eq. (11.25) by writing $e = bx_r$, where the value of x_r is generated by a random number generator (RNG) in the interval $(-1, +1)$.

11.5.3 Correlation at the Process Level

Compact model parameters can be generated by means of an integrated chain of simulation programs, consisting of process and device simulation, followed by parameter extraction (see chapter 1). Introduction of statistical variations in the parameters of the process simulation ultimately leads to properly correlated model parameters. To save CPU time and make Monte-Carlo methods applicable, the FABRICS package [11.22, 11.23] uses analytical models for the process and device simulation. Random number generators introduce the process variations or "disturbances" as they are called, at two levels: local and global. A local disturbance has a mean value D and a variance σ_D^2; each of them is submitted to global variations, again with a mean value and a variance: $(\bar{D})$, $\sigma_{\bar{D}}^2$, $\bar{\sigma}_D^2$, $\sigma_{\sigma_D^2}^2$.

Local variations may be within one chip (matching), global over one wafer, but we can also consider a wafer for local variations as distinct from a batch for global variations, or batch variations (local) against process variations (global).

In this way the spread in model parameters is calculated from the spread of the real causes, the process disturbances. Drawbacks are that the inaccuracy of the analytical models influences the results and that the statistics of the process disturbances cannot be measured directly, but must be derived indirectly from measurements (PCM structures, sheet resistances, model parameters).

The advantage of the model parameter correlation method is that it does not require any knowledge of the process; it can be measured on the device as it is. Drawbacks are that not all parameters can be always correlated with one or two others and that extrapolation to other geometries and/or slight process modifications are not allowed.

The process block approach with PCM data takes an intermediate position and might be considered as a good compromise between the process disturbance method and the model parameter method.

11.6 MOS Statistical Modelling

Generally in MOS IC's two types of statistical variations have been considered: intradie and interdie fluctuations. For VLSI logic the major design constraints stem from interdie spread including variation from batch to batch. Since the design of precision analog circuits is based on component ratios rather than absolute values, for this application local variation or mismatch is important. Usually in both cases the spread in the compact model parameters has been correlated with electrical process parameters. However, since sheet resistances have little effect on MOSFET properties, these process parameters differ from those in a bipolar approach.

Generally the following have been considered as physical causes of parameter spread in MOS devices: the statistical distribution of fixed oxide charge and implanted ions, the fluctuations in geometrical quantities such as channel length and channel width, and gate oxide thickness owing to edge non-uniformity or surface roughness. Since most of the above quantities are determined by different process steps, this approach has the advantage that the basic variables are statistically uncorrelated. However, a disadvantage is that the variance of most variables cannot be obtained separately in an easy way, but only in combination. Although in principle the oxide thickness and the flatband voltage can be obtained from separate capacitance measurements, owing to inherent accuracy problems DC measurements of the transistor characteristics are preferred. However, in this case for instance, the threshold voltage of a large-size device provides information on the gate insulator thickness only in combination with that on fixed oxide charge. Alternatively, the spread in length and width can be obtained from measuring transistors with different aspect ratios, but only in combination with the spread in gate oxide thickness.

Since all model parameters can be expressed in terms of the independent variables (compare the results of section 11.2), their spread can be calculated from the measured values of the variance of the elementary quantities, which are typical for a given process. By substituting the above information in an arbitrary circuit simulator, the expected spread of circuit properties can be calculated. In this way a statistical parametric yield estimator has been developed [11.4].

11.6.1 Mismatch in MOSFETs

Assuming that the total mismatch of a parameter P is composed of many singular physical events and that the events have a correlation distance much smaller than the transistor dimensions, mathematical analysis [11.19] shows that the values of the parameter P are normally distributed, and that the variance is related to transistor length and width, according to

$$\sigma_P^2 \sim 1/ZL.$$

11.6 MOS Statistical Modelling

Examples are the fixed oxide charge (Q_{ox}), the number of implanted ions and the gate insulator thickness [11.24]. The assumption of a short correlation distance implies that no correlation exists within the distance d between two transistors. However, long distance substrate doping and gate oxide thickness variations may lead to enhanced parameter variation with increasing distances. Therefore the variance of the difference in parameters between a pair of transistors is given by

$$\sigma_P^2 = A_P(ZL)^{-1} + S_P d^2, \tag{11.27}$$

where A_P is the area proportionality constant, and S_P describes the variance of P with the spacing d. Experimentally it has been found that for devices which are not of too large a size, the latter correlation is at least an order of magnitude smaller [11.25].

An exception is formed by the geometries L and W. Some authors consider the variance in L and W as a constant [11.4, 11.21]. This assumption is reasonable from the viewpoint that lithographic mismatch is the only source. However, if edge roughness is considered as a physical cause of mismatch between two closely spaced devices, it can be argued [11.2] that

$$\sigma_L^2 = A_L Z^{-1} \tag{11.28}$$

and

$$\sigma_Z^2 = A_Z L^{-1}. \tag{11.29}$$

Applying the above scheme to the difference in threshold voltage and keeping only major sources of spread, it can be shown that [11.21]

$$\sigma_{V_T}^2 = \frac{2}{ZLC_{ox}^2} \{q(Q_{0x} + Q_B + qD_i) + A_{ox}(Q_{0x}^2 + q^2 D_i^2 + Q_B^2)\}, \tag{11.30}$$

where D_i is the implanted threshold adjustment dose, Q_B is the depletion charge outside the implanted layer and A_{ox} is a proportionality constant for the variance of the gate oxide thickness. Note that in the above equation use has been made of the well-known relation for a Poisson distribution, e.g.

$$\sigma_{D_i}^2 = D_i(ZL)^{-1}. \tag{11.31}$$

Naturally, since effects of width and length on V_T have been omitted, the above relations only apply to transistors with dimensions well above a few microns. For σ_y^2 an equation similar to (11.30) can be obtained, but in this case Q_{0x} and D_i are lacking. Fig. 11.17 gives some experimental results [11.25] for σ_{V_T} and σ_y. A reasonable fit with the predicted geometry relations is found. From an analysis of the σ_y-results and the additional data $C_{ox} = 7 * 10^{-8}$ F/cm², $Q_B = 3 * 10^{-8}$ C/cm² and $qD_i = 4 * 10^{-8}$ C/cm² a value

$$A_{ox} = 0.5 * 10^{-12} \text{ cm}^2$$

is obtained. Somewhat higher values have been reported earlier [11.24]. Using this result it can be concluded from Eq. (11.30) that the oxide thickness

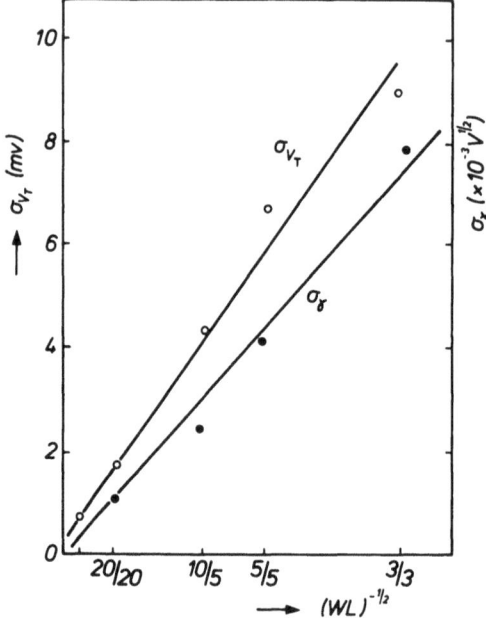

Fig. 11.17 a Standard deviation of the difference in threshold voltage and body coefficient for a MOSFET pair as a function of the inverse transistor area

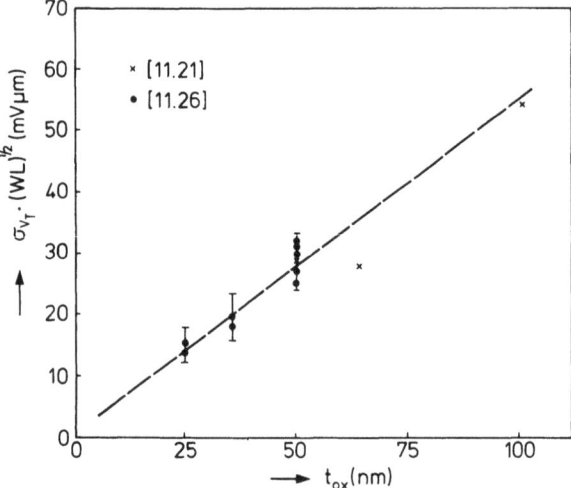

Fig. 11.17 b Standard deviation of threshold voltage difference for a MOSFET pair as a function of gate oxide thickness

11.6 MOS Statistical Modelling

spread contributes less than 10% to the threshold spread. In addition it follows from the interpretation of the data of Fig. 11.17 with Eq. (11.30) that

$$Q_{0x} \approx 1 * 10^{-8} \, C/cm^2.$$

The latter result is found too from the average value of the threshold voltage. Following the previous approach it is easily derived from (11.27), (11.28) and (11.29) that

$$\frac{\sigma_\beta^2}{\beta^2} = \frac{A_L}{ZL^2} + \frac{A_Z}{Z^2L} + \frac{A_{0x}}{ZL} + \frac{A_\mu}{ZL}, \tag{11.32}$$

where A_μ is a constant characteristic for surface mobility variations. The latter quantity may vary over a wafer owing to variations in the oxide thickness and the various charges, which both affect the normal field. Note that in (11.32) owing to edge roughness the terms depend on the geometry according to a higher power. This implies that their contribution is gaining importance at smaller dimensions. Unfortunately no data for micron-size devices has yet been reported. Fig. 11.18 gives the standard deviation as a function of the square-root of the area for larger devices only [11.25]. From the fair result it can be concluded that for not too small devices only the last two terms of (11.32) dominate. However, since the value of σ_β strongly varies with rotated placement of transistors, and additionally the correlation between threshold voltage variations and gain constant variations is almost

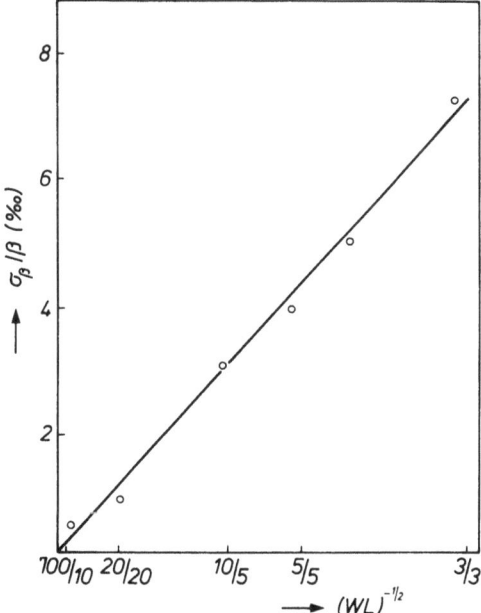

Fig. 11.18 Relative standard deviation of the difference in gain constant versus the inverse transistor area

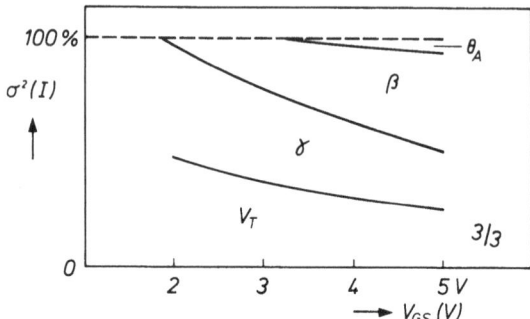

Fig. 11.19 Relative contribution of the model parameters to the mismatch of the drain current

zero, it can be concluded that the mobility term of Eq. (11.32) is dominant. This is in agreement with the resulting value

$$A_\mu = 6 * 10^{-12} \text{ cm}^2,$$

which is found from the result of Fig. 11.18.

Finally Fig. 11.19 gives the relative contribution of variations of threshold voltage, body-effect coefficient, gain constant and mobility reduction factor (θ_A) to the spread in drain current as a function of gate drive. All data have been obtained experimentally [11.26]. This result agrees with physical intuition. While the variation of gain constant dominates in the entire bias region, the threshold spread contribution diminishes at increasing gate bias, whereas the contribution of θ_A increases owing to the effect of source series resistance.

The above results have been applied to calculate the expected standard deviation of electrical circuit properties, like the output voltage of a bandgap circuit [11.25] and current sources [11.24]. A fair agreement with experimental values has been obtained. From the above results it is also clear that often the obtainable circuit accuracy can be improved by increasing the size of the appropriate devices. However, since the standard deviation only improves with the square root of the area, this has to be handled with care.

11.6.2 Parametric Yield Estimation in MOS VLSI

For logic design, interdie variation of parameters has been found to be the dominant source of spread. Since in this case control of the manufacturing process over a long period has to be taken into account as well, the resulting interdie variations are much larger than in the previous paragraph. From the limited literature available [11.5, 11.7, 11.27] it can be concluded that most authors consider the variations in length, width, gate oxide capacitance and flatband voltage to be the major factors responsible for the statistical

11.6 MOS Statistical Modelling

variation of device characteristics. This is not unreasonable. The sensitivity to other parameters, such as changes in the doping profile are estimated to be at least an order of magnitude smaller. For instance, ion implantation can be much better controlled in volume production than etching. Since length, width, oxide thickness and oxide charge are determined at different steps in the manufacturing process, they should be uncorrelated, independent statistical variables. Despite some published results [11.5], unfortunately no convincing proof of this statement has been given.

On the other hand, since most model parameters can be expressed in simple terms of these four statistical variables, e.g.

$$P = A\Delta L + B\Delta Z + Ct_{0x} + DV_{FB} + E, \tag{11.33}$$

geometrical statistical yield analysis becomes feasible.

Using the above relations in a preprocessor, the statistical analysis of circuit performance parameters can be carried out by means of circuit simulation. Since process variations are assumed to make small perturbations on the considered statistical variables, the performance constraints, which define the yield boundaries, can be approximated by a linear function of these variables. Therefore for each circuit performance parameter F we have the relation

$$\sigma_F^2 = a_L \sigma_{\Delta L}^2 + a_Z \sigma_{\Delta Z}^2 + a_C \sigma_{t_{0x}}^2 + a_V^2 \sigma_{V_{FB}}^2. \tag{11.34}$$

Consequently, if the value of the variances of ΔL, ΔZ, etc. have been measured for a given process, only 5 circuit simulations are required to calculate the parametric yield [11.4].

Experimental verification of the method has been made by a comparison of the measured delay and average power dissipation for an n-MOS inverter chain with that predicted by the statistical model [11.4]. The variance of ΔL, ΔZ, t_{0x} and V_{FB} has been determined by measuring the gain constant β and the threshold voltage of a number of test transistors with different nominal values of length and width. For instance, ΔL can be measured by comparing large-width devices with varying channel length. Alternatively, t_{0x} can be determined from β_0 and flat-band voltage from the residual value of threshold voltage after removal of length, width and oxide capacitance dependence. In addition to the parameters extracted from the linear region, complete compact model parameters were obtained for the mean devices in one slice selected as representative of the lot.

A comparison of the mean delay of the inverter chain on each of the nine slices with the calculated delay using the model parameters and statistical data is shown in Fig. 11.20. Although no mention has been made of the effect of parasitic capacitances on the result, a fair correlation is observed.

In addition to the process dependence of circuit performance, the effect of both temperature range of operation and variation in supply voltage have to be considered in parameteric yield estimation. Regarding the latter effect, one additional simulation is required to determine the change in performance parameters at the specification limits of supply voltage. As with

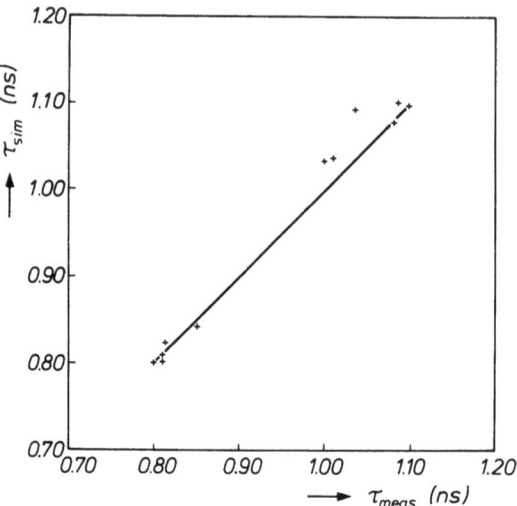

Fig. 11.20 Simulated spread in circuit delay time versus measured values

statistical variables, a linear dependence of performance parameters on supply voltage is assumed. Since the performance constants must be guaranteed for the full specification range of the supply voltage, the voltage dependence is used to define the worst-case limits for each of the performance parameters. Usually the sign of the linear coefficients of the parameters on voltage is used to determine which extreme will produce worst-case dependence. The temperature dependence of circuit properties follows from the sensitivity of individual device parameters. The principal temperature-dependent device parameters are the threshold voltage and the mobility (compare paragraph 6.3.4 and 6.4.2). Using the appropriate temperature scales in the model preprocessor, one additional simulation is required to obtain the temperature effect on the parametric yield.

References

[11.1] R. Berry: Correlation of Diffusion Process Variations with Variations in Electrical Parameters of Bipolar Transistors. Proc. IEEE 57, 1513 (1969).
[11.2] R. W. Dutton, D. A. Divekar: Bipolar Models for Statistical IC Design. In: Process and Device Modeling for Integrated Circuit Design (F. van de Wiele, W. L. Engl, P. G. Jespers, eds.). Noordhoff, Leyden (1977).
[11.3] C. J. B. Spanos, S. W. Director: Parameter Extraction for Statistical IC Process Characterization. IEEE Trans. Comp.-Aid. Des. *CAD-5*, 66 (1986).
[11.4] P. Yang, et al.: An Integrated and Efficient Approach for MOS VLSI Statistical Circuit Design. IEEE Trans. Comp.-Aid. Des. *CAD-5*, 5 (1986).
[11.5] P. Cox, et al.: Statistical Modeling for Efficient Parametric Yield Estimation of MOS VLSI Circuits. IEEE Trans. Electr. Dev. *ED-32*, 471 (1985).
[11.6] P. J. Rankin: Statistical Modelling for Integrated Circuits. IEEE Proc. *129*, 186 (1982).

[11.7] N. Herr, B. Garbs, J. J. Barnes: A Statistical Modeling Approach for Simulation of MOS VLSI Circuit Designs. IEDM Techn. Digest (1982), p. 290 (paper 11.5).

[11.8] Ph. Balaban, J. J. Golembeski: Statistical Analysis for Practical Circuit Design. IEEE Trans. Circ. Syst. *CAS-22*, 100 (1975).

[11.9] S. Inohira, et al.: Statistical Model Including Parameter Matching for Analog Integrated Circuits Simulation. Trans. Electr. Dev. *ED-32*, 2177 (1985).

[11.10] P. M. Solomon, D. D. Tang: Bipolar Circuit Scaling. IEEE Int. Solid-St. Circ. Conf. (1979), p. 86 (paper WPM 8.4).

[11.11] P. A. H. Hart, T. v. 't Hof, F. M. Klaassen: Device Down Scaling and Expected Circuit Performance. IEEE J. Solid-St. Circ. *SC-14*, 343 (1979).

[11.12] T. Smedes: Optimization and Down Scaling of Processes for ECL Circuits. Master Thesis, Technical University, Eindhoven (1986).

[11.13] H. C. de Graaff, W. J. Kloosterman: New Formulation of the Current and Charge Relations in Bipolar Transistor Modeling for CACD Purposes. IEEE Trans. Electr. Dev. *ED-32*, 2415 (1985).

[11.14] J. W. Slotboom: Computer-Aided Two-Dimensional Analysis of Bipolar Transistors. IEEE Trans. Electr. Dev. *ED-20*, 669 (1973).

[11.15] N. Shiono: Emitter Perimeter-to-Area Ratio Effects on High-Frequency Transistor Current Gain and Its Degradation. Jap. J. A. P. *18*, 1097 (1979).

[11.16] G. A. M. Hurkx: On the Sidewall Effects in Submicrometer Bipolar Transistors. IEEE Trans. Electr. Dev. *ED-34*, 1939 (1987).

[11.17] R. K. Brayton, G. D. Hachtel, A. L. Sangiovanni-Vincentelli: A Survey of Optimization Techniques for Integrated-Circuit Design. Proc. IEEE *69*, 1334 (1981).

[11.18] A. M. Mood, F. A. Graybill, D. C. Boes: Introduction to the Theory of Statistics, 3rd Ed. McGraw-Hill, Tokyo (1974).

[11.19] A Papoulis: Probability, Random Variables and Stochastic Processes. McGraw-Hill, Kogagusha, Tokyo (1965).

[11.20] P. Yang, P. Chatterjee: Statistical Modelling of Small Geometry MOSFETs. Techn. Digest IEDM 286 (1982), (paper 11.4).

[11.21] K. R. Lakshmikumar, R. A. Hadaway, M. A. Copeland: Characterization and Modeling of Mismatch in MOS Transistors for Precision Analog Design. IEEE J. Solid-State Circ. *SC-21*, 1057 (1986).

[11.22] W. Maly, A. Strojwas: Statistical Simulation of the IC Manufacturing Process. IEEE Trans. Comp.-Aid. Des. *CAD-1*, 120 (1982).

[11.23] S. R. Nassif, A. Strojwas, S. W. Director: Fabrics II, A Statistically Based IC Fabrication Process Simulator. IEEE Trans. Comp.-Aid. Des. *CAD-3*, 40 (1984).

[11.24] J. B. Shyu, G. Temos, F. Krummenacher: Random Error Effects in Matched MOS Capacitors and Current Sources. IEEE Journal of Solid-State Circuits *SC-19*, 948–955 (1984).

[11.25] M. J. M. Pelgrom, A. C. J. Duinmayer: Matching Properties of MOS Transistors. Digest ESSCIRC (1988).

[11.26] M. J. M. Pelgrom: Delay Lines with Surface Channel Charge-Coupled Devices. Ph.D. Thesis, University of Technology, Twente, The Netherlands (1988).

[11.27] M. J. B. Bolt, A. Trip, H. J. Verhagen: Statistical Worst-Case MOS Parameter Extraction. Proceedings on Microelectronic Test Structures. Edinburgh (1989).

Subject Index

Acceptors 11
Accumulation 148, 184, 203, 256
Auger recombination 26, 28
Avalanche breakdown 29
Avalanche multiplication 27–30, 72–75, 233
Avalanche current 29, 72
Average deviation 302–308

Bandgap 8, 164
Bandgap narrowing 19, 20, 37
Bandgap voltage 19, 93, 95, 105, 112
Barrier lowering (drain-induced) 168, 258
Base charge 45, 107, 109, 320
Base charge at zero bias 44, 45
Base current 56, 104, 124
Base current, ideal 56, 95, 101, 109, 124, 135
Base current, non-ideal 56, 95, 105, 109, 124, 136
Base push-out 65, 70, 109
Base push-out factor 109
Base resistance 76–79, 109, 286, 291, 319
Base transit times 46, 47, 51, 70, 89
Base widening 65, 70, 79, 109
Body (substrate) effect 156, 159, 162, 166, 197, 200, 214, 216, 325
Boltzmann distribution 9, 10, 16, 25
Boltzmann transport equation 13
Breadboarding 1
Breakdown voltage 29, 73, 75, 319
Built-in electric field 47, 49, 82, 117, 158
Built-in junction voltage 58, 67
Bulk charge 181, 206, 265
Bulk (bottom) components 123, 316
Buried-channel 161, 172, 251, 259
Buried layer 79

Capacitances 181, 183, 208–211, 228, 241
Capture cross sections 25
Channel conductivity 243, 260, 262

Channel-length modulation 196, 222, 238, 271, 275
Channel-stop implantation 169
Channel charge 154, 181, 252
Charge control 35, 40–42, 47, 80, 87, 88, 102, 108, 111
Charging current 41, 179, 182
Charge sheet approximation 149, 152
Charge storage 70, 108, 134
Chynoweth's formula 28, 73
Circuit optimization 323
Circuit simulation 2, 4, 346
Collector series resistance 79, 110, 287, 288, 290
Common emitter 80, 87, 289, 292
Common base 81
Compact models 2
Conduction band 8
Conductivity modulation 78, 109, 117, 129, 137
Continuity equations 17, 35, 40, 85
Correlation between model parameters 3, 332, 337
Correlation coefficient 245, 332
Covariance 332, 338
Current crowding 77, 117, 126, 132
Current gain 38, 81, 87, 103, 134, 135, 292, 322
Cut-off frequency 81, 87, 95, 102, 292, 297, 298

Density of states 7, 8
Depletion capacitance 58–62, 93, 102, 109, 134, 137, 293, 320, 336
Depletion charge 45, 58–62, 123, 124, 147, 149, 152, 157, 165, 169, 183, 202, 203, 206, 226, 229, 295,
Depletion-mode 148, 163, 251, 255, 259, 308
Depletion region 49, 144, 163, 201, 269
Detailed balance 25
Device simulation 2, 4, 316, 323, 339
Diffusion constant 16, 37
Diffusion current 16, 152, 153

Subject Index 349

Diffusion noise 31, 244
Diffusion-recombination length 36
Diffusion voltage 58, 67, 93, 105, 164, 200
Distribution function trapped electrons 25
Donors 11
Doping profile 48, 157, 319
Drain Current 152, 154, 196, 200, 211, 216, 238, 263, 268, 276
Drain charge 182, 216, 226
Drain conductance 173, 237, 239, 275, 300, 304, 306
Drain-induced barrier lowering 168, 215, 258, 325
Drift current 152, 153
Drift velocity 41

Early effect 62–65, 81, 102, 104, 107, 129, 140, 294, 295
Early factor 62, 107, 136
Early voltage 62, 63, 102, 108, 129, 134, 141, 322
Ebers-Moll model 44, 99–107, 128, 133, 338
ECL gate 323
Effective mass tensor 8
Einstein relation 16, 49, 66, 82, 94
Electrical dimensions 314, 316
Electrostatic potential 18, 144
Emitter series resistance 75, 286, 287, 291
Empirical models 4, 273, 277
Energy density 14
Energy relaxation 18
Enhancement mode 161, 163, 195, 299
Epitaxial collector layer 61, 65, 110, 293
Epilayer resistance 79
Equivalent circuit 39, 81, 88, 92, 100, 108, 116, 134, 136, 230, 241, 245, 278, 291
Error, overall 299
Excess carrier concentration 36
excess phase shift 84, 117, 126
Expectation 331
Extrinsic part 125

Fermi-Dirac distribution 9, 14
Fermi integral 9
Fermi level 9, 10
Fermi level, intrinsic 10
Fermi potentials 15, 147
Flat-band voltage 146, 211, 262
Flicker noise 32, 92
Forward current 50, 52
Fringing field 188
$1/f$ noise 92, 246

Gain constant 197, 325, 326, 343
Gate charge 181, 206, 265, 277
Gate current 179, 277
Gate delay 323, 324, 346

Gate driving voltage 176, 199, 216, 222, 237, 260, 263
Gate-induced noise 245
Gate insulator 144
Gate overlap capacitances 188, 211
Generation 43, 85
Generation rate 17, 28
Geometrical dimensions 315
Geometrical scaling 3, 299, 316, 325
Grading coefficient 58, 184, 294
Gummel numbers 37, 40, 43
Gummel plot 105, 286, 295, 296
Gummel-Poon model 47, 99, 107–114, 129, 136
Gummel-Poon parameters 109, 130, 337

Hard saturation 112, 287, 288
Hole lifetime 27
Hot carriers 67, 69, 110, 117, 119, 234
Harmonic distortion (3rd order) 290

Inactive region 125
Initial guess 284
Injection, high 45, 52–53, 57, 91, 96, 107, 129, 135, 136, 137
Injection, low 36, 44, 50–52, 67, 85
Injection model 35, 38, 39, 99
Injection region 66
Input admittance 289
Integral charge control relation 45–47, 107
Intrinsic concentration 10
Intrinsic Fermi level 10, 18
Inverse mode of operation 71–72, 206, 207, 227
Inversion (layer charge) 23, 145, 148, 181, 202, 204
Ion implantation (threshold implant) 157, 160, 200, 218, 229
Ionized impurities 93
Iteration, Gauss-Newton 283

Jacobian matrix 282
Junction, abrupt 58, 61, 62, 89
Junction, linear 59, 62
Junction capacitance 184
Junction field-effect transistor 267–273

Kirchhoff current law 230
Kirk effect 65, 67–70, 74, 117

Lateral *pnp* transistor 132–142
LDD MOSFET 152, 188, 190, 230, 234, 308, 310
Least-squares method 282, 301
Levenberg-Marquardt approximation 283
Lightly doped drain 152
Linear parameter determination 284, 285
Linear regression 332, 337, 338
LOCOS (bird's beak) 169

Mask dimensions 316, 323
Matrix batch 318, 319, 323
Maxwell-Boltzmann distribution 9, 147
Mean square noise 245, 291
Mean value 331
MESFET 267, 273
Mextram model 104, 114–128, 129, 293
Mextram parameters 126, 130, 293, 337
Miller's formula 73
Mismatch (parameter), 340–344
Mobility 15, 16, 21–25, 94, 170
Moll-Ross relation 45, 107
Monte Carlo simulations 339
Multiplication factor 29, 73

Narrow-Channel effect 169, 213, 214
Neutral emitter 119
Noise 30–32, 243, 246, 291
Noise Figure 92
Non-ideality coefficients 103
Normal distribution 331, 335
Nyquist's formula 32, 243, 291

Oxide charges 146, 343
Oxide sidewall 316

Parameters for compact models 2, 3, 103, 104, 109, 126, 138, 197, 212, 223, 224, 240, 256, 264, 273, 277
Parameter extraction 3, 281–312, 339
Parasitic capacitances 179, 184, 241
Parasitic pnp transistor 100, 129
Pinch-off 177, 220, 269
pn product 10, 19, 53, 112
Poisson's equation 18, 58, 147, 167, 176
Process block 4, 314, 323, 335
Process control module (PCM) 315, 316, 323, 335, 336, 337, 339
Process simulation 4, 316, 323, 339

Quasi-Fermi levels 42, 66
Quasi-Fermi potentials 20
Quasi-saturation 57, 65–71, 91, 96, 109, 111–112, 119, 129, 290, 293, 297
Quasi-static approximation 40–42, 80, 86, 102, 178, 182, 243

Random number generator 339
Recombination 25–27, 35, 38, 43, 85, 132
Recombination current 41
Recombination rate 17
Recombination, Shockly-Read-Hall 25, 56
Recombination velocity 27, 37
Recursion formula 283
Residual vector 282

Resistivity 21, 24
Reverse current 51, 52

Saturated drift velocity 24, 36, 61, 65, 112, 172
Saturation current bipolar 100, 134, 319
Saturation current MOS 211, 217, 231, 237, 269
Saturation voltage 155, 197, 201, 211, 218, 231, 237, 238, 254, 264, 269, 276
Scaling rules, geometrical 315, 316–318
Schrödinger wave equation 7
Series resistance 75–80, 102, 109, 125, 185, 229, 270, 278, 286, 309
Sheet resistance 24, 94, 186, 319, 339
Shifted Maxwellian 9, 14
Shockly-Read-Hall mechanism 25, 56
Short-Channel effects 165, 213
Shot noise 30, 91, 291
Sidewall components 123, 294, 318, 319
Silicon sidewall 316
Small-signal a.c. 40, 241
Smoothing factor 111, 198, 238
Source charge 182, 206, 226, 278
Source resistance 230, 270, 276, 309
Space charge 18, 144, 148, 252
Specific resistance 24
SPICE, circuit simulator 110, 112
Spin degeneracy 8
Spread 337, 340, 344
Spreading resistance 185, 187
Statistics 331
Statistical modelling 3, 315, 333, 337, 340
Standard deviation 332, 342, 343
Stored charge 45, 47, 102, 117–119
Stored charge in collector epilayer 65–67, 110, 111, 119
Strong inversion 148, 154, 196, 204, 225, 243, 300
Substrate-collector junction 100
Substrate current 100, 105, 125, 233, 308
Substrate effect 156, 300
Subthreshold current 199, 244, 300, 306, 307
Subthreshold slope 224, 237
Surface mobility 171, 216, 256
Surface potential 148, 152, 236
Sustaining voltage 75

Table models 3
Temperature dependence mobility 23, 173
Temperature dependence n_i 93, 164
Temperature dependence parameters 104–105, 164
Thermal noise 31, 91, 243, 291
Thermal voltage 100, 147
Threshold voltage 149, 156, 161, 197, 201, 214, 254, 263, 325, 330, 341
Threshold voltage shift 159, 168, 201, 214, 325

Time delay 84
Transconductance 80, 239, 272
Transfer admittance 80
Transit time 41, 87–91, 95, 102, 103, 108, 110, 129, 134, 178
Transparency 41
Transport model 35, 38–40, 100
Trap density 25

Unity parameters 316, 318, 322, 334

Valence band 8
Variance 332, 338, 341–345

Velocity overshoot 267
Velocity saturation 23, 172, 211, 220, 263, 270
Vertial *npn* transistor 100, 101

Wave equation 7
Wave vector 7, 8
Weak avalanche 29, 72
Weak inversion 148, 150, 155, 183, 223, 239, 244
White noise 32, 91, 291
Work function difference 145, 161, 164
Worst case 315, 346

Y-parameters (admittance parameters) 80–85

 Springer-Verlag Wien New York

Computational Microelectronics
Editor:
S. Selberherr

P. A. Markowich

The Stationary Semiconductor Device Equations

1986. 40 figs. IX, 193 pages.
Cloth DM 98,—, öS 686,—
ISBN 3-211-81892-8
Prices are subject to change without notice.

The static semiconductor device problem is treated in an "applied mathematics" way. The device equations are derived from physical principles; qualitative properties like existence, uniqueness, continuous dependence-on-data, and regularity of solutions are analysed by means of the modern theory of elliptic boundary value problems, and singular perturbation methods are employed to investigate the structure of solutions. This analysis describes depletion layer phenomena qualitatively and quantitatively. Physical interpretations of the mathematical results are given and pitfalls of the physical model are discussed. The structural results obtained by the singular perturbation analysis serve as a basis for the derivation and convergence analysis of numerical discretisation techniques, i.e. of finite element and finite difference methods.

The monograph provides device modelers with the basic mathematical results and techniques necessary for designing efficient simulation programs and for a profound interpretation of simulation results. Also, it serves as a source of mathematically challenging problems for researchers in numerical analysis, singular perturbation theory, and p. d. e. analysis.

Springer-Verlag Wien New York
Moelkerbastei 5, P.O. Box 367, A-1011 Wien
Heidelberger Platz 3, D-1000 Berlin 33
175 Fifth Avenue, New York, NY 10010, USA
37-3, Hongo 3-chome, Bunkyo-ku, Tokyo 113, Japan

Springer-Verlag Wien New York

Computational Microelectronics
Editor:
S. Selberherr

C. Jacoboni, P. Lugli

The Monte-Carlo Method for Semiconductor Device Simulation

1989. 228 figures. X, 356 pages.
Cloth DM 186,—, öS 1300,—
ISBN 3-211-82110-4

Prices are subject to change without notice

The subject of the book is the application of the Monte Carlo method to the simulation of semiconductor devices. It introduces the reader to the Monte Carlo technique as applied to the study of transport in semiconductors, and to the modelling of semiconductor devices. The book is at a tutorial level, where all the details of a Monte Carlo algorithm are discussed. Since the use of the Monte Carlo technique requires an accurate knowledge of the physical system under investigation, a general overview of the basis of the physics of transport in semiconductors is also provided. A review of the Monte Carlo simulation of actual devices is also presented, together with possible vectorization schemes.

The Monte Carlo technique is a fairly new tool in the area of device modeling, traditionally dominated by simulators based on drift-diffusion or on balance-equation models. A comparison of the characteristics of the different methods is presented, pointing out the areas and limits of applicability of each of them.

The book allows the reader (even with a limited knowledge of transport in semiconductor devices) to become accostumed to the Monte Carlo simulation and to quickly set up a simulator for a specific problem.

Springer Verlag Wien New York
Moelkerbastei 5, P.O. Box 367, A-1011 Wien
Heidelberger Platz 3, D-1000 Berlin 33
175 Fifth Avenue, New York, NY 10010, USA
37-3, Hongo 3-chome, Bunkyo-ku, Tokyo 113, Japan

Springer-Verlag Wien New York

S. Selberherr

Analysis and Simulation of Semiconductor Devices

1984. 126 figures. XIV, 294 pages.
Cloth DM 146,—, öS 1022,—
ISBN 3-211-81800-6

Prices are subject to change without notice

Contents: Introduction — Some Fundamental Properties — Process Modeling — The Physical Parameters — Analytical Investigations About the Basic Semiconductor Equations — The Discretization of the Basic Semiconductor Equations — The Solution of Systems of Nonlinear Algebraic Equations — The Solution of Sparse Systems of Linear Equations — A Glimpse on Results — Author Index — Subject Index.

Numerical analysis and simulation has become a basic methodology in device research and development. This book satisfies the demand for a thorough review and judgement of the various physical and mathematical models which are in use all over the world today. A compact and critical reference with many citations is provided, which is particularly relevant to authors of device simulation programs. The physical properties of carrier transport in semiconductors are explained, great emphasis being laid on the direct applicability of all considerations. An introduction to the mathematical background of semiconductor device simulation clarifies the basis of all device simulation programs. Semiconductor device engineers will gain a more fundamental understanding of the applicability of device simulation programs. A very detailed treatment of the state-of-the-art and highly specialized numerical methods for device simulation serves in an hierarchical manner both as an introduction for newcomers and a worthwhile reference for the experienced reader.

Springer-Verlag Wien New York

Moelkerbastei 5, P.O. Box 367, A-1011 Wien
Heidelberger Platz 3, D-1000 Berlin 33
175 Fifth Avenue, New York, NY 10010, USA
37-3, Hongo 3-chome, Bunkyo-ku, Tokyo 113, Japan

GPSR Compliance

The European Union's (EU) General Product Safety Regulation (GPSR) is a set of rules that requires consumer products to be safe and our obligations to ensure this.

If you have any concerns about our products, you can contact us on

ProductSafety@springernature.com

In case Publisher is established outside the EU, the EU authorized representative is:

Springer Nature Customer Service Center GmbH
Europaplatz 3
69115 Heidelberg, Germany

www.ingramcontent.com/pod-product-compliance
Ingram Content Group UK Ltd.
Pitfield, Milton Keynes, MK11 3LW, UK
UKHW060012240426

470314UK00007B/52